# Tourism and Development in the Himalaya

This book examines the unique characteristics of the Himalaya that mark them as a special region among other orographic regions of the world. The Himalayan range is an important global asset for ecological, climatic, cultural, spiritual, and economic reasons. Its diversity of landscapes, climates, and biotic systems makes the Himalaya an extremely attractive region for tourism. The book examines tourism and development in the Himalaya region, exploring its sociocultural, environmental, and economic dimensions.

The contributors address Himalayan issues from a holistic perspective, emphasizing the uniqueness of the region, together with concerns it shares with other montane, developing parts of the world. With a framework of sustainable development, this book elucidates interdisciplinary perspectives on nature, society, economic development, poverty, justice, health, social and environmental vulnerability, faith and culture, Indigenous rights, women, conflict, heritage and living culture, and many other concepts that broaden our understanding of tourism and development in mountain areas. Many contributors are from the Himalaya region, or have worked there extensively, lending strength through native and insider perspectives.

This work will be useful for advanced undergraduate and graduate students, research and teaching scholars, policy makers, practitioners, and anyone interested in the Himalaya and their distinctive tourism and development-related potential and challenges.

**Gyan P. Nyaupane** is Professor of Community Resources and Development and Associate Dean of Research in the Watts College of Public Service and Community Solutions at Arizona State University.

**Dallen J. Timothy** is Professor of Community Resources and Development at Arizona State University and Visiting Professor at universities in China, Spain, and South Africa.

## Routledge Cultural Heritage and Tourism Series

Series editor: Dallen J. Timothy

*Arizona State University, USA*

The Routledge Cultural Heritage and Tourism Series offers an interdisciplinary social science forum for original, innovative, and cutting-edge research about all aspects of cultural heritage-based tourism. This series encourages new and theoretical perspectives and showcases ground-breaking work that reflects the dynamism and vibrancy of heritage, tourism, and cultural studies. It aims to foster discussions about both tangible and intangible heritages, and all of their management, conservation, interpretation, political, conflict, consumption, and identity challenges, opportunities, and implications. This series interprets heritage broadly and caters to the needs of upper-level students, academic researchers, and policy makers.

**Resilience, Authenticity and Digital Heritage Tourism**
*Deepak Chhabra*

**Cultural Heritage and Tourism in Japan**
*Takamitsu Jimura*

**Medieval Imaginaries in Tourism, Heritage and the Media**
*Jennifer Frost and Warwick Frost*

**Children, Young People and Dark Tourism**
*Edited by Mary Margaret Kerr, Philip Stone and Rebecca H. Price*

**Tourism and Development in the Himalaya**
Social, Environmental, and Economic Forces
*Edited by Gyan P. Nyaupane and Dallen J. Timothy*

**Cultural Tourism and Cantonese Opera**
*Jian Ming Luo*

For more information about this series, please visit: www.routledge.com/Routledge-Cultural-Heritage-and-Tourism-Series/book-series/RCHT

# Tourism and Development in the Himalaya

Social, Environmental, and Economic Forces

**Edited by Gyan P. Nyaupane and Dallen J. Timothy**

Routledge
Taylor & Francis Group

LONDON AND NEW YORK

First published 2022
by Routledge
4 Park Square, Milton Park, Abingdon, Oxon OX14 4RN

and by Routledge
605 Third Avenue, New York, NY 10158

*Routledge is an imprint of the Taylor & Francis Group, an informa business*

*British Library Cataloguing-in-Publication Data*
A catalogue record for this book is available from the British Library

*Library of Congress Cataloging-in-Publication Data*
Names: Nyaupane, Gyan P., 1968– editor. | Timothy, Dallen J., editor.
Title: Tourism and development in the Himalaya : social, environmental and economic forces / edited by Gyan P. Nyaupane and Dallen J. Timothy.
Description: Abingdon, Oxon ; New York, NY : Routledge, 2022. | Series: Routledge cultural heritage and tourism | Includes bibliographical references and index.
Identifiers: LCCN 2021060663 (print) | LCCN 2021060664 (ebook) | ISBN 9780367466275 (hardback) | ISBN 9781032272610 (paperback) | ISBN 9781003030126 (ebook)
Subjects: LCSH: Tourism—Himalaya Mountains Region. | Economic development—Himalaya Mountains Region. | Himalaya Mountains Region—Economic conditions.
Classification: LCC G155.H54 T678 2022 (print) | LCC G155.H54 (ebook) | DDC 306.4/819095496—dc23/eng20220412
LC record available at https://lccn.loc.gov/2021060663
LC ebook record available at https://lccn.loc.gov/2021060664

ISBN: 978-0-367-46627-5 (hbk)
ISBN: 978-1-032-27261-0 (pbk)
ISBN: 978-1-003-03012-6 (ebk)

DOI: 10.4324/9781003030126

Typeset in Times New Roman
by Apex CoVantage, LLC

# Contents

# Figures

# Tables

# Contributors

## The editors

**Gyan P. Nyaupane** is Professor of Community Resources and Development and Associate Dean of Research in the Watts College of Public Service and Community Solutions at Arizona State University. His research interests include sustainable tourism, natural resource management, and human-environment interactions.

**Dallen J. Timothy** is Professor of Community Resources and Development at Arizona State University and Visiting Professor at universities in China, Spain, and South Africa. His tourism research interests include geopolitics, heritage, pilgrimage, and issues pertaining to the Global South.

## The contributors

**Julia N. Albrecht** is a senior lecturer in the Department of Tourism, University of Otago, Dunedin, New Zealand. Her research interests are in tourism and visitor management for sustainable tourism.

**Viacheslav Andreychouk** is a full professor and the head of the Geoecology Department at the Faculty of Geography and Regional Studies, University of Warsaw, Poland. His research is concentrated on holistic concepts of geography and geotourism. He has also been recognized as a speleologist, researcher, and explorer of numerous caves.

**Michal Apollo** is an assistant professor at the Institute of Earth Sciences, Faculty of Natural Sciences, University of Silesia in Katowice, Poland. Michal is an enthusiastic researcher, traveler, mountaineer, ultra-runner, diver, photographer, science popularizer, and NGO activist. His areas of expertise are tourism management, consumer behaviors, and environmental and socioeconomical issues.

**Rojan Baniya** is a graduate fellow and PhD student in the Department of Tourism, Hospitality and Event Management at the University of Florida. His research interests are corporate social responsibility and sustainability within the tourism and hospitality industry.

**Buddha Basnyat** is a renowned physician working in the field of mountain medicine/rescue and infectious disease. He is the founder and president of the Mountain Medicine Society of Nepal and is currently the medical director of the Himalayan Rescue Association. He is also the past president of the international society of mountain medicine and has been a member of the International Commission for Alpine Rescue, medical commission (ICAR-Medcom).

**Sanjeeb S. Bhandari** is an emergency medicine physician from Nepal currently working in Roanoke, Virginia. He is passionate about mountains, has a deep interest in wilderness medicine, and has worked as a volunteer physician in HRA Everest ER at Mt. Everest Base Camp at an altitude of 5,300 meters, in 2014 and 2017.

**Alton C. Byers** is a mountain geographer, conservationist, and mountaineer specializing in interdisciplinary and applied research, high mountain environments, climate change, glacier hazards, and community-based conservation and development programs.

**Neil Carr** is a professor in the Department of Tourism, University of Otago, New Zealand. His research focuses on understanding behavior within tourism and leisure experiences, with a particular emphasis on children and families, sex, and animals (especially dogs).

**Neel Kamal Chapagain** is an associate professor and the Director of the Centre for Heritage Management at Ahmedabad University in India. An architect by education, he is interested in context-responsive design, and integrated approaches to heritage, bridging the binary perceptions of natural and cultural, as well as the tangible and intangible.

**Kishor Chitrakar** is a PhD candidate in the Department of Tourism, the University of Otago, New Zealand. His research area of interest includes community-based tourism and gastronomic tourism.

**Lisa Choegyal** is a writer and sustainable tourism specialist based in Kathmandu, Nepal, since 1974. She consults throughout the Asia Pacific region, is the director of Tiger Mountain Pokhara Lodge, and serves as New Zealand's Honorary Consul to Nepal.

**Jonathon Day** is an associate professor at Purdue University. His research interests include sustainable tourism and responsible travel.

**Mariela Fernandez** is an associate professor in the Parks, Recreation, and Tourism Management Department at Clemson University. Mariela's research examines environmental injustices affecting Latinx urban communities, with primary focus on their limited access to community-based parks. She is also interested in how Latinx communities organize against environmental injustices.

**Agnieszka Gawlik** is an assistant professor at the WSB University in Opole, Poland. Agnieszka conducts research on the effectiveness of markets and market failures. She is the author of several dozen publications, passionate lecturer, and consultant who combines an economic approach with care for the natural environment and social development.

**Lee Jolliffe** recently retired from the University of New Brunswick, Canada, and is currently Visiting Professor in the Business School at the University of Ulster, Northern Ireland. Her research interests include cultural tourism, art and tourism, and tea and coffee tourism. She has worked extensively in Asia, the Caribbean, and Europe.

**Jharna Joshi** is an architect with a master's degree in historic preservation and cultural landscapes, and a PhD in tourism management. Her areas of specialization include architecture, conservation, cultural landscapes, and cultural tourism. Currently, she is involved in designing and restoring traditional buildings, as well as teaching at Kathmandu University, Nepal.

**Anup K C** is an assistant professor at Tribhuvan University, Nepal, and a PhD candidate at Clemson University, USA. His research work focuses on ecotourism and socioeconomic and environmental issues in Nepal.

**Sanjay K. Nepal** is a professor in the Department of Geography and Environmental Management at the University of Waterloo, Canada. He is the Past President of the Canadian Association of Geographers (CAG). His current research focuses on wildlife conservation, and on social and cultural capital, tourism and post-disaster recovery.

**Matiram Pun** is a high-altitude researcher with extensive field experience in the Himalaya. After completing his graduate work in mountain medicine and high-altitude physiology at the University of Calgary, Canada, he worked as an assistant professor in the Department of Clinical Physiology, Institute of Medicine, Kathmandu, Nepal. He also worked as a researcher at the Institute of Mountain Emergency Medicine, European Academy of Bolzano (Eurac Research) and currently serves in an advisory capacity for the Mountain Medicine Society of Nepal.

**Milan Shrestha** is a senior lecturer in the School of Sustainability at Arizona State University. A broadly trained environmental anthropologist, Milan's recent research focus has been on the sociocultural intersectionality of smallholder agriculture, environmental change, and disaster risks in high mountains.

**Brijesh Thapa** is a professor and head of the School of Hospitality and Tourism at Oklahoma State University, USA. His research theme is within the nexus of tourism, conservation, and sustainability.

**Yana Wengel** is an associate professor at the Hainan University-Arizona State University Joint International Tourism College, China. Yana takes a human geography lens on social sustainability and international development. Her research interests include volunteer tourism, creative methodologies (like LEGO® Serious Play® and Ketso), mountaineering, and nature-based tourism.

# Part 1

# Background and Contextual Setting

# 1 Introduction

## Tourism and Development in the Himalaya

*Gyan P. Nyaupane and Dallen J. Timothy*

## Introduction

The great Himalaya, one of the longest orographic ranges in the world with the highest mountains on the planet, is home to more than half a billion people in Nepal, Bhutan, Pakistan, China (Tibet), India, Afghanistan, Bangladesh, and Myanmar. The Himalaya provide a life-support base for an additional 1.9 billion people who live in downstream basins, and three billion people rely on food produced in these basins, making the Himalaya an important resource for half of the world's population. The Himalaya are a critical global asset for economic, sociocultural, and environmental reasons. The diversity of their physiographic landscapes, climates, and biotic systems makes the Himalaya an extremely attractive region for tourism. In addition, the region is very rich in cultural diversity, with multitudes of cultural mosaics, ancient civilizations, archeological assets, and other historical sites. The region is also known for its hospitable people, who embody the mantra *Atithi Devo Bhava*, meaning the guest is God (from *Upanishad*, an ancient holy writ). Despite all these attractions and resources, tourism in the Himalayan region is still in its infancy and its contribution to the economy is below the global average because of several physical, social, economic, and political challenges.

Being one of the youngest mountain ranges on the planet and surrounded by densely populated areas with extreme poverty, the Himalayan region is highly vulnerable to both natural and human-induced environmental and social changes occurring at local, national, regional, and global scales. These changes include, but are not limited to, natural disasters, population growth, climate change, and political conflict. Their age and unique and dynamic geophysical processes as well as their cultural diversity, underdevelopment, and remoteness make the Himalaya different from other mountains. Fragility, marginality, and diversity are unique characteristics of the Himalaya (Jodha, 2001). This uniqueness has major implications for socioeconomic development in general and tourism development in particular, which warrants concerted attention in the area of tourism and sustainable development.

The Himalaya offer critically important geo-ecological, climatic, and cultural assets to the world (Wester et al., 2019). The Himalaya are known for highlands

DOI: 10.4324/9781003030126-2

and majestic mountains. The region is home to 400 mountains over 7,000 meters in elevation and 14 mountains over 8,000 meters and includes the Tibetan plateau, the largest and highest plateau on the planet (Douglas, 2021). The Himalaya are the source of 10 major rivers whose basins cover an area of over 4.2 million $km^2$ (Wester et al., 2019). The region is also known as the 'Third Pole', having the largest area of permanent ice cover outside of the North and South Poles, and the Himalaya play an important role in global hydrological cycles. The region is also diverse in terms of climate, with some of the wettest (over 10 meters of precipitation annually) to the driest (less than 20 cm annually) places on earth. The temperature ranges from over 45°C in the foothills and plains to the some of the coldest places on earth. The area's topography and climate create a perfect environment for diverse flora and fauna, making the Himalaya one of most biologically diverse regions in the world, hosting four global biodiversity hotspots (Chettri et al., 2008; Wester et al., 2019).

Beyond astonishing geophysical and climatic wonders, the Himalaya are full of human stories. There is a deep spiritual and moral connection between the people and the landscape, as many mountains, rivers, animals, and plants are considered sacred. Culturally, the Himalaya are the site of convergence of three major religions. Hinduism and Buddhism originated in the Himalaya, and Islam has flourished in the Western Himalaya. The region is a linguistic mega center, where one-sixth of all human languages are spoken (Turin, 2019). Despite the region's cultural and linguistic diversity, one thing that binds the Himalaya together culturally is its role as the root of the Sanskrit language, one of the most complex and ancient languages in the world. Additionally, mountain communities have developed and adapted various complex systems, including languages, arts and architecture, agriculture, and institutions for resource management. These great inventions were not accidental; they were developed through trial and error over hundreds of generations. The Himalaya have been a place of discovery for millennia for yogic sciences, including yoga, meditation, and Ayurveda, and more recently they have been a living laboratory for geologists, paleontologists, archaeologists, biologists, and biomedical scientists. There is no place on earth where one can study human adaptations to high altitude better than in the Himalaya.

Despite the worldwide importance of the Himalaya, the region has been largely ignored by the global community and its initiatives (Wester, 2019). The Himalaya were explored and exploited by the British who colonized most of the land south of the Himalaya, including present-day India, Pakistan, and Bangladesh. The Himalaya were further introduced to the Western world as a mysterious and adventurous place by various novels and movies. In his novel *Lost Horizon*, James Hilton (1937) led readers to believe that the fictional Shangri-La, a utopian paradise, a peaceful place without war, dispute, and illness, exists somewhere in the Himalaya.

There were two parallel forces that introduced the Himalaya to the Western world in the 20th century. First, mountain explorers and expedition teams tried to conquer the highest peaks, which was done successfully in 1953 when Tenjing Norgay Sherpa and Sir Edmund Hillary climbed Mount Everest. The global media coverage and news of this event drew Westerners' attention to the Himalaya. Second, the counterculture revolution in the West brought people to the region for meditation and spirituality. For example, the Beatles' transcendental meditation in the foothills of the Himalaya inspired millions of young Westerners to adopt eastern spiritual practices (Douglas, 2021). The first one was inspired by geophysical awe and the second one by outward spirituality, although the former is more prevalent in the Western media. Today, both of these positionalities drive millions of tourists to the region. The Himalaya have seen many changes, not just by Western tourism, but also by the Western paradigm of nature and development.

The utilitarian value of nature, guided by Judeo-Christian traditions and the commodification of nature—a neoliberal approach to resource governance and privatization—has heavily influenced the symbolic, ancestral, and moral relationships between native people, nature, and landscape. As a result, the story of spirituality, sacred spaces, human connections, struggles, and adaptations, and their connections to tourism and development are less prevalent in the literature. This book analyzes these and other issues from a critical perspective. The Himalaya are unique from both biophysical and socioeconomic perspectives. Thus, the book aims to make the case that these unique characteristics ought to be considered while proposing any development and policy intervention. This chapter begins with an introduction to the region and its tourism, followed by a brief summary of the chapters included in this book.

## The Himalaya

*Himalay* (हिमालयः) is a compound Sanskrit word made of two words. '*Hima*' means snow and '*alay*' means house or dwelling, so the literal meaning of the *Himalay* is a place with a lot of snow. Within the entire Indian subcontinent, snow is found primarily in the high mountain chain on the north, which are called 'Himalay'. In Sanskrit, the plural form of the term, Himalay is हिमालयाः 'Himalaya', so there is no need to add an 's' to make the term plural. Further, the range is known as Himalaya, although with a slight variation in pronunciation, in every language spoken in the region, including Nepali, Hindi, Urdu, and Tibetan. This book therefore uses the term Himalaya instead of Himalayas. '*Himali*' and '*hima*' are used as adjectives, such as himali plants, himali animals, himali herbs, himchituwa 'snow leopard', himtal 'glacial lakes', himbristi 'snowfall', and himnadi 'glaciers'. Some specific mountains, such as Gaurishankar Himal, have Himal added in their names. A single mountain peak is also known as Himal in Nepal. Mountains have several names in Sanskrit, including Parvat, Giri, Naga, Adri, Shaila, and Maheebhurut. These names are reflected in the names of various high mountains in the Himalaya, for example, Dhaulagiri and Nangaparvat. In ancient religious writings, such as *Mahabharat* and *Upanishad*, the Himalaya have been personified as Himavat (ruler of the Himalayan Kingdom), Himraj (king of snow), or Pravateshor (lord of mountains). Himavat has been explained as the father of Ganga (the Ganges), the river goddess, and Parvati, wife of Shiva. Mountains, rivers, and other natural features are also personified and considered sacred in the

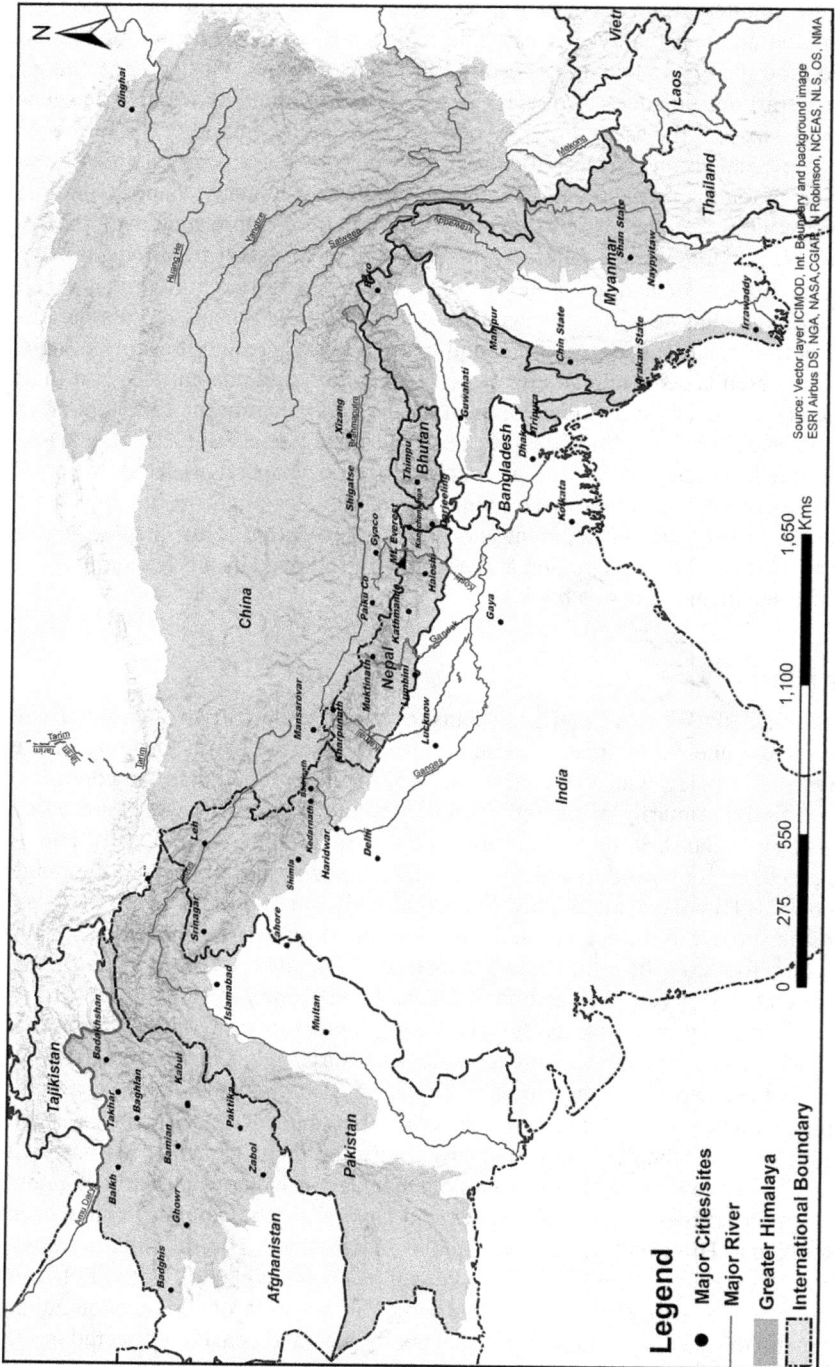

*Figure 1.1* Physical map of the Himalayan region (map drawn by Gokarna Jung Thapa)

Himalaya. The *Puranas* (sacred Hindu writings) mention the Himalayan kingdom as a mountainous country that was ruled by Himavat. The *Mahabharat* also mentions many kingdoms in the Himalaya, including Kulinda, Parvata, Nepa, Kirata, Kimpurusha, and Kinnara.

The Himalayan range passes through six countries: Afghanistan, Pakistan, India, Nepal, Bhutan, and China (Tibet). The extended region, or the Greater Himalaya, is also known as the Hindu Kush Himalaya (HKH) and includes the Pamir, Hindu Kush, and Karakoram ranges that cover about 4,000 km from east to west (Figure 1.1). Some reports (Wester et al., 2019) incorporate eight countries, including Bangladesh and Myanmar, as part of the Hindu Kush Himalayan region as these countries have some low mountain areas connected to the Himalayan range and are also impacted by the Himalaya directly. Among the eight countries, the entire geographic areas of Nepal and Bhutan fall into the Himalayan region. The Tibetan Plateau of China and parts of Yunnan province and Xingjian Uygur autonomous region cover a long stretch of the Greater Himalaya on the northern side. Most of Afghanistan (29 of 34 provinces) and Pakistan's Khyber Pakhtunkhwa and Balochistan provinces, Azad Jammu and Kashmir, Gilgit-Baltistan administrative territories, and federally administered Tribal Areas occupy the western part of the Greater Himalaya. Eleven states of India are located on both the western and eastern sides of the Himalaya. The Chittagong Hills, a mountain division of Bangladesh, and Chin, Shan, Rakhine, and Kachin states of Myanmar are also often considered extensions of the Himalaya on the southeastern side (see Table 1.1).

*Table 1.1* Area, population, and Human Index in the Himalayan region

| Countries | Total area (km²) | Regions in the Himalaya | Population in 2017 (m) | Human Dev Index (ranking, 1- 185) |
|---|---|---|---|---|
| Afghanistan | 652,230 | All provinces except Kandahar, Helmand, Nimroz, Farah, and Herat | 22.85 | 0.511 (169) |
| Bhutan | 38,394 | Entire country | .78 | .654 (129) |
| China | 9,596,960 | Parts of the provinces of Yunnan and Sichuan, the Xinjiang autonomous region, and all of Tibet | 33.29 | .761 (85) |
| India | 2,387,590 | 11 states, including Assam, Uttarakhand, Himachal Pradesh, Manipur, Jammu and Kashmir (Indian-administered area), Meghalaya, Mizoram, Nagaland, Sikkim, Tripura, Arunachal Pradesh, and Darjeeling and Kalimpong districts of the West Bengal state | 86.27 | .645 (131) |

*(Continued)*

Table 1.1 (Continued)

| Countries | Total area (km²) | Regions in the Himalaya | Population in 2017 (m) | Human Dev Index (ranking, 1- 185) |
|---|---|---|---|---|
| Nepal | 147,181 | Entire country | 28.75 | .602 (142) |
| Pakistan | 882,000 | Parts of Khyber Pakhtunkhwa province, Balochistan province, Azad Jammu and Kashmir (AJK), Gilgit-Baltistan and federally administered Tribal Areas (FATA) | 51.47 | .557 (154) |
| Bangladesh* | 144,000 | Chittagong hills | 1.78 | .632 (133) |
| Myanmar* | 676,577 | Chin, Shan, Rakhine, and Kachin states | 11.70 | .583 (147) |
| Total | 14,524,932 | | 236.90 | |

Source: Wester et al., 2019; UNDP, 2020

*More extended region of the Himalaya

Although most of the High Himalaya are uninhabitable or have sparse populations, some of the lowlands and valleys are among the most densely populated areas in the world. According to the Human Development Index, some Himalayan countries are among the world's most underprivileged, in terms of health, education, and per capita income (UNDP, 2020). All Himalayan countries except China have Human Development Index scores below the world average of .737 (Table 1.1).

## Tourism

Tourism is not new to the region, as the Himalaya have been explored by saints and seers as a source of peace, enlightenment, and sacred thoughts for millennia. This is evidenced by the location of major pilgrimage sites in the High Himalaya, including Hindu, Buddhist, and Sikh shrines, such as Badrinath, Kedarnath, Gangotri, Yamunotri and Hemkund in Uttarakhanda; Vaisno Devi and Amarnath cave sanctuaries in Jammu and Kashmir; Pashupatinath and Muktinath temples in Nepal; and Mount Kailash in southern Tibet. These sacred sites are located in remote parts of the Himalaya and for centuries were not easily accessible, so the number of pilgrims was limited. A rough estimate shows that only five thousand to ten thousand pilgrims visited Badrinath each year in the mid-19th century, which was accessible only by a 30-day hike (Spaltenberger, 2002). However, after the extended network of roads, hotels, lodges, and tourism services, the number of pilgrims has increased significantly within the last few decades (Kala, 2014). Grotzbach (1994) estimated that the total number of pilgrims to Uttarakhand more than doubled from 355,000 in 1975 to 751,000 in 1989. People used to travel in the Himalaya for trade through high mountain passes. Tibet was a major source of salt, and people from the southern side of the Himalaya used to trade food and spices for salt. Regional and domestic pilgrimage has existed for centuries. Even today, regional and domestic pilgrimage is the most popular form of tourism in the region.

The Himalayan region was visited by outsiders when most of South Asia, including present-day India, Pakistan, Bangladesh, and Myanmar, was colonized by the British Empire. Some regions and countries in the Himalaya have been more accessible than others, although this has changed from time to time. For example, during colonial times, Tibet and the Indian Himalaya were open to Westerners but the Nepalese Himalaya were closed. As a result, many Everest expeditions prior to 1951, when the country was liberated from the 104-year reign of the notorious Rana regime, were accessed through Tibet, as Tibet was not directly controlled by China at that time. Many Indian hill stations, including Shilong, Darjeeling, Nainital, Almora, Ranikhet, Dehra Dun, Musoorie, Chakrata, Shimla, Srinagar, Murree, and Abbotabad, were developed as recreational spaces for the British who were living in India during the colonial era. These hill stations are now very popular among domestic tourists to escape the hot summer months in India. For example, Himachal Pradesh alone received 16,829,231 domestic tourists in 2019, and the majority of them visited the Shimla district (Lohumi, 2020).

Modern mass tourism in the Himalaya started in the mid-20th century after a series of major global and regional changes, including the end of World War II in 1945, India's independence in 1947, the opening of Nepal to foreigners after the end of the Rana Regime in 1951, the successful Mount Everest expedition in 1953, and the opening of roads and airports in the mountains. During the past five decades, tourism has grown exponentially in the Himalayan region. For example, Nepal's tourism grew from 6,000 arrivals in 1962 to 1,197,000 arrivals in 2019, contributing to 6.7% of the total GDP in 2019 (MCTCA, 2020) (Table 1.2). Tourism is the single biggest foreign exchange earner for Nepal, and the industry

*Table 1.2* Tourism in the Greater Himalayan countries*

| Countries | International tourist arrivals in 2019 (000) | Total GDP contribution (2019) | Top four markets | Travel and Tourism Competitive Index (TTCI) | TTCI global ranking |
|---|---|---|---|---|---|
| Bhutan | 316 | Not available | India, Bangladesh, Maldives, China | Insufficient data | Insufficient data |
| China | 65,700 | 11.6% | Hong Kong, Macau, Myanmar, South Korea | 4.9 | 13 |
| India | 17,910 | 6.9% | Bangladesh, US, UK, Sri Lanka | 4.4 | 34 |

(*Continued*)

*Table 1.2* (Continued)

| Countries | International tourist arrivals in 2019 (000) | Total GDP contribution (2019) | Top four markets | Travel and Tourism Competitive Index (TTCI) | TTCI global ranking |
|---|---|---|---|---|---|
| Nepal | 1,197 | 6.7% | India, China, US, UK | 3.3 | 102 |
| Pakistan | 1,800 | 5.7% | UK, US, India, China | 3.1 | 121 |
| Bangladesh* | 1,026 | 2.7% | India, China, Pakistan, US | 3.1 | 120 |
| Myanmar* | 4,364 | 5.9% | Thailand, China, Japan, India | Not listed | Not listed |

Source: compiled by the authors from various sources including UNWTO, 2021; WTTC, 2021; WEF, 2020, and others.

* Afghanistan data are unavailable.

provides over one million jobs (6.6% of total employment) in the country. Bhutan's tourism grew from 287 tourists in 1974 to 315,599 tourists in 2019 (Tourism Council of Bhutan, 2020) (Table 1.2). According to the World Travel and Tourism Council (WTTC, 2021), this is the most improved tourism region in the world since 2017 and is home to several of the fastest growing tourism destinations on the planet.

Despite tourism's exponential growth, according to the World Economic Forum (WEF, 2020), the Himalayan countries except China and India rank toward the bottom in the Travel and Tourism Competitive Index (TTCI). All Himalayan countries listed on the WEF TTCI index have poor tourism service infrastructures and lack environmental sustainability. Bangladesh and Pakistan have done poorly in international openness and in prioritizing tourism and the human resource labor market. Nepal lacks adequate ground transportation and has a poor business environment, while Pakistan and India face many safety and security concerns.

Despite these challenges and barriers, domestic tourism is very popular in the region. The number of domestic tourists is several times higher than that of international tourists. For example, over 95% of tourists to Himachal Pradesh are from India, and Pakistan's domestic tourism was estimated to be 43 million in 1995 (Sharma, 2000). Globally, domestic tourism represents around 80% of all travel, and in many Himalayan countries, including China, the share of domestic tourism is higher than the global average. More importantly, domestic tourism is proven to be more sustainable and resilient than international tourism as can be seen during the COVID-19 crisis (Nyaupane et al., 2020). Domestic tourism is playing a critical role in tourism recovery during the current pandemic as international tourism is still uncertain, many international borders remain closed, and new variants of the virus are emerging.

Along with economic opportunities, tourism has brought many challenges, from environmental degradation to cultural clashes. The impacts of tourism in the Himalaya are widely reported, including deforestation as a result of increasing demand for firewood and construction, land degradation, solid waste disposal challenges, and sanitation problems. The way environmental impacts are presented and researched has some fundamental incongruities. First, the way impacts are reported usually focuses on a global audience rather than emphasizing local priorities. Reports focus on how the impacts are seen by tourists and other outsiders along major trekking trails, mostly in aesthetic terms rather than how these impacts affect local livelihoods. Second, although the environmental, social, and economic changes in the Himalaya are not only a result of tourism, tourism receives much of the blame. The negative effects of human activities are also caused by other development endeavors, including modernized agriculture and infrastructure development, including roads and hydroelectricity dams. Hence, it is important to recognize that tourism is not the only culprit and does not exist in a vacuum; other phenomena contribute to these diminished conditions. Third, the measurement and assessment of impacts are discrete and discipline based. However, the impacts should be analyzed with a holistic approach. For example, long-term environmental degradation is frequently justified locally for short-term profits and revenue generation, which is the antithesis of sustainable development and an example of understanding tourism narrowly and apart from the greater socioecological whole. Beyond the physical environment, tourism has also brought modernity and technologies to Himalayan destinations, which have modified traditional values and cultures (Nyaupane et al., 2014).

Sustainable development has been proposed as a tool to ameliorate the challenges created by modernity and development. However, the approach has fallen short of recognizing that the current economic and development paradigm of overconsumption and overproduction, which generates revenue and jobs in the short term, is unsustainable. Himalayan destinations are also facing the issue of unmanaged growth for the sake of national, provincial, and local economic progress. Oftentimes, growth-driven planning and policies that promote unsustainable consumption and production borrowed from other regions are not conducive to the Himalayan socioecological system.

Himalayan countries have been exploring various options for their economic development, including commercial agriculture, hydropower, manufacturing, and tourism. The agricultural sector is important for food security, but in terms of employment globally, as agriculture is mechanized, its contribution to job creation has declined significantly during the past century. The Himalaya are rich in water resources, but the dams and reservoirs developed for hydropower are extremely costly and risky given the region's fragile and tectonically active geomorphology. Because of inaccess, environmental sensitivity, and global competitiveness, manufacturing is not the best option for the region either. Thus, many of the Himalayan countries have targeted tourism as an economic driver. The tourism industry, if properly managed, is far more sustainable than some of the industries mentioned earlier and should be seriously considered a mainstream economic sector for the Himalayan communities.

## About the Book

This book aims to examine many of the characteristics of the Himalaya that mark them as a special and unique region among other orographic regions of the world. With the underlying thread of sustainable development, this book brings together interdisciplinary perspectives on nature, society, economic development, poverty, social and environmental vulnerability, faith and culture, Indigenous people, women, geopolitical conflicts, heritage and living culture, attachment to place, and many other concepts to broaden our understanding of tourism and development in mountain areas. Although the Himalayan mountains are one mega socioecological region from both biophysical and social perspectives, the region is politically fragmented, which has complicated how people generally understand this part of the world and makes conducting research and implementing broad regional policies difficult at best. These aforementioned issues have been more researched and written about in the context of some countries more than others. There is a dearth of literature that takes a holistic approach to tourism and development. Within the limited scope of the extant literature, most work, including books and journal articles, about the Himalayan region are isolated case studies or collections of cases, which undermines the common challenges and socioecological interactions between destinations in the region. This volume is a major departure from the existing case study-based approach. Instead, it addresses a more holistic and interdisciplinary perspective. The majority of the authors are from the region, which brings a native perspective of the Himalaya, and the contributors who are not from the region have worked there extensively and understand it well.

The book is conceptually structured into four successive parts. The first part of the book (Chapters 1 and 2) provides a comprehensive background of the region. Chapter 2, authored by the editors, Dallen Timothy and Gyan P. Nyaupane, provides a broad regional overview of both the physical and cultural environments of the region, including the topographic and abiotic systems, and a description of the cultural landscape. The chapter sets the locational context for the concepts and empirical examples that follow throughout the book. The chapter highlights unique geographic features of the Himalayan region, including different belts of the Himalayan range from the Outer Himalaya in the south to the Lesser Himalaya, the Great Himalaya, and the Tibetan/Trans-Himalaya, as well as the region's climatic conditions, glacial landscapes, and biodiversity. The chapter then delves into the region's tangible and intangible heritage and culture, including religions, ethnicities, and languages, and the relationships between these features and tourism.

The second part of the book (Chapters 3–7) presents the unique issues and challenges the Himalayan region faces and their implications for tourism and development. Within this part, the third chapter discusses vulnerabilities, resilience, hazards, and adaptation as they relate to the Himalayan climate and socioecological systems. Here, Nyaupane examines these issues conceptually, followed by Himalaya-specific characteristics that contribute to understanding both vulnerability and resilience. The chapter connects these concepts and discusses various types of disasters and crises that have impacted the Himalaya in general and are more specific to tourism. The contribution concludes with some short-term and long-term policy implications that could be helpful in addressing the region's key issues and challenges.

The High Himalaya have been remote and inaccessible until recently, but the advent of high-mountain tourism, such as trekking, mountaineering, and pilgrimage, has put a lot of pressure on sensitive alpine ecosystems since the mid-20th century. In Chapter 4, Alton C. Byers and Milan Shrestha examine the impacts of high-altitude tourism on the alpine ecosystems of the Himalaya, including excessive shrub harvesting for fuelwood and lodge construction, solid waste management, sanitation, and water contamination. Through their extensive review of the literature and the authors' own research and observations in the Himalaya mountains, the chapter provides a detailed account of the environmental damage caused by tourism in the Himalayan region. Byers and Shrestha's chapter, however, does not harp only on the downside of tourism; it also discusses several success stories of community-based conservation and restoration, and offers some actionable solutions.

The next chapter, by medical scientists Matiram Pun, Sanjeeb Bhandari, and Buddha Basnyat, examines the physiological effects of high-altitude travel on the human body caused by low-level oxygen and the medical complications associated with it. When people living in the lowlands ascend to high altitudes, typically above 2,500 meters, they experience high-altitude effects, such as an increased breathing frequency, tiredness on exertion, increased heart rate, and increased frequency of urination, which can be lethal if ignored. Home to the highest mountains in the world, the Himalaya have exacted some severe outcomes on people's travel to the region. People living and traveling in the Himalaya face various manifestations of altitude sickness, or 'hypoxia'. The chapter offers important insights into the causes, trends, medical diagnoses, and treatments of altitude sicknesses, as well as the importance of awareness, education, and preparation for travelers at high elevations. The chapter fills a major gap in knowledge about high-altitude sickness that is lacking in the social sciences in general and is more specific to the human mobility and tourism literature as it pertains to mountain tourism.

Geopolitical disputes and political instability are among the major barriers to tourism development in the Himalayan region, especially in transfrontier areas, which otherwise could have significant development potential. Chapter 6 examines these conflicts and their effects on resource management and tourism development. In this chapter, Dallen Timothy describes tourism, conflict, and geopolitics from a general perspective and provides an overview of both historical and current situations with regard to the major conflicts in the Himalayan region, including, among others, the India-Pakistan border conflict, the Sino-Indian border dispute, the China-Bhutan border conflict, Nepal's Maoist insurgency, and Bhutan's refugee crisis, and their implications for tourism.

Chapter 7 examines tourism in the Himalaya from a moral geographies perspective. Although porters provide critical services to trekking and mountaineering tourists, being placed at the lowest level in the labor hierarchy, they are the most disenfranchised and exploited group. With erudite examples from the Mount Everest region, Sanjay K. Nepal scrutinizes the role of porters in Himalayan mountaineering and trekking, porters' perspectives of self and others, and the social exclusion and discrimination of these key tourism players. To contextualize the issue, the chapter describes moral geographies in a tourism context, bringing in the broader social science literature on social justice, morality, and ethics, and ties them together with tourism. The chapter provides conceptually rich yet socially important issues of the mistreatment of mountain porters, which might be faced by other mountain workers in orographic regions throughout the world.

The third part of the book (Chapters 8–13) discusses the uniqueness of various forms of tourism in the Himalayan region. Being home to majestic mountains, beautiful natural and cultural landscapes, and unique biophysical systems, the Himalaya are the leading destination for ecotourism, a form of nature-based tourism that embraces the principles of sustainability, education, and host community well-being. In Chapter 8, Lisa Choegyal observes the intricacies of ecotourism in the Himalaya, with special reference to the efforts currently being undertaken in the countries of the region to develop more ecologically and socially sensitive forms of mountain tourism. The chapter provides an overview of the ecotourism concept, its growth, and its trends from a global perspective and then hones in on the ecotourism issues of the Himalaya, including its challenges, constraints, and opportunities in the countries of the region. She utilizes a strong empirical foundation based upon a number of ecotourism projects in the region.

Indigenous people comprise a significant portion of the Himalayan population. They are among the most vulnerable and economically, socially, and politically marginalized of any people in the region. Tourism provides opportunities for the Himalayan Indigenous communities to improve their livelihoods and become socially and economically empowered. Chapter 9, by Anup K C and Mariela Fernandez, connects Indigenous people with tourism. The chapter discusses various forms for tourism adopted by Himalayan Indigenous communities, and the impacts of tourism on these communities and major tourism issues they have been facing, including natural disaster risks, changing climate, political instability, cultural changes, and infrastructure-related challenges.

Chapter 10 examines the role of volunteer tourism, a form of tourism in which someone travels to participate in voluntary, humanitarian, and charity work, in the Himalaya. Although community and family volunteering has a long history in the Himalaya, this chapter focuses mainly on the Western notion of Himalayan voluntourism, highlighting its most common and recent manifestations, such as orphanage work and disaster relief, and the benefits of these activities in the region. In this chapter, Dallen Timothy also turns readers' attention to many of the major criticisms and pitfalls of volunteer tourism, including the exploitation of children, the voyeuristic nature of philanthropic travel, and the complications associated with transitory medical clinics.

Pilgrimage has been one of the most prominent travel phenomena in the Himalaya for centuries. Chapter 11, by Agnieszka Gawlik, Michal Apollo, Viacheslav Andreychouk and Yana Wengel, assesses pilgrimage tourism in the Himalaya. They discuss the unique characteristics of pilgrimage tourism in the Himalayan region and its socioeconomic and environmental impacts. Through an extensive and exhaustive review of current literature, extant debates, and ethnographic experience in this region, the authors highlight the impacts of contemporary mass pilgrimage in the Himalaya, including overcrowding, congestion, and environmental degradation in mountain sacred spaces.

Kishor Chitrakar, Neil Carr, and Julia N. Albrecht's chapter (Chapter 12) examines the role of community-based homestay tourism in the evolution of local cuisine and culinary traditions. The chapter broadly discusses changing food cultures and the impacts of homestay on regional cuisine in the Himalaya and delves deeper into the nuisances of culinary changes as a result of homestay tourism using a community-based homestay example in Nepal. The authors demonstrate how community homestays can expand the boundaries of local foodscapes by preserving and revitalizing indigenous foodways, while at the same time accepting food and gustatory traditions from other cultures. The chapter also discusses the challenges of balancing between authentic local food and guests' food preferences.

Agriculture is an important part of people's heritage. In several parts of South Asia, tea cultivation, spurred largely through colonialism, is a salient manifestation of agricultural heritage. Chapter 13, by Lee Jolliffe, discusses tea-based agritourism as a sustainable development tool in the Eastern Himalaya. The chapter presents tea tourism as a growing niche form of tourism with growing potential in the Eastern Himalaya, extending from eastern Nepal across northeastern India, Bhutan, Tibet, Yunnan Province (China), and northern Myanmar. This part of the Himalaya holds great potential for promoting the teascapes of rolling cultivated tea plantations or forest grown tea, tea factories and production centers, small tea holdings, and tea shops or tea bungalows—all of which comprise an unmistakable tea agriheritage landscape that appeals to domestic and international tourists. Besides describing the tea heritage and its tourism potential, Jolliffe examines some of the issues and challenges facing the development of a viable and sustainable tea tourism product in the region.

The fourth part (Chapters 14–16) takes a solutions perspective with regard to tourism in the Himalaya. The region has witnessed many of tourism's adverse impacts on its fragile ecology and rich culture, as the previous chapters have well indicated. In Chapter 14, Rojan Baniya and Brijesh Thapa examine corporate social responsibility (CSR) as a response strategy in promoting sustainable tourism in the region. The chapter begins with an overview of the impacts of tourism in light of ecological and sociocultural sustainability, and the role tourism businesses can play in mitigating the (un)sustainability issues endemic to the region. Some of the CSR practices by the tourism industry in general and more specific to the region are also discussed. Despite the potential of CSR to help guide sustainable tourism, the authors conclude that CSR practices by tourism businesses have not been very effective in the region. The chapter provides some recommendations that tourism businesses and governments can consider while implementing CSR in the Himalaya.

In Chapter 15, Jonathan Day positions tourism as a tool for achieving sustainable development goals. His chapter explores some tourism-related business models that are commonly applied to achieve the United Nations Sustainable Development Goals (SDGs) in general and goals specific to the Himalaya. Through two examples in Nepal, the chapter demonstrates the power of tourism to support community-building activities. Among the 17 UN SDGs, the chapter make the case that tourism-related programs encourage gender equality (SDG5), improve health and well-being (SDG3), reduce inequalities (SDG10), and create decent work (SDG8). The chapter further examines the power of tourism to overcome some of the rampant social inequalities in the region, such as the caste system, which are rooted in most of the Himalayan communities.

In Chapter 16, Neel Kamal Chapagain and Jharna Joshi examine cultural heritage as a crucial tourism asset in the Himalaya from a management and policy perspective. Using empirical fodder from Upper Mustang, a Trans-Himalayan region of Nepal, and other localities, the chapter discusses the issues heritage tourism destinations are facing, including cultural heritage and nature protection, community concerns, and poor management. From a more balanced perspective, the chapter offers heritage tourism as a tool for heritage protection and community development through a culturally sensitive and collaborative approach.

The final chapter by Dallen Timothy and Gyan P. Nyaupane draws together many of the main issues and concepts threaded throughout the book and points to future directions for additional research that need scholarly attention in a dynamic region of both tradition and modernity.

The Himalayan region is a dynamic province of geophysical and cultural change, yet contradictorily much of the mountains' appeal lies in the traditional lifestyles of their people and the permanence of their natural landscapes. This mountainous region of Asia is one of the most mysterious and inspirational places on the earth. It is simultaneously powerful and strong, yet vulnerable and fragile. It exudes a powerful appeal that derives from ancient travelers' tales and accounts of modern expeditions that are broadcasted worldwide through social media and other online platforms. Thus, while much of the region remains relatively isolated and inaccessible, it is becoming increasingly consumable metaphorically through modern technology and literally through improved infrastructure, widening transportation networks, and a higher global awareness of what the Himalaya have to offer. Through carefully planned development, this region, like many others, can reap the fiscal and social rewards of tourism while avoiding many of its pitfalls. The chapters in this book draw attention to both the benefits and costs of tourism and provide sound advice for the region in going forward in its efforts to develop an industry that capitalizes on special people, places, and experiences—the very essence of the Himalaya.

# References

Chettri, N., Shakya, B., Thapa, R., & Sharma, E. (2008). Status of a protected area system in the Hindu Kush-Himalayas: An analysis of PA coverage. *The International Journal of Biodiversity Science and Management*, 4(3), 164–178.

Douglas, E. (2021). *Himalaya: A Human History*. New York: W.W. Norton.

Grotzbach, E. (1994). Hindu-Heiligtümer als pilgerziele im Hochhimalaya. *Erdkunde*, 48(3), 181–193.

Hilton, J. (1937). *Lost Horizon*. New York: William Morrow and Company, Inc.

Jodha, N.S. (2001). *Life on the Edge: Sustaining Agriculture and Community Resources in Fragile Environments*. London: Oxford University Press.

Kala, C.P. (2014). Deluge, disaster and development in Uttarakhand Himalayan region of India: Challenges and lessons for disaster management. *International Journal of Disaster Risk Reduction*, 8, 143–152.

Lohumi, B.P. (2020). Tourist flow to Himachal stayed sluggish in 2019. Retrieved from www.tribuneindia.com/

MCTCA (2020). *Nepal Tourism Statistics 2019*. Kathmandu: Ministry of Culture, Tourism & Civil Aviation.

Nyaupane, G.P., Lew, A.A., & Tatsugawa, K. (2014). Perceptions of trekking tourism and social and environmental change in Nepal's Himalayas. *Tourism Geographies*, 16(3), 415–437.

Nyaupane, G.P., Paris, C.M., & Li, X. (2020). Introduction: Special issue of domestic tourism in Asia. *Tourism Review International*, 24(1), 1–4.

Sharma, P. (2000). Tourism and livelihood in the mountains: Regional overview and the experience of Nepal. In M. Banskota, T.S. Papola, & J. Richter (Eds.), Growth, Poverty Alleviation and Sustainable Resource Management in the Mountain Areas of South Asia (pp. 349–376). Feldafing, Germany: German Foundation for International Development.

Spaltenberger, T. (2002). Tourism in the Himalayas. Retrieved from www.spaltenberger.de/usa/himalayantourism.pdf

Tourism Council of Bhutan (2020). Bhutan tourism monitor 2019. Retrieved from www.tourism.gov.bt/uploads/attachment_files/tcb_K11a_BTM%202019.pdf

Turin, M. (2019). Concluding thoughts on language shift and linguistic diversity in the Himalayas: The case of Nepal. In S.K. Sonnatag & M. Turin (Eds.), *The Politics of Language Contact in the Himalaya* (pp. 163–176). Cambridge: Open Book Publishers.

UNDP (2020). *Human Development Reports*. United National Development Programme. Retrieved from http://hdr.undp.org/sites/default/files/hdr2020.pdf

UNWTO (2021). *International Tourism Highlights, 2020 Edition*. Madrid: World Tourism Organization.

WEF (2020). *The Travel & Tourism Competitiveness Report 2019*. Geneva: World Economic Forum.

Wester, P., Mishra, A., Mukherji, A., & Shrestha, A.B. (Eds.). (2019). *The Hindu Kush Himalaya Assessment: Mountains, Climate Change, Sustainability and People*. Cham, Switzerland: Springer.

WTTC (2021). *2021 Annual Research: Key Highlights*. London: World Travel and Tourism Council.

# 2 The Himalaya
## Physiographic and Cultural Perspectives

*Dallen J. Timothy and Gyan P. Nyaupane*

## Introduction

The Himalaya are a vast landmass that stretches uninterruptedly about 2,500 km from east to west, and entails very narrow and steep terrain from south to north (200–400 km), with a total area of about 595,000 sq km (Encylopaedia Britannica, 2021). The variation in topography, altitude, precipitation, wind, and slope creates diversity in flora and fauna, and cultures, which makes the Himalayan region one of the most intriguing parts of the world, wielding considerable appeal in the global tourism marketplace. Tourism is an increasingly important part of Himalayan socioeconomic life, yet the physical geography presents significant challenges to tourism development, and the cultural environment is beset with strengths and weaknesses that have to be carefully planned and developed for the benefit of the Himalayan communities.

This chapter provides a broad overview of both the physical and cultural environments of the region, including the topographic and biotic systems, and a description of the cultural geography and cultural landscapes of the region. Its aim is to set the geographical, cultural, and tourism contexts for the concepts and cases that will follow throughout the book.

## Himalayan Physical Geography and the Natural Environment

The Himalaya are the highest mountains on the planet, home to more than 110 peaks that extend higher than 7,300 meters (24,000 feet) in elevation. The most famous of these, of course, is Mount Everest, which, according to the latest estimates (remeasured in 2020), is 8,848.86 meters (29,032 feet) in elevation, slightly higher than the last measures calculated by both China and Nepal in the 1950s (Khadka, 2020). Some scientists believed the world's tallest mountain might have shrunk slightly as a result of the 2015 Nepal earthquake, which was measured at 7.8 on the Richter scale and exacted severe damage and much loss of life throughout central Nepal (Ghose, 2017; Howard, 2015). However, the 2020 measurements suggest that the mountain is slightly taller than previously estimated. The reasons for this change could be both the advancement of more

DOI: 10.4324/9781003030126-3

precise measurement methods and geological phenomena. Since the Himalayan range formed as a result of the Indian tectonic plate penetrating into the Asian plate, and this movement still continues, Mount Everest and other mountains are still growing a few millimeters per year.

The world's top 50 highest mountains are all located in the Himalaya or in the ranges that comprise the Pan-Himalayan region (see Table 2.1 for information on the ten highest peaks). Many of the highest peaks are located on international borders, divided between adjacent states. This is an issue that has significant potential to raise sovereignty and environmental concerns, but because of their exceeding height and largely inaccessible locations, their binational divided status has not been a significant problem yet.

The Himalaya lie in the northern part of South Asia, bordering the Tibetan Plateau in the north (over 4,570 meters) and the plains of the Indian subcontinent in the south. In the west, the Sulaiman Mountains are located along the border of Pakistan and Afghanistan and extend north to the Pamir Knot, where the Himalaya begin in the west. In the east, the Himalaya extend to the Chin Hills and the Arakan Range, where India merges with Myanmar and Southeast Asia (Lew et al., 2015). The Himalayan range comprises four belts with distinct physiographic characteristics and geologic histories: the Outer Himalaya (900–1,200 meters), also known as Shiwalik (India) or Churia Hills (Nepal); the Lesser Himalaya (3,700–4,500 meters); the Great Himalaya (6,100–8,849 meters); and the Tibetan/ Trans-Himalaya (5,300–5,900 meters), which run parallel from east to west. Likewise, the range is often divided into western, central, and eastern sections (Alter, 2019; Encyclopaedia Britannica, 2021; Rajwar, 1993). The Great Himalaya range is the backbone of the entire system, with many of its peaks extending above the zone of perpetual snow.

Some 19 major rivers drain the Himalaya, including the Brahmaputra and the Indus, which are the largest in the region, each with a catchment basin of around 260,000 square kilometers. Other important drainage rivers include the Jhelum, Ravi, Beas, Chenab, and Sutlej in the western region in India and Pakistan—all

*Table 2.1* The world's ten highest peaks

| Mountain | Location | Elevation (meters) | Elevation (feet) |
|---|---|---|---|
| Mount Everest | Nepal/China | 8,849 | 29,032 |
| K2 | Pakistan/China | 8,611 | 28,251 |
| Kangchenjunga | Nepal/India | 8,586 | 28,169 |
| Lhotse | Nepal/China | 8,516 | 27,940 |
| Makalu | Nepal/China | 8,485 | 27,838 |
| Cho Oyu | Nepal/China | 8,188 | 26,864 |
| Dhaulagiri | Nepal | 8,167 | 26,795 |
| Manaslu | Nepal | 8,163 | 26,781 |
| Nanga Parbat | Pakistan | 8,126 | 26,660 |
| Annapurna | Nepal | 8,091 | 26,545 |

Source: compiled from multiple sources

part of the Indus system. The Ganges, Yamuna, Mahakali, Karnali, Rapti, Gandaki, Bagmati, and Koshi, rivers are part of the Ganges river system, while the Raidak, Manas, and Tisa rivers form part of the Brahmaputra fluvial system. Most major rivers flow through deep gorges; the Ganges-Brahmaputra system flows eastward, whereas the Indus system follows a northwesterly route (Encyclopaedia Britannica, 2021). Many of the rivers that drain the montane landscapes and feed the valleys are older than the mountains they pass through. Many scientists believe that the Himalaya were formed so slowly that the oldest rivers continued to flow in their channels and were able to carve their valleys relatively fast (Encyclopaedia Britannica, 2021).

The Himalayan glaciered landscapes are remarkable and one of their most notable features. The Khumbu Glacier drains the Everest region southward and is one of the most popular trekking routes to Everest. Other large glaciers feed the headwaters of the region's many rivers, but climate change appears to be shrinking them quickly. For example, the Khumbu moves approximately 30 cm per day, whereas the Baltoro Glacier is estimated to move approximately two meters per day (Encyclopaedia Britannica, 2021).

The geographical extent of the Himalaya has been a point of discussion in the West for more than a century (Bose, 1972; Burrard & Hayden, 1907; Douglas, 2021; Hund & Wren, 2018), but it is generally accepted that the mountain ranges abutting the Himalaya, such as the Pamirs, the Hindu Kush, and the Karakorum Mountains, are quasi-separate entities. Nonetheless, they form what is known as the broader Himalayan region, Hindu Kush Himalaya, or the Pan-Himalaya, because they are adjacent ranges that connect to the Himalaya, share common features, and have similar geological roots. As noted in the previous chapter, this book focuses on the Himalaya and their adjacent mountain regions in the west, to the range's easternmost extent in the Indian state of Arunachal Pradesh. Thus, the Pan-Himalaya form a natural region from Afghanistan's Wakhan Corridor to the northeastern states of India. Bangladesh and Myanmar are not normally included in the Himalayan region, but their hill provinces are connected with the Himalayan range, and both of these countries are part of the Hindu Kush Himalayan river system.

Geologically, the Himalaya are still relatively young (50–60 million years) and are continuing to form with the collision of the Eurasian and Indian tectonic plates (Searle, 2013). They are characterized by steep, jagged peaks, deep valleys and river gorges, and multitudes of alpine glaciers. Many different elevation zones comprise unique ecosystems and biomes with varying climates and vastly different flora and fauna regimes. While many of the highest peaks lie above the permanent snow line, the majority of the range, including the Outer and the Lesser Himalayan ranges, is below the snow line (Encyclopaedia Britannica, 2021). The tectonic movement, the extent of the snow line, the glacier equilibrium line, and the microclimates determined by local relief and topography determine the climate, the ecosystems and biomes, and therefore the human activities that take place within them.

Precipitation in the mountains varies by elevation, orographic lift, and location. The massive height and width of the Himalaya create a barrier against cold air masses from the north, protecting the southern plain, and through its orographic effect, it forces the moist summer monsoon clouds and winds to drop most of their moisture load before heading northward, providing plenteous rain and snowfall in northeast India, Bhutan, Nepal, northwest India, and Pakistan. However, it results in arid conditions in the Trans-Himalayan region, including the Mustang and Dolpa districts of Nepal, Ladakh union territory of India, and the Tibetan Plateau (Encyclopaedia Britannica, 2021). The average yearly rainfall on the southern part of the mountains ranges from 1,530 mm at Uttarakhand and Himachal Pradesh to 3,050 mm in parts of West Bengal, 5,000 mm at some stations in Sikkim, and in Arunachal Pradesh, the average is around 3,000 mm per year (Dhar & Nandargi, 2004; Shrestha et al., 2019). The Trans-Himalayan region, including Ladakh and the arid regions of Pakistan and Afghanistan, usually receive only 75–150 mm of precipitation per annum. Precipitation on the Tibetan Plateau ranges between 100 and 300 mm per year (Jiang & Ting, 2017).

Numerous fertile valleys pockmark the Himalayan ranges. This is where most of the population lives and where their cultures developed. It is also where most of the agriculture and subsistence lifestyles are carried out. Most Himalayan valleys are long and narrow, having been shaped by the fluvial systems that pass through them. The Tibetan Plateau, sometimes referred to as the 'Roof of the World', is an immense elevated plateau of more than 4,500 meters on the northern side of the mountains. Its geographical reach (2.5 million km²) includes most of China's Tibet Autonomous Region (TAR), as well as most of Qinghai province and marginal parts of the provinces of Sichuan and Yunnan. It also extends slightly into northwestern Nepal, northern Bhutan, and the Indian regions of Himachal Pradesh and Ladakh (Kolås & Thowsen, 2005; van Schaik, 2011). The plateau is home to many fluvial headwaters, including thousands of glaciers, and is nicknamed the 'third pole' because it retains the world's third largest accumulation of ice and snow after the South and North Poles (Yao et al., 2007). Climate change is a serious concern throughout the Himalaya, no more so than on the Tibetan Plateau. The region is becoming an increasingly busy laboratory for researching the effects of global warming (Gao et al., 2019; Qiao et al., 2019; Wester et al., 2019).

The Himalayan vegetation zones are determined by elevation, levels of precipitation, exposure to sun, and soil composition. These zones are usually categorized into alpine (including subalpine) (3,000 meters to the snow line), temperate (2,000–3,000), subtropical (1,000–2,000 meters), and tropical (less than 1,000 meters). The foothills in the eastern and central areas are home to tropical rainforests in Nepal, Bhutan, and India's Arunachal Pradesh (Mizono, 2016; Rau, 1974; Singh & Singh, 1987). The lower-elevation foothills of Bhutan, Nepal, and the Indian states of Sikkim, Uttarakhand, and Himachal Pradesh are home to subtropical vegetation (Banerjee & Bandopadhyay, 2016). The temperate zone in Nepal, Bhutan, and northwest India is home to various broadleaf deciduous and evergreen coniferous forests. There are relatively few trees in the alpine zone, as it lies primarily above the tree line. Small scrub vegetation, grasses, mosses, and lichens characterize the flora landscapes of the higher elevations. On the northern flanks of the Himalaya, in parts of Pakistan, India (Arunachal Pradesh, Ladakh, and

Jammu and Kashmir), and the Tibetan Plateau, cold deserts dominate with montane grasslands and alpine tundra-types of vegetation, including mosses, lichens, and alpine shrubs. Coniferous forests flank the outer margins of the Tibetan Plateau, but the largest portion of its territory is comprised of alpine deserts and grasslands (Chang, 1981).

Animal life is abundant and has played a central role in Himalayan livelihoods since the beginning of the Anthropocene. Several elevation-withstanding and cold-tolerant endemic species live above the tree line, but these are largely limited to spiders and other insects. The best-known native endangered species include one-horned rhinoceroses (*Rhinoceros unicornis*), Asian elephants (*Elephas maximus*), red pandas (*Ailurus fulgens*), Bengal tigers (*Panthera tigris tigris*), snow leopards (*Uncia uncia*), Himalayan *tahr* (*Heritragus jemlahicus*, Himalayan musk deer (*Moschus leucogaster*), gray wolf (*Canis lupus lupus*), black bears (*Ursus arctos*), Ganges River dolphins (*Platanista gangetica*), and various birds, fish, amphibians, and reptiles (Qamar et al., 2011; Rajeev et al., 2020). Different species occupy different climatic zones and ecosystems, including the rainforests and temperate zones.

Some animals have been domesticated over millennia for use as work animals and sources of nutrition. The Yak is perhaps the most pervasive and important domestic animal in the region. Yaks are a type of cattle that was domesticated from wild yaks native to the Himalaya and the Tibetan Plateau. They are commonly used as beasts of burden and pack animals, as well as a source of meat, milk, and fiber (Tamang, 2001, 2021; Wangchuk et al., 2013). Their dung is also used as fuel for heating and cooking, especially above the tree line where wood is unavailable. For thousands of years, humans have relied on yaks for survival, and the animals play an important role in regional folklore, sport, and celebratory events (Bhasin, 2011; Jina, 2009; Schuler, 1979; Wangchuk et al., 2013). Yaks are strong and hardy, and can navigate the difficult montane terrain. Their expanded lung capacity enables them to survive in low-oxygen elevations, and they are capable of efficiently processing nutrients from a scant diet of scrub and alpine grasses.

## *Tourism and the Physical Geography of the Himalaya*

The most obvious natural feature of the region is the mountain peaks. The Himalaya have long exuded a sense of awe and wonderment among the people of the world for their massive heights, unique cultures, and in some cases the 'off-limits' nature of regional geopolitics, which kept tourists at bay for many years until relatively recently (e.g., Bhutan, Nepal, and Tibet).

The mountaineering industry began in the early 1950s with the ascent of the world's highest peaks during the 1950s and into the 1960s, which brought notoriety to the men and women who scaled them and made the peaks a major target for global tourism. Of the ten highest peaks, Annapurna was the first to be scaled on June 3, 1950, by Maurice Herzog and Louis Lachenal. This was followed by the most famous climb (Mount Everest) by Edmund Hillary and Tenzing

Norgay in May 1953. Nanga Parbat was conquered on July 3, 1953, followed by K2 (July 31, 1954), Cho Oyu (October 19, 1954), Makalu (May 15, 1955), Kangchenjunga (May 25, 1955), Manaslu (May 9, 1956), Lhotse (May 18, 1956), and Dhaulagiri on May 13, 1960 (Sale et al., 2012). Other climbers challenged the natural elements in the 1980s and 2000s by ascending the peaks during the winter months, with the most recent winter ascent taking place on January 16, 2021, by a team of Nepali mountaineers, led by Mingma Gyalje Sherpa and Nirmal Purja, on K2 (Skolnick & Sharma, 2021).

Mount Everest continues to hold a prominent place in the Himalayan lore, and many people visit Nepal just to get a glimpse of the world's highest peak, if not attempt to try climbing it (Zurick, 2006). Everest has played a prominent role in the mystique of the Himalayan region and is the centerpiece of the trekking and mountaineering sectors. Nonetheless, mountaineering has become successful outside the Everest region in many other areas of the Himalaya where hiking, glacier climbing, rappelling, and trekking may be less onerous (Apollo, 2017). The Himalaya were pivotal in the development of mountaineering tourism, especially with regard to mountain collecting, or peak-bagging or high-pointing, as it is known in the industry (Apollo et al., 2021). Many adventurers seek to conquer as many summits as possible or achieve a goal associated with industry-recognized assemblages of peaks, such as the 'Eight Thousanders' (the 14 mountains greater than 8,000 meters in elevation) or the Seven Summits: the highest peaks on each continent. The Himalayan highlands of Bhutan, China, India, Nepal, and Pakistan have become a popular venue for undertaking these activities and a critical laboratory for studying them.

Mountain peaks, caves, lakes, and river headwaters are deemed extremely sacred among many Himalayan religions and have been revered as places of worship for thousands of years. Today, these natural features remain auspicious destinations for pilgrims of many faiths. Mount Kailash (6,638 meters), TAR, China, is revered by adherents of Jainism, Bon, Hinduism, and Buddhism, and has different symbolic spiritual values for each group. Tibetan Buddhists revere it as the center of the universe. In Hindu tradition, Kailash is the abode of Lord Shiva and a direct line to heaven. For Jains, it is venerated as the location where the faith's founder, Rishabhanatha, broke free from the cycle of life, death, and rebirth (Rubin Museum of Art, 2015; Sarao, 2014). Several thousand pilgrims of these faiths visit Mount Kailash each year, but its high elevation and remote location limit the number of pilgrim tourists who are able to visit (Chakrabarty & Sadhukhan, 2020). Several other montane pilgrimages venerate mountains as sacred abodes of deity or are otherwise closely connected to the physiographic landscapes of the mountain, including pilgrimages to Badrinath Temple (3,100 meters, Vishnu) and Kedarnath Temple (3,500 meters, Shiva), both in Uttarakhand, India. Likewise, the Vishnu temple of Muktinath in Nepal (3,800 meters) is one of the world's highest temples. It is devoted to Vishnu and is considered sacrosanct by Hindus and Buddhists and is visited by 30,000–40,000 pilgrims each summer (Basnyat, 2014; de Jong & Grit, 2019).

The Amarnath Cave Temple, associated with Lord Shiva, in India's Jammu and Kashmir is one of Hinduism's holiest shrines. Owing to its high altitude (3,888 meters), the locality is covered in snow most of the year and is only accessible for a short time each summer (Shah, 2013). During the brief six-week visitor season in July and August, several hundred thousand pilgrims visit en masse. Despite its popularity, this is known to be one of the most dangerous pilgrimages in the Himalaya because of altitude sickness, poor health, and lack of physical preparation among older pilgrims in general, and traffic accidents on dangerous montane roads (Basnyat, 2014).

Gosainkunda (4,380 meters) is an alpine freshwater lake in central Nepal, which draws several thousand pilgrims and other tourists each year for ritual bathing its healing waters. It is associated with Lord Shiva and Parvati, and is especially popular among Hindu pilgrims from Nepal and India during certain festivals, including *Janai Purnima*, which falls sometime in August. The lake and its surrounds are also very popular among trekkers in the Langtang Valley and along the Helambu-Dhunche hiking trail (Basnyat, 2014; Zafren, 2007).

Besides mountains, rivers have garnered the most tourism attention in this region. Rivers are worshipped as goddesses and are extremely sacred in Himalayan religious dogma and are therefore important pilgrimage destinations. Of India's seven holy rivers, four of them have Himalayan origins: the Ganges, Yamuna, Indus, and Saraswati rivers. The Saraswati is sometimes called the 'invisible river', because its location and course are currently unknown, but many believe it was an ancient river recorded in holy writ, which might presently exist underground or might have changed course. Others believe it to be a mythical or allegorical waterway (Prasad, 2017). The Ganges is the holiest river for Hindus and is venerated all along its 2,525-km course, from its Himalayan glacial origins to its confluences with other rivers and its delta in the Bay of Bengal. Rishikesh, India, is a sacred Hindu Himalayan destination because of its location on the upper Ganges River, where the river flows from the mountains into the north Indian plains. The city's original appeal was its sacred river location, but its reach is now far wider than its sacrosanct origins (Pinkney & Whalen-Bridge, 2018). Rishikesh has also become an important fitness/yoga and river-rafting destination. Many Himalayan rivers now serve a general tourism function beyond their original spiritual connotations.

## The Cultural Environment

The Himalaya are home to numerous ethnic and cultural groups, defined largely by the languages they speak (Barfield, 2010; Berreman, 1963; Bista, 2004; Dahal, 2020; Kalman, 2010; Malik, 2006) (Table 2.2). The Tibetan Himalaya and Great Himalaya are home primarily to different Tibetan people who speak Tibeto-Burman languages, whereas Kashmiri, Gujari, Gaddi, and other Indo-European cultures inhabit the Lesser Himalaya. Many diverse groups occupy different parts of the region, with some having been acclimatized to the highest elevations (e.g., Sherpas in Nepal), while others have remained in the lower altitudes as farmers and pastoralists (Berreman, 1963; Eraly, 2008).

*Table 2.2* Major ethnic groups in the Himalaya

| Country | Approximate Number of Ethnic Groups in the Country | Sample Ethnic Groups Populating the Pan-Himalaya | Estimated Population in the Entire Country |
|---|---|---|---|
| Afghanistan | 14 recognized groups | Uzbek<br>Pashai<br>Wakhi<br>Ishkashimi<br>Kyrgyz | 2 million<br>500,000<br>21,000<br>3,100<br>1,300 |
| Bhutan | 3 major, many minor | Ngalop<br>Nepali<br>Sharchop<br>Indigenous people<br>(e.g., Kheng, Monpa,<br>Lepcha, Drokpa) | 428,700<br>300,100<br>111,465<br>17,158 |
| China | 56 official | Uyghur<br>Tibetan<br>Tajik<br>Monba<br>Lhoba | 12.8 million<br>6.3 million<br>50,200<br>7,500<br>3,682 |
| India | More than 2,000 | Kashmiris<br>Dogra<br>Rajput<br>Nyishi<br>Bhutia | 6.8 million<br>2.5 million<br>2.2 million<br>250,000<br>60,300 |
| Nepal | 126 | Chhetri<br>Khas Brahmin/Bahun<br>Magar<br>Tharu<br>Tamang | 5.05 million<br>3.7 million<br>2.2 million<br>2.0 million<br>1.8 million |
| Pakistan | 6 major, many minor | Punjabi<br>Pashtun<br>Kashmiri<br>Kho<br>Balti | 106.5 million<br>36.7 million<br>353,064<br>332,200<br>200,000 |

Source: compiled from multiple sources

The primary traditional livelihoods of the region are agriculture and pastoralism, which entails herding domesticated livestock (e.g., cattle, sheep, goats, yaks, horses, and camels) on lands that possess enough vegetation for animal grazing. Some groups were traditionally more nomadic in character, moving from place to place without permanent settlements, whereas others lived in settlements and practiced farming and transhumance: the custom of moving livestock from their home base to grassy, higher elevations during the summer months.

Today, in the Himalaya, there is a wide mix of livelihoods that derive from the pastoral and agricultural heritage of the region. People in the lowlands and many valleys engage in growing vegetables, fruits, grains, and livestock. In the higher elevations, usually above 1,500 meters, transhumance continues to be a chief sub-sistence practice as villagers trek seasonally to the higher altitudes to graze their animals during the summer, returning to their villages for the winter (Chakraborty, 2017; Mitra et al., 2013; Namgay et al., 2013). These pastoral traditions are still the foundations of sustenance throughout the entire region and are better suited to high elevations, harsh environments, and rugged topography than settled agricul-ture. These transhumant communities and their livelihoods are at especially high risk from climate change and its effects on alpine vegetation, access, and water surplus or water scarcity (Aryal et al., 2013; Pandit, 2017).

The population of this region is highly diverse. As of the 2011 census, India was home to 705 scheduled tribes, which are the country's Indigenous people, whose status is formally acknowledged by national law (Timothy, 2021). The Indian states that are part of its Himalayan reach include Arunachal Pradesh, Assam, Himachal Pradesh, Jammu and Kashmir (Union Territory), Ladakh (Union Ter-ritory), Nagaland, Manipur, Meghalaya, Mizoram, Tripura, Sikkim, Uttarakhand, and West Bengal. All of these states and union territories are home to numerous tribal groups (see Table 2.3). The Indigenous people of India's Himalayan states are manifold and diverse (Ghurye, 1980; Sundar, 2016). There are 26 major tribes and over 100 sub-tribes in the state of Arunachal Pradesh alone, including the Singpo, Aka, Adi, Dafla, Nyishi, Mishmi, and many others.

Hinduism, Islam, and Buddhism are the dominant religions of the Himalaya, but others such as Christianity, Sikhism, and Jainism are also well represented (Table 2.4). The people in the Himalayan reaches of Pakistan and Afghanistan are predominantly Muslims, by more than a 95 percent majority. There is also a large Muslim population in India's Jammu and Kashmir (68.8%) and Ladakh (46.4%) union territories, as well as in the eastern states of Assam (34.2%) and West Ben-gal (27%). Islam is represented in all of India's Himalayan regions, as well as in Tibet. Nepal also has a sizable Muslim populace, which constitutes approximately 4.4 percent of the total population, particularly in the Terai region. Most Nepali

*Table 2.3* Scheduled tribal populations of the Himalayan states of India, 2011

| State | Scheduled Tribe Populations | Percentage of State Total Population |
|---|---|---|
| Arunachal Pradesh | 951,821 | 68.8% |
| Assam | 3,885,094 | 12.5% |
| Himachal Pradesh | 391,968 | 5.7% |
| Jammu & Kashmir (incl. Ladakh) | 1,492,414 | 11.9% |
| Nagaland | 1,710,612 | 86.5% |
| Manipur | 903,235 | 35.1% |
| Meghalaya | 2,555,974 | 86.2% |
| Mizoram | 1,036,201 | 94.4% |
| Tripura | 1,710,612 | 86.5% |
| Sikkim | 205,886 | 33.7% |
| Uttarakhand | 292,502 | 2.9% |
| West Bengal | 5,294,014 | 5.8% |

Source: Census of India (2011)

*Table 2.4* Main religions in the Himalayan countries

| Country | Primary Religion(s) | Percentage of Total National Population |
|---------|---------------------|------------------------------------------|
| Afghanistan | Islam | 99.7 |
| | Other (Sikhism, Hinduism, Bahá'i, etc.) | 0.3 |
| Bhutan | Buddhism | 75.3 |
| | Hinduism | 22.1 |
| | Other (Bon, Christianity, and Islam) | 2.6 |
| China | No religion | 52.2 |
| | Folk religion (including Taoism) | 21.9 |
| | Buddhism | 18.2 |
| | Christianity | 5.1 |
| | Islam | 1.8 |
| | Other (Judaism, Bon, Hinduism, etc.) | 0.8 |
| India | Hinduism | 79.8 |
| | Islam | 14.2 |
| | Christianity | 2.3 |
| | Sikhism | 1.7 |
| | Buddhism | 0.7 |
| | Jainism | 0.4 |
| | Other (Judaism, Bon, Zoroastrianism, etc.) | 0.9 |
| Nepal | Hinduism | 81.3 |
| | Buddhism | 9.0 |
| | Islam | 4.4 |
| | Kiratism | 3.0 |
| | Christianity | 1.4 |
| | Other (Prakriti, Bon, Jainism, etc.) | 0.9 |
| Pakistan | Islam | 96.3 |
| | Hinduism | 2.1 |
| | Christianity | 1.3 |
| | Other (Sikhism, Buddhism, Bahá'i, Zoroastrianism, etc.) | 0.3 |

Source: Central Intelligence Agency (2021)

Muslims arrived in the area during the 9th- and 12th-century Muslim invasions from what is now India (Nyaupane, 2009).

The regions and countries with the largest number of Buddhist adherents are Tibet (78.5), Bhutan (75.3%), Ladakh (39.7%), and Sikkim (27.4%). Nepal and Arunachal Pradesh have large Buddhist populations, and Buddhism is represented in all of the other Indian Himalayan states in smaller degrees. Hinduism is the dominant faith in Nepal (81.3%) and India's Assam (61.5%), Himachal Pradesh (95.2%), Manipur (41.4%), Tripura (83.4%), Sikkim (57.8%), Uttarakhand (83%), and West Bengal (70.5%), with sizable cohorts in Bhutan (25%), Arunachal Pradesh (29%), Jammu and Kashmir (28.8%), Ladakh (12.1%), and Meghalaya (11.5%). Hinduism has a significant minority presence in Pakistan and all of India's northeastern states but is negligible in Tibet and Afghanistan.

Although Christianity is represented in all Himalayan countries (practiced secretly in Afghanistan), it is most prominent in India. Christianity is the dominant religion in several of India's northeastern states, including Arunachal Pradesh (30.3%), Nagaland (87.9%), Manipur (41.3%—on par with Hinduism), Meghalaya (74.6%), and Mizoram (87.2%) (Census of India, 2011). The spread of Christianity in India's northeast occurred primarily during the mid-20th century around the time of India's independence and has continued unabated into the 21st century. The tribal communities of several northeastern Himalayan states have almost entirely converted to Christianity, a trend that has accelerated since the 1980s. The change from animist practices to Christianity resulted from missionary efforts, the British colonizers turning much of the education system over to the Catholic Church, and widespread person-to-person conversions within the tribal communities in recent decades (Frykenberg, 2008).

Besides the four mainstream faiths, other religions are represented, including Jainism, Sikhism, Bon, and Kirat Mundhum. There is a substantial Sikh presence in India's Uttarakhand, Jammu and Kashmir, and Ladakh. Bon, an indigenous pre-Buddhist set of animist beliefs, is practiced by 12.5 percent of Tibet's population, 0.04 percent of Nepal's population, and a significant number of people in India and Bhutan (Phuntsho, 2014; Singh, 2011). All four countries are home to hundreds of Bon monasteries. Kirat Mundhum is an Indigenous religion practiced by Kirat ethnic groups in Nepal, Sikkim, and the Darjeeling area of West Bengal which, like other animist and shamanist belief systems, focuses on the oneness of humans and their natural environments, the deification of nature, and ancestor worship (Lungeli, 2020).

Together with religion, language is one of the primary markers of cultural identity and is often a surrogate measure of ethnicity and nationhood. Over 300 languages, one-sixth of all human languages, are spoken in the Himalayan region (Turin, 2019). Many of these languages are endangered. Table 2.5 lists the major languages spoken in the Himalayan countries, most of which fall within the Tibeto-Burman and Indo-European language families, although there are significant pockets of Altaic (Turkic), Austro-Asiatic, Dravidian, and Daic (Kra-Dai) languages. Sinitic (e.g., Mandarin) and non-Kurukh Dravidian (e.g., Tamil) languages are also widely spoken by non-indigenous populations in the area (Guneratne, 2010; Savada, 1993). At least two language isolates, or languages not related to any other language, are also spoken. These are Burushaski, spoken by approximately 112,000 in far northern Pakistan and a few hundred in India's Jammu and Kashmir, and Kusunda, which is spoken by fewer than 100 people in central and western Nepal (Tikkanen, 1995; Yadava, 2014). Most of the languages spoken in the Himalayan region are derived from Sanskrit, one of the most ancient and sophisticated languages in the world that has been used in writing ancient Hindu and Buddhist scriptures and is still being used in linguistic, arts, philosophy, and religious ceremonies. Buddhist and Hindu manuscripts, such as Vedas, Upanishad, Gita, and Hitopadesh, were written in Sanskrit over 2,500 years ago (Eagle, 1999). These scriptures are important documents not just for religious purposes but also to learn about the arts, music, archeology, history, medicine, yoga and meditation, and the lives of people of who lived in the region thousands of years ago. There is a new language revival movement to preserve Sanskrit and

*Table 2.5* Examples of languages and dialects spoken in the Himalayan countries

| Country | Approximate Number of Languages in the Country | Languages Spoken in the Himalaya | Language Family | Estimated Number of Native Speakers* |
|---|---|---|---|---|
| Afghanistan | 74 | Dari | Indo-European | 5 million |
| | | Ishkashimi | Indo-European | 2,500 |
| | | Kati | Indo-European | 37,000 |
| | | Pashai | Indo-European | 100,000 |
| | | Sanglechi | Indo-European | 2,200 |
| | | Shughni | Indo-European | 50,000 |
| | | Tajik | Indo-European | 8 million |
| | | Wakhi | Indo-European | 30,000 |
| Bhutan | 25 | Brokpa | Sino-Tibetan | 5,000 |
| | | Bumthangkha | Sino-Tibetan | 30,000 |
| | | Chocangacakha | Sino-Tibetan | 20,000 |
| | | Dzalakha | Sino-Tibetan | 15,000 |
| | | Dzongkha | Sino-Tibetan | 171,000 |
| | | Khengkha | Sino-Tibetan | 50,000 |
| | | Lakha | Sino-Tibetan | 8,000 |
| | | Lepcha | Sino-Tibetan | 15,000 |
| | | Nepali | Indo-European | 650,000 |
| | | Nyenkha | Sino-Tibetan | 8,700 |
| | | Tshangla | Sino-Tibetan | 157,000 |
| China | 302 | Amdo (Tibetan) | Sino-Tibetan | 101,000 |
| | | Dbus Tsang Tajik | Sino-Tibetan | 1.2 million |
| | | Dbus Tsang (Tibetan) | Sino-Tibetan | 1.4 million |
| | | Groma | Sino-Tibetan | 27,000 |
| | | Khams (Tibetan) | Sino-Tibetan | 39,600 |
| | | Mandarin Chinese | Sinitic | 10.1 million |
| | | Uyghur | Turkic | |
| India | 19,000–20,000 | Adi | Sino-Tibetan | 1 million |
| | | Balti | Sino-Tibetan | 1.3 million |
| | | Bhotiya | Sino-Tibetan | 70,000 |
| | | Chakma | Indo-European | 230,000 |
| | | Dogri | Indo-European | 2.6 million |
| | | Garhwali | Indo-European | 2.5 million |
| | | Hindi | Indo-European | 425 million |
| | | Kashmiri | Indo-European | 6.8 million |
| | | Kumaoni | Indo-European | 2 million |
| | | Ladakhi | Sino-Tibetan | 120,000 |
| | | Lepcha | Sino-Tibetan | 50,000 |
| | | Nepali | Indo-European | 2.9 million |
| | | Nishi | Sino-Tibetan | 299,000 |
| | | Punjabi | Indo-European | 31.1 million |
| | | Purgi | Sino-Tibetan | 94,000 |
| | | Sikkimese | Sino-Tibetan | 70,000 |

*(Continued)*

*Table 2.5* (Continued)

| Country | Approximate Number of Languages in the Country | Languages Spoken in the Himalaya | Language Family | Estimated Number of Native Speakers* |
|---------|------------------------------------------------|-----------------------------------|-----------------|--------------------------------------|
| Nepal | 123 | Bajjika (Western Maithili) | Indo-European | 800,000 |
| | | Bhojpuri | Indo-European | 1.6 million |
| | | Gurung | Sino-Tibetan | 325,600 |
| | | Limbu | Sino-Tibetan | 343,600 |
| | | Magar | Sino-Tibetan | 788,500 |
| | | Maithili | Indo-European | 3.1 million |
| | | Nepali | Indo-European | 11.8 million |
| | | Newari | Sino-Tibetan | 846,600 |
| | | Sherpa | Sino-Tibetan | 114,830 |
| | | Tamang | Sino-Tibetan | 1.4 million |
| | | Thakali | Sino-Tibetan | 5,242 |
| | | Tharu | Indo-European | 1.5 million |
| Pakistan | 74 | Balti | Sino-Tibetan | 379,000 |
| | | Burushaski | Language isolate | 112,000 |
| | | Dogri | Indo-European | 1.6 million |
| | | Kalami | Indo-European | 100,000 |
| | | Kashmiri | Indo-European | 353,000 |
| | | Khowar | Indo-European | 290,000 |
| | | Kohistani | Indo-European | 200,000 |
| | | Lahnda | Indo-European | 42 million |
| | | Pahari-Potwari | Indo-European | 3.8 million |
| | | Shina | Indo-European | 600,000 |
| | | Wakhi | Indo-European | 20,000 |
| | | Yidgha | Indo-European | 6,200 |

Source: compiled from multiple sources

* Note: source dates range from 1995 to 2019

other languages in the region. Uttarakhand and Himachal Pradesh, two Himalayan states of India, recently recognized Sanskrit as their second official language.

Most of the languages spoken in the region use the scripts derived from the Brahmi script. Even the Tibetan language uses a derived written version of the Brahmi script. The most common form of the script in the Central and Eastern Himalaya now is Devnagari, the fourth most widely used writing system in the world. The Devnagari script, which is used for over 120 languages in the Himalayan region, was developed from the Brahmi script in the 1st to 4th centuries. The Western Himalaya, including parts of India, Pakistan, and Afghanistan currently use Arabic script. Urdu is the official language of Pakistan, which was derived from Arabic, Persian, Sanskrit, and Hindi languages in northern India in the 12th century.

### Tourism and the Cultural Heritage of the Himalaya

Owing to the largely nomadic lifestyles of many High Himalayan peoples, there are relatively few tangible remains of certain ethnic groups, while for others, the material culture is rather prominent with ancient homes and villages remaining home to today's descendent populations. In these cases, settlement patterns and vernacular architecture are among the most visible and pervasive elements of cultural heritage in the region and form the foundations of much of the area's cultural tourism appeal. Intangible heritage is equally important as a tourism asset. Religion is one of the most powerful forces that mold a society and culture. The varied religious traditions of the Himalaya have resulted in many unique funerary rituals, social customs, holidays, celebrations, and festivals, which all lend color and diversity to the intangible cultural landscape of the region.

The tangible and intangible religious heritage of Buddhism, Hinduism, Islam, and multifarious other faiths coalesce into a remarkable permanent imprint in the Himalaya, some of which creates the region's unique and most prevalent cultural tourism appeal (Timothy, 2021). In Bhutan, for instance, the cultural focus of tourism are the ever-present *dzongs*—Buddhist fortress-monasteries that serve both religious and governance functions. These are an important component of tour packages in Bhutan (Nyaupane & Timothy, 2010), and similar structures are found in Tibet, including the world-famous Potala Palace in Lhasa, the traditional home of the Dalai Lama, the holiest site of Tibetan Buddhism, and a UNESCO World Heritage Site (Amundsen, 2001; Shepherd, 2006). Likewise, the birthplace of Buddha in Lumbini, Nepal, located on the edge of the Outer Himalaya, is one of Buddhism's most venerated locations and draws hundreds of thousands of Buddhist pilgrims and non-pilgrim tourists each year, and is also a World Heritage Site (Nyaupane, 2009; Nyaupane et al., 2015; Rai, 2020).

As noted earlier in the physical geography section, the region is famous for its highland pilgrimages. Although mountains and rivers form the geophysical foundations of many pilgrimages because these are the physical manifestations of sacred space, multitudes of temples built along the rivers, in caves, and in mountain valleys are a crucial part of the cultural landscape of the region. Famous Hindu and Buddhist temples in Kathmandu lend much tourism appeal to the city and are part of the Kathmandu Word Heritage Site. Badrinath Temple in Uttarakhand, India, one of the four Dhamas (most important pilgrimage sites for the Hindus) draws hundreds of thousands of devotees and curiosity-seekers each year. The Tiger's Nest monastery in Bhutan is one of that country's most prominent cultural sites. Lamayaru is one of the oldest Buddhist sites in Ladakh, and the Jokhang is the main temple in Lhasa, Tibet (Rubin Museum of Art, 2015).

Indigenous architecture and settlement patterns are extremely unique and strongly reflect a shortage of timber, an abundance of stone, harsh climates, ancient social hierarchies, and family clan relationships (Handa, 2009). These vernacular landscapes and the religious sites are among the region's most important cultural heritage attractions. Some valleys, such as Kathmandu, on the other hand, had a shortage of stone; so residents built wooden structures instead, which allowed them to create a unique Pagoda architecture that has been widely used in the Himalayan region, including Tibet, today.

From an intangible perspective, faith and life cycle commemorations, harvest abundances, the annual return of transhumant herders and their flocks, and hunting successes have spurred celebratory events and festivals throughout the region

(Sharma & Sharma, 1998). Folk religion is the foundation of many celebrations of localized deity, in addition to the holidays associated with faith-wide celebrations (Chophy, 2019; Sinha, 2006). Harvest, hunting, and pastoral blessings are celebrated throughout the region (Sharma et al., 2016). Himalayan food traditions are becoming more common in other parts of the Indian subcontinent (Negi et al., 2020; Tanwar et al., 2018; Timothy, 2021).

## Conclusion

The Himalaya Mountains are home to some of the most dramatic montane topography on the earth and some of the most fascinating living cultures. The region is vast, diverse, and a crucial laboratory for understanding many global forces affecting mountain ecosystems and livelihoods today. The Himalayan environment and its inhabitants are sensitive to processes such as climate change, natural disasters, globalization, transportation and infrastructure development, and technological change.

The Himalayan region provides a great diversity of topography, climate, fauna, flora, and culture from northeastern India to Afghanistan in the west, from the Outer Himalayan fluvial and delta plains in the south to the highest mountains and the Tibetan plateau in the north. The region has some of the most densely populated urban areas and the sparsest human settlements on the planet. However, the physical and cultural diversity of the region is often ignored by oversimplification, generalizations, and Western-centric development (Wester et al., 2019).

The current COVID-19 pandemic has brought significant challenges to the region, particularly in terms of tourism, but it has also raised danger for the protection of the region's fragile ecosystems, economies, and cultures. From a tourism perspective, the pandemic has reduced the economic well-being of a region that was already on the global margins. Tourism was completely shut down at the onset of the COVID-19 pandemic, but there is an opportunity to enhance regional and domestic tourism to revive the tourism industry. However, most of the tourism infrastructure is built to meet the needs of Western tourists. The pandemic therefore may also be used as an opportunity to develop more sustainable and regenerative tourism that is less dependent on Western tourists.

The Himalaya are shared by five countries. Nepal and India control most of the southern part of Himalaya. Bhutan, a small kingdom, controls a small portion in the Eastern Himalaya, and Pakistan controls a part of the Western Himalaya. China controls the Tibetan Plateau in the north of the Himalaya. Their governance and policies, unfortunately, are not consistent in terms of managing natural and cultural resources and tourism. For example, just as the 2021 climbing season was set to begin in May 2021, and against increasing numbers of COVID-19 cases in the Everest Base Camp in Nepal, China announced its intensions to erect a 'separation line' or border barrier/fence across the small summit of the mountain as a means of preventing climbers from the Chinese and Nepali sides from mingling on the mountaintop to prevent spreading the coronavirus from Nepal into China. A team of Tibetan climbing guides has been assigned to set up the barrier. Climbers on the north side (China) of Everest would be prohibited from crossing the borderline or coming into contact with people or items from the Nepalese side (Reuters, 2021). This type of unilaterally created barrier ruins the sensitive and unique landscape of the Himalaya. Further, there is also a race among powerful countries, including China and India, to develop the Himalaya through transportation, hydropower, and tourism for economic prosperity without any cross-border collaboration, or environmental and cultural considerations.

## References

Alter, S. (2019). *Wild Himalaya: A Natural History of the Greatest Mountain Range on Earth*. New Delhi: Aleph.

Amundsen, I.B. (2001). On Bhutanese and Tibetan dzongs. *Journal of Bhutan Studies*, 5, 8–41.

Apollo, M. (2017). The true accessibility of mountaineering: The case of the high Himalaya. *Journal of Outdoor Recreation and Tourism*, 17, 29–43.

Apollo, M., Mostowska, J., Maciuk, K., Wengel, Y., Jones, T.E., & Cheer, J.M. (2021). Peak-bagging and cartographic misrepresentations: A call to correction. *Current Issues in Tourism*, 24(4), 1970–1975.

Aryal, S., Cockfield, G., & Maraseni, T.N. (2013). Vulnerability of Himalayan transhumant communities to climate change. *Climatic Change*, 125, 193–208.

Banerjee, A., & Bandopadhyay, R. (2016). Biodiversity hotspot of Bhutan and its sustainability. *Current Science*, 110(4), 521–527.

Barfield, T. (2010). *Afghanistan: A Cultural and Political History*. Princeton: Princeton University Press.

Basnyat, B. (2014). High altitude pilgrimage medicine. *High Altitude Medicine & Biology*, 15(4), 434–439.

Berreman, G.D. (1963). Peoples and cultures of the Himalayas. *Asian Survey*, 3(6), 289–304.

Bhasin, V. (2011). Pastoralists of Himalayas. *Journal of Human Ecology*, 33(3), 147–177.

Bista, D.B. (2004). *People of Nepal*. Kathmandu: Ratna Pustak Bhandar.

Bose, S.C. (1972). *Geography of the Himalaya*. Ann Arbor: University of Michigan.

Burrard, S.G., & Hayden, H.H. (1907). *A Sketch of the Geography and Geology of the Himalaya Mountains and Tibet*. Calcutta: Government of India.

Census of India (2011). *Census of India, 2011*. New Delhi: Ministry of Home Affairs. Retrieved from https://censusindia.gov.in/

Central Intelligence Agency (2021). The World Fact Book. Retrieved August 12, 2021, from https://www.cia.gov/the-world-factbook/

Chakrabarty, P., & Sadhukhan, S.K. (2020). Destination image for pilgrimage and tourism: A study in Mount Kailash region of Tibet. *Folia Geographica*, 62(2), 71–86.

Chakraborty, M. (2017). Transhumance of Indian Himalayas in transition: A prospect for social research. *Journal of the Anthropological Survey of India*, 66(1/2), 117–123.

Chang, D.H.S. (1981). The vegetation zonation of the Tibetan Plateau. *Mountain Research and Development*, 1(1), 29–48.

Chophy, G.K. (2019). Agents of the godlings: An ethnographic account of folk Hinduism in Himachal Pradesh. *Journal of the Anthropological Survey of India*, 68(1), 123–134.

Dahal, B.P. (2020). *Culture in Nepal—Himalaya*. Riga, Latvia: Omniscriptum.

de Jong, M., & Grit, A. (2019). Implications for managed visitor experiences at Muktinath temple (Chumig Gyatsa) in Nepal: A netnography. In M. Griffiths & P. Wiltshier (Eds.), *Managing Religious Tourism* (pp. 135–143). Wallingford: CABI.

Dhar, O.N., & Nandargi, S. (2004). Rainfall distribution over the Arunachal Pradesh Himalayas. *Weather*, 59(6), 155–157.

Douglas, E. (2021). *Himalaya: A Human History*. New York: W.W. Norton.

Eagle, S. (1999). The language situation in Nepal. *Journal of Multilingual and Multicultural Development*, 20(4–5), 272–327.

Encyclopaedia Britannica (2021). Himalayas. Retrieved April 30, 2021, from www.britannica.com/place/Himalayas

Eraly, A. (2008). *India: People, Place, Culture, History*. London: Dorling Kindersley.

Frykenberg, R.E. (2008). *Christianity in India: From Beginnings to the Present*. Oxford: Oxford University Press.

Gao, J., Yao, T., Masson-Delmotte, V., Steen-Larsen, H.C., & Wang, W. (2019). Collapsing glaciers threaten Asia's water supplies. *Nature*, 565, 19–21.

Ghose, T. (2017). Did mount Everest really shrink? Scientists measure peak again. *Live Science*, January 25, 2018. Retrieved May 3, 2021, from www.livescience.com/57621-scientists-measure-mount-everest-again.html

Ghurye, G.S. (1980). *The Scheduled Tribes of India*. New Brunswick, NJ: Transaction Books.

Guneratne, A. (Ed.). (2010). *Culture and the Environment in the Himalaya*. London: Routledge.

Handa, O. (2009). *Himalayan Traditional Architecture with Special Reference to the Western Himalayan Region*. New Delhi: Rupa & Company.

Howard, B.C. (2015). Did Nepal earthquake change mount Everest's height? *National Geographic*, April 28. Retrieved April 5, 2021, from www.nationalgeographic.com/science/article/150428-everest-height-nepal-earthquake-geology-science

Hund, A.J., & Wren, J.A. (Eds.). (2018). *The Himalayas: An Encyclopedia of Geography, History, and Culture*. Santa Barbara, CA: ABC-CLIO.

Jiang, X., & Ting, M. (2017). A dipole pattern of summertime rainfall across the Indian subcontinent and the Tibetan Plateau. *Journal of Climate*, 30(23), 9607–9620.

Jina, P.S. (2009). *Cultural Heritage of Ladakh Himalaya*. New Delhi: Kalpaz Publications.

Kalman, B. (2010). *India: The People*. New York: Crabtree.

Khadka, N.S. (2020). Mt Everest grows by nearly a metre to new height. *BBC News*, 8 December. Retrieved April 3, 2021, from www.bbc.com/news/world-asia-55218443

Kolås, Å., & Thowsen, M.P. (2005). *On the Margins of Tibet: Cultural Survival on the Sino-Tibetan Frontier*. Seattle: University of Washington Press.

Lew, A.A., Hall, C.M., & Timothy, D.J. (2015). *World Regional Geography: Human Mobilities, Tourism Destinations, Sustainable Environments* (2nd ed.). Dubuque, IA: Kendall-Hunt.

Lungeli, D. (2020). Human embeddedness with nature: An ecocritical reading of some Mundhums in Limbu culture. *The Outlook*, 11, 14–23.

Malik, I.H. (2006). *Culture and Customs of Pakistan*. Westport, CT: Greenwood Press.

Mitra, M., Kumar, A., Adhikari, B.S., & Rawat, G.S. (2013). A note on transhumant pastoralism in Niti Valley, Western Himalaya, India. *Pastoralism: Research, Policy and Practice*, 3, article 29.

Mizono, K. (2016). The distribution and management of forests in Arunachal Pradesh, India. In R.B. Singh & P. Prokop (Eds.), *Environmental Geography of South Asia: Contributions Toward a Future Earth Initiative* (pp. 189–208). Tokyo: Springer.

Namgay, K., Millar, J., Black, R., & Samdup, T. (2013). Transhumant agro-pastoralism in Bhutan: Exploring contemporary practices and socio-cultural traditions. *Pastoralism: Research, Policy and Practice*, 3, article 26.

Negi, A., Kumari, D., Kukreti, R., & Dani, R. (2020). An analytical study of Uttarakhand cuisine. *Sodh Sanchar Bulletin*, 10, 49–53.

Nyaupane, G.P. (2009). Heritage complexity and tourism: The case of Lumbini, Nepal. *Journal of Heritage Tourism*, 4(2), 157–172.

Nyaupane, G.P., & Timothy, D.J. (2010). Power, regionalism and tourism policy in Bhutan. *Annals of Tourism Research*, 37(4), 969–988.

Nyaupane, G.P., Timothy, D.J., & Poudel, S. (2015). Understanding tourists in religious destinations: A social distance perspective. *Tourism Management*, 48, 343–353.

Pandit, M.K. (2017). *Life in the Himalaya: An Ecosystem at Risk*. Cambridge, MA: Harvard University Press.

Phuntsho, K. (2014). *The History of Bhutan*. London: Haus.

Pinkney, A.M., & Whalen-Bridge, J. (Eds.). (2018). *Religious Journeys in India: Pilgrims, Tourists, and Travelers*. Albany: State University of New York Press.

Prasad, R.U.S. (2017). *River and Goddess Worship in India: Changing Perceptions and Manifestations of Sarasvati*. London: Routledge.

Qamar, F.M., Ali, H., Ashraf, S., Daud, A., Gillani, H., Mirza, H., & Rehman, H.U. (2011). Distribution and habitat mapping of key fauna species in selected areas of western Himalaya, Pakistan. *Journal of Animal and Plant Sciences*, 21(2), 396–399.

Qiao, B., Zhu, L., & Yang, R. (2019). Temporal-spatial differences in lake water storage changes and their links to climate change throughout the Tibetan Plateau. *Remote Sensing of Environment*, 222, 232–243.

Rai, H.D. (2020). Buddhism and tourism: A study of Lumbini, Nepal. *Journal of Tourism & Hospitality Education*, 10, 22–52.

Rajeev, S., Pandey, V.V., & Abhishek, Y. (2020). Nag Tibba the beauty trail of lesser Himalayas of Uttarakhand with special reference to fauna and flora conservation. *Flora and Fauna*, 26(1), 3–10.

Rajwar, G.S. (1993). *Garhwal Himalaya: Ecology and Environment*. New Delhi: Ashish.

Rau, M.A. (1974). Vegetation and phytogeography of the Himalaya. In M.S. Mani (Ed.), *Ecology and Biogeography in India* (pp. 247–280). Dordrecht: Springer.

Reuters (2021). China to create 'line of separation' at Everest summit on COVID fears. Retrieved May 9, 2021, from www.reuters.com/world/asia-pacific/china-create-line-separation-everest-summit-covid-fears-2021-05-09/.

Rubin Museum of Art (2015). Eight beautiful sacred places in the Himalayas. Retrieved May 9, 2021, from https://rubinmuseum.org/blog/sacred-spaces-in-the-himalayas

Sale, R., Jurgalski, E., & Rodway, G. (2012). *On Top of the World: The New Millennium*. Gloucester: Snowfinch.

Sarao, K.T.S. (2014). Pilgrimage to Mt Kailash: The abode of Lord Shiva. *Dialogue*, 16(2), 214–231.

Savada, A.M. (1993). *Nepal and Bhutan: Country Studies*. Washington, DC: Library of Congress.

Schuler, S. (1979). Yaks, cows and status in the Himalayas. *Contributions to Nepalese Studies*, 6(2), 65–72.

Searle, M. (2013). *Colliding Continents: A Geological Exploration of the Himalaya, Karakorum, & Tibet*. Oxford: Oxford University Press.

Shah, A.R. (2013). A sociology lens of pilgrimage tourism in Kashmir Valley: A case of holy Amarnath pilgrimage. *The Tibet Journal*, 38(3/4), 57–85.

Sharma, S.K., Gupta, N., Johnson, J.A., & Sivakumar, K. (2016). 'Fish festivals' in the Garhwal Himalaya: Conservation options amidst age-old practices. *Current Science*, 110(7), 1155–1156.

Sharma, S.K., & Sharma, U. (Eds.). (1998). *Social and Cultural Heritage of Sikkim and Bhutan*. New Delhi: Anmol.

Shepherd, R. (2006). UNESCO and the politics of cultural heritage in Tibet. *Journal of Contemporary Asia*, 36(2), 243–257.

Shrestha, S., Yao, T., & Adhikari, T.R. (2019). Analysis of rainfall trends of two complex mountain river basins on the southern slopes of the Central Himalayas. *Atmospheric Research*, 215, 99–115.

Singh, C. (Ed.). (2011). *Recognizing Diversity: Society and Culture in the Himalaya*. Oxford: Oxford University Press.

Singh, J.S., & Singh, S.P. (1987). Forest vegetation of the Himalaya. *The Botanical Review*, 53(1), 80–192.

Sinha, V. (2006). Problematizing received categories: Revisiting 'folk Hinduism' and 'Sanskritization'. *Current Sociology*, 54(1), 98–111.

Skolnick, A., & Sharma, B. (2021). How climbers reached the summit of K2 in winter for the first time. *New York Times*, January 19. Retrieved March 30, 2021, from www.nytimes.com/2021/01/19/sports/summit-k2-nepalese-sherpas.html

Sundar, N. (2016). *The Scheduled Tribes and Their India: Politics, Identities, Policies, and Work*. Oxford: Oxford University Press.

Tamang, J.P. (2001). Food culture in the Eastern Himalayas. *Journal of Himalayan Research and Cultural Foundation*, 5(3–4), 107–118.

Tamang, J.P. (2021). Himalayan food culture. In N. Kumar, G. van Driem, & P. Stobdan (Eds.), *Himalayan Bridge* (pp. 277–298). London: Routledge.

Tanwar, M., Tanwar, B., Tanwar, R.S., Kumar, V., & Goyal, A. (2018). Himachali dham: Food, culture, and heritage. *Journal of Ethnic Foods*, 5(2), 99–104.

Tikkanen, B. (1995). Burushaski converbs in their South and Central Asian areal context. In M. Haspelmath & E. König (Eds.), *Converbs in Cross-Linguistic Perspective* (pp. 487–528). Berlin: De Gruyter.

Timothy, D.J. (2021). Heritage and tourism: Alternative perspectives from South Asia. *South Asian Journal of Tourism and Hospitality*, 1(1), 35–57.

Turin, M. (2019). Concluding thoughts on language shift and linguistic diversity in the Himalayas: The case of Nepal. In S.K. Sonnatag & M. Turin (Eds.), *The Politics of Language Contact in the Himalaya* (pp. 163–176). Cambridge, UK: Open Book Publishers.

van Schaik, S. (2011). *Tibet: A History*. New Haven, CT: Yale University Press.

Wangchuk, D., Dhammasaccakarn, W., Tepsing, P., & Sakolnakarn, T. (2013). The yaks: Heart and soul of the Himalayan tribes of Bhutan. *Journal of Environmental Research and Management*, 4(2), 189–196.

Wester, P., Mishra, A., Mukherji, A., & Shrestha, A.B. (Eds.). (2019). *The Hindu Kush Himalaya Assessment: Mountains, Climate Change, Sustainability and People*. Cham, Switzerland: Springer.

Yadava, Y.P. (2014). Language use in Nepal. In National Planning Commission Secretariat (Ed.), *Population Monograph of Nepal, Volume 2: Social Demography* (pp. 51–72). Kathmandu: Central Bureau of Statistics.

Yao, T., Pu, J., Lu, A., Wang, Y., & Yu, W. (2007). Recent glacial retreat and its impact on hydrological processes on the Tibetan Plateau, China, and surrounding regions. *Arctic, Antarctic, and Alpine Research*, 39(4), 642–650.

Zafren, K. (2007). Pilgrimages in the high Himalayas. In A. Wilder-Smith, E. Schwartz, & M. Shaw (Eds.), *Travel Medicine* (pp. 279–283). Oxford: Elsevier.

Zurick, D. (2006). Tourism on top of the world. *Focus on Geography*, 48(4), 9–16.

# Part 2

# Tourism Issues and Challenges in the Himalayan Region

# 3 Himalayan Vulnerabilities and Resilience

## Climate Change and Natural Disasters, Shifting Physical Landscapes, and Development

*Gyan P. Nyaupane*

## Introduction

Home to unique physiography, geology, steep terrain, heavy monsoon rain, and high seismic activity, the Himalaya are vulnerable to both anthropogenic (including climate change) and natural disasters, such as glacier retreat, avalanches, flooding, landslides, droughts, earthquakes, and the destruction of ecosystems (Nyaupane & Chhetri, 2009; Vaidya et al., 2019). The region's disasters are escalated by climate change, globalization and rapid economic transformations, geopolitical issues, and more recently the COVID-19 pandemic. Other internal and external factors, such as population growth, overconsumption and overproduction, and human-induced environmental and social changes occurring at local, national, regional, and global levels make the people of the Himalaya and surrounding regions vulnerable to both natural and human forces (Nyaupane et al., 2014). The interactions between these complex drivers of change not only add uncertainty to the region but also have major implications globally (Wester et al., 2019). These changes directly impact the life-support base of more than half a billion people who consider the Himalaya their home, an additional 1.9 billion people who live in downstream basins, and over three billion people who rely on food produced in the river basins.

The impacts of disasters and changes on the lives and livelihoods of the Himalayan people have been sporadically reported, and each disaster is taken as a single event. Although the region is divided by national borders, the Himalaya are a grand mountain system, and the challenges faced by communities in the region are similar. This chapter takes a comprehensive approach to analyzing vulnerabilities and highlights various approaches to addressing them through more resilient and adaptive mechanisms. The chapter begins with a review of the literature on vulnerabilities, hazards, resilience, and adaptability, followed by Himalaya-specific characteristics that contribute to both vulnerability and resilience. The chapter connects these concepts and discusses various types of disasters and crises that have impacted the region in general and are more specific to tourism. Finally, the chapter proposes some short-term and long-term policy solutions that could be helpful in addressing the region's key issues.

DOI: 10.4324/9781003030126-5

## Vulnerability, Resilience, and Adaptability

Vulnerability, resilience, hazards, and adaptation are related concepts. Vulnerability is a function of the exposure and sensitivity to hazardous conditions, and the ability or capacity (resilience) to cope, adapt, or recover (Smit & Wandel, 2006). Many exposure and sensitivity factors make the communities of the Himalaya vulnerable. These include socioeconomic conditions, such as poverty, inequality, population growth, and political instability; human-induced physical conditions, such as unplanned development and poor infrastructure; and biophysical conditions, such as challenging physiography, seismic activity, steep topography, and remoteness, coupled with climate change. Exposure, sensitivity, and vulnerability lead to disastrous outcomes from hazards, such as floods, droughts, and avalanches.

The literature outlines four dimensions of vulnerability: physical, social, economic, and environmental (Vaidya et al., 2019). Physical vulnerabilities are related to physical characteristics, including remoteness, difficult topography, proximity to hazards, and design and quality of infrastructure. People in the Himalaya, in general, face poor quality infrastructure each day. According to the World Economic Forum (2014), the overall quality of infrastructure, such as roads, air transport, and electrical services in Bangladesh, Nepal, Pakistan, and India is lower than the global average. Social vulnerability refers to institutions, cultural values, human capacities, social capital, and other human characteristics. Himalayan communities have many social vulnerabilities, including economic inequality, social exclusion and marginalization, discrimination based on religion, caste, gender, age, disability, and sexual orientation (Bergstrand et al., 2015; Vaidya et al., 2019). Vulnerability is also dependent on the economic status of individuals, communities, and nations (Vaidya et al., 2019). Those with a stronger economic base can invest more in building infrastructure and can access emergency services more quickly (Vaidya et al., 2019). This is particularly important for the Himalayan countries, which are less resourceful economically and classified as Least Developed Countries (LDCs). There is even inequality within the countries between communities in urban centers and rural areas, mountains and plains, and distance or nearness from political and economic hubs. In general, the poverty rate is higher in the mountains and hills of the Himalaya than in the plains (Gioli et al., 2019). The mountain communities, particularly Indigenous communities, who are economically, socially, and politically marginalized, are even more vulnerable. Notably, Pandey et al. (2016) found that urban households have better access to available resources, including additional sources of income, compared to those in rural areas. Environmental aspects refer to the exploitation of natural resources, such as deforestation, mining, and air and water pollution, impacts of global environmental change, environmental regulations, and the works of natural resource management institutions (UNISDR & UNEP, 2007).

The concept of resilience, on the other hand, emerged as a potential development paradigm that allows individuals and communities to adapt to changing environments (Lew, 2014). Resilience is helpful to achieve sustainable development goals through the strengths that can help overcome vulnerability to change. The main concept of resilience refers to a system's ability to resist disturbance and maintain its basic function (Folke, 2006). Resilience is not just an outcome, but it is a process or ability that can be built through repeated pathways and mechanisms (Guo et al., 2018). Although vulnerability and resilience are conceptually related, high levels of vulnerability do not necessarily imply low levels of resiliency (Espiner & Becken, 2014). The research on resilience has been focused primarily on three paradigms, including ecological resilience, social resilience, and more integrated social-ecological systems (SESs) resilience (Wilson, 2013), the latter having gained more traction from a systems viewpoint.

For operationalization, resilience can be broken down into several domains, including social, institutional, economic, and ecological (Holladay & Powell, 2013). Social resilience can be gained from factors such as social capital, including bonding, bridging, and linking, which can be obtained from trust, knowledge sharing, and social networks (Folke et al., 2005). Institutional resilience depends on factors such as flexibility, adaptability, self-organization, collaborative learning, funding, and institutional mission, vision, and goals (Nyaupane et al., 2020). Economic resilience is built upon a range of economic opportunities to access revenue streams, alternative livelihoods, prevention of leakage, and economic growth or stability (Adger, 2000). Lastly, ecological resilience refers to an ecological system's ability to maintain itself by absorbing disturbance without changing its behavior (Gunderson, 2000; Holladay & Powell, 2013). Community-based resource management through community forestry and buffer zone programs in Nepal, for example, have promoted more resilient communities by supporting sustainable livelihoods (UNISDR & UNEP, 2007).

The adaptive capacity of an individual, community, or institution enhances its resilience. Adaptation refers to a process, action, or outcome in a system, ranging from a household to the global level, to better cope with, manage, or adjust to some changing conditions (Smit & Wandel, 2006). The essence of adaptation, therefore, is coping with change. Adaptation can be anticipatory, autonomous, or reactive based on timing and spontaneity (Smit et al., 2000; Smit & Wandel, 2006). Adaptation is often assessed using five factors: the community's social network, livelihood strategies, adjustment strategies, resource availability, and resource accessibility. Although these concepts can be applied in any environment, the Himalaya are different from other mountains, which requires concerted efforts to understand the issues the region is facing.

## Himalayan Characteristics

The Himalaya are different from plain environments and other regions of the world because of their inaccessibility, fragility, marginality, and diversity or heterogeneity (Jodha, 2001). Inaccessibility is caused by steep slopes, high altitudes, overall terrain conditions, and seasonal precipitation. The major manifestations of inaccessibility are poor communications, inadequate transportation, and limited mobility, which contribute to the remoteness of the Himalayan region (Jodha, 2001). For centuries, isolation and inaccessibility have contributed to cultural diversity and helped protect the natural environment, but it has become a major vulnerability factor today.

Limited carrying capacity as a result of sensitive biophysical and social-ecological systems adds to the fragility of the Himalayan region (Jodha, 2001; Nyaupane & Chhetri, 2009). The degree and scale of impacts of any disturbances in the Himalaya caused by fragility make the outcomes more damaging and irreversible. Fragility is a major limitation of development projects and certain types of tourism, as some activities can be undertaken only at a small scale, which reduces the carrying capacity of the tourism industry (Nyaupane & Chhetri, 2009). The scale of infrastructure development such as multiple-lane highways and big airports, dams, and large-scale resorts is not environmentally conducive in the Himalayan region. Fragility, therefore, makes the Himalayan environment and the communities' livelihood more vulnerable to changes, including natural and human-made disasters and climate change.

The Himalayan region is considered marginal geopolitically, socially, and economically. Being marginal to mainstream society, politics, and the economy, the livelihoods, culture, and traditions of Himalayan communities have been influenced and impacted by other dominant societies of the region and the West. A poor resource base and weak institutional foundations are among the main causes of marginality in the Himalaya (Nyaupane & Chhetri, 2009). Other factors contributing to marginality are physical isolation and remoteness, fragile and low-productivity resources, and human factors that include weak institutions that prevent participation in mainstream development activities (Jodha, 1991).

Extreme heterogeneity in terms of elevation, geology, edaphic conditions, orientation, and steepness of slopes has created diversity in biotic conditions. Mountains in general facilitate species isolation and the creation of new species (Körner & Ohsawa, 2005). The Himalayan region is one of the most diverse ecosystems among the global mountain biomes (Wester et al., 2019). Endemic flora and fauna have survived in the Himalaya by adapting to various conditions; particularly, altitude has caused both plant and animal species to migrate for adaptation for their survival. The biodiversity of the Himalaya is also an important asset for maintaining the gene pool and ethnobotany (Nepal & Chipeniuk, 2005). The biodiversity contributes to sustaining food systems, nutrition, health, and livelihoods (Xu et al., 2019). The region has ten major river systems which connect very high mountains to lower elevations, creating a diverse terrestrial and aquatic ecosystem. The physiographic diversity has also created cultural diversity, in terms of religion, language, traditions, and institutions. Heterogeneity or diversity itself is an important asset of the Himalaya for local community livelihoods, including tourism.

The Himalayan characteristics, both biophysical and social, also provide various opportunities, such as niches for specific products, activities, and services that offer a comparative advantage over plains and other regions of the world (Jodha, 2001). Mountains have long been considered wasteland and barriers to economic development, but now the very characteristics that were considered liabilities have made the region very lucrative for tourism, medicinal plant-based products, and hydropower. Some of the features that can be further harnessed are the production of off-season fruits and vegetables, medicinal plants, ornamental gardens, and niche tourism, such as dark-sky astrotourism and wellness tourism. These characteristics could potentially enhance the Himalayan communities' economic conditions and quality of life through the promotion of high-value and specialized tourism products, such as climbing, hiking, whitewater rafting, birdwatching, stargazing, cultural tourism, and pilgrimage.

## Disasters and Climate Change in the Himalaya

Disasters do not occur by accident; they are functions of hazards, exposure, and vulnerabilities (IPCC, 2012; UNISDR & UNEP, 2007). Although the two terms, 'disasters' and 'crises' are often used interchangeably to explain the disruption originated by natural or manmade disasters, they are distinct. Disasters are defined as "those situations where the event which disrupts the routine of the community concerned, and in response to which adjustments have to be made, is triggered externally" (Faulkner, 2001, p. 138), whereas crises are caused by the failure to adapt to ongoing changes or to the problems created by disasters (Faulkner, 2001). The human toll, property damage, and disruption of daily life are exacerbated by other factors such as institutions and other social systems, economy, environmental characteristics from the local to the global level (Ritchie & Jiang, 2019).

A recent report shows that between 2000 and 2010, 749 hazard events were reported in the Himalayan region, killing almost 400,000 people. These events reflected a 143% increase from 1990 to 2000 (Vaidya et al., 2019). Tourism destinations and amenities are often located in vulnerable areas, such as along rivers and lakes, and on mountain tops, which increases exposure and induces hazards (Faulkner, 2001). In general, tourism is more prone to disasters than other industries, as evidenced by major calamities including the current pandemic. The Himalayan region is particularly prone to earthquakes, and climate change-induced disasters, including floods, landslides, glacial lake outbursts, and avalanches.

### Earthquakes

The Himalayan range was formed from the collision of Indian and Asian plates. The collision continues and the region is still tectonically very active, which is the main reason for seismic activities in the Himalaya. The Indian plate is penetrating underneath the Eurasian plate at a rate of about 45 mm per annum and also rotating anticlockwise (Sella et al., 2002), which creates an arc shape as the Asian plate rises. The rate of rise is approximately 9 mm per year at Mount Everest. Being located in tectonically active zones, the region experiences far more earthquakes than lowland areas do (Wang et al., 2019).

Although there is little by way of seismic record in the Himalaya before 1800, the region is believed to have experienced one of the planet's largest earthquakes

(8.2Mw) in the Kumaon region of India and western Nepal in 1505 (Bilham, 2004). Within the last 100 years, the region suffered four major earthquakes, including the Nepal-India border earthquake in 1938 (8Mw), the Kashmir earthquake in Pakistan in 2005 (7.6Mw), the Sichuan earthquake in China in 2008 (7.9 Mw), and the Gorkha earthquake in Nepal in 2015 (7.8Mw). These earthquakes led to losses of hundreds of thousands of lives, destroyed property and infrastructure, and triggered other mountain hazards, including landslides, debris, and mudflow (Cui et al., 2008; Wang et al., 2019).

The 2015 earthquakes in Nepal cost over 9,000 lives, injured an additional 25,000 individuals, and directly impacted 8 million people. Tourist numbers decreased by 40% following the earthquake, and the total loss of direct tourism revenue was around USD 456 million for one year following the earthquake (Goldberg, 2015; ICIMOD, 2020). The 2015 earthquake also killed 89 international tourists from 18 countries (Beirman et al., 2018). The event severely impacted the tourism industry, as many tourists were either evacuated or forced to cancel their trips following the catastrophic events, and bookings dropped sharply for several months. Critical tourism infrastructures and attractions such as hotels, restaurants, museums, historic buildings, shopping centers, and roads were destroyed. Following the earthquake, the tourism industry faced several challenges in the long term, including search and rescue, security risk mitigation, public perceptions of risk and insecurity, travel advisories from key tourist origin countries, lack of travel insurance availability, and increased liability insurance coverage for tourism companies (Beirman et al., 2018). Similarly, China's Sichuan earthquake in 2017 resulted in the loss of millions of dollars in tourism revenue.

### *Climate Change and Its Effects*

According to Wester et al. (2019), although the region has very low per capita fossil fuel $CO_2$ emissions (one-sixth of the global average), the Himalaya have suffered considerably from climate change (Bürki et al., 2005; Harrison & Wallace, 2005). The vulnerability and hazards in the montane region have increased and intensified as a result of climate change (Wang et al., 2019). The region has experienced a steady increase in annual average temperatures since the mid-1970s. Studies show that the warming rate in the Himalaya over the last five decades has been 0.2°C per decade (Krishnan et al., 2019). Even if global warming is limited to 1.5°C by the end of the century, according to the Paris Agreement, the region will likely continue warming at least 0.3°C higher than the global average. The warming trend will be much higher (0.7°C higher than global average) in the northwest Himalaya (Krishnan et al., 2019).

Some of the major threats associated with climate change in the Himalaya are altering the hydrological cycle, mostly monsoon patterns with increased severity and frequency of storms, which directly trigger mountain hazards (Wang et al., 2019). For example, floods and flood-induced dangers such as landslides directly threaten the livelihoods of people, including loss of agricultural lands and livestock, and the destruction of tourism and other infrastructure.

The monsoon starts from the northeastern parts of South Asia and moves gradually westward. Most floods take place during monsoon season and directly impact tourism as the industry's seasonal changes in demand depend on monsoonal changes in a number of ways. First, fewer tourists arrive during the wet phase of monsoon between June and September. Heavy rains and cloud cover make nature-based tourism activities inconducive and decrease the aesthetic value of the landscapes by blocking the views of the mountains and affecting the visibility of the scenery (Gomez Martin, 2005). Physical changes in certain attractions make some activities impossible, such as sudden changes in fluvial water levels or floods, making rafting and other water-based adventure activities risky. Second, wet conditions cause a high level of discomfort in certain activities, such as hiking, which is not enjoyable in the rain (Gomez Martin, 2005). Third, access to destinations is limited during the monsoon season. Most of the region's small airstrips remain unpaved, and thus during monsoon season they become muddy or prone to erosion. Similarly, many mountain roads are washed away by landslides, erosion, and other mass wasting because of high and intense levels of rainfall (Nyaupane & Chhetri, 2009). Fourth, extreme events, such as floods, glacier lake outburst floods (GLOFs), and avalanches take the lives of tourists and their support staff, damage property, destroy ancient heritage sites and artifacts, destroy critical tourism infrastructure (e.g., resorts, viewpoints, airports), and public infrastructure (e.g., reservoirs, roads, bridges), and even trigger additional mountain hazards, including the destruction of montane ecosystems (Nyaupane & Chhetri, 2009).

Floods are the most common threats in the Himalaya; they account for 17% of casualties and 51% of property damage (Vaidya et al., 2019). Floods also induce landslides, which frequently block river courses, causing massive outbursts of water that wash away human settlements downstream. For example, in 1893, a landslide in the Western Himalaya blocked the Biraganga River, a tributary of the Ganges, for 10 months. Another landslide in India in 1969 blocked the Alkananda River for three months (Kala, 2014). Two of the major causes of human casualties from disasters in the state of Uttarakhand, India, are excessive rain and its accompanying landslides (Kala, 2014). Between 2010 and 2014, floods and landslides killed 182,721 people in Bangladesh, 158,974 people in China, 140,292 in India, 139,641 in Myanmar, 90,850 in Pakistan, 17,465 in Afghanistan, and 17,390 in Nepal. Altogether, 747,333 lives were lost in the Himalaya and their river basins during that period (Vaidya et al., 2019). In Uttarakhand, during the monsoon season of 2013, landslides and flooding caused by heavy rainfall killed more than 4,120 people, most of whom were pilgrims visiting the Badrinath and Kedarnatha temples (Kala, 2014). The area received over 340 mm of rain in a single day, which is almost four times above the daily normal monsoon rainfall (Kala, 2014). The event also impacted 1,800 villages; 2,500 families became homeless, 150 bridges were destroyed and 17,000 sq km road was damaged (Kala, 2014). Although heavy rainfall caused the event, unplanned and haphazard construction, mismanaged tourism, and related activities including intensive mining in this fragile ecosystem, were some of the reasons this natural disaster became a man-made crisis, increasing the intensity and magnitude of damage.

Global warming in the High Himalaya has begun to reduce snow cover, melt glaciers, and degrade permafrost (Wang et al., 2019). This is a source of concern since the greater Himalaya hold the largest mass of glaciers outside the polar regions (Xu et al., 2009). Mountain glaciers are highly sensitive to even minor changes in atmospheric temperatures. A time-series analysis of satellite images shows that, owing to the rapid melting caused by global warming, the region had lost almost a quarter of its total glacial area over the past 30 years (Bajracharya et al., 2020). As a result, the number of glacial lakes increased by 11% (Bajracharya et al., 2020). Alam and Regmi (2004) estimate that an increase of temperature by 3–4°C could result in the loss of 60–70% of snow from the Himalaya. Melting ice and snow expand glacial runoff, increasing the potential for GLOFs (Rai & Gurung, 2005).

Although there is no accurate record of GLOF events before the 1980s, one recent study shows that the region has experienced over 50 GLOFs, with more than half of them being recorded in Nepal (Bajracharya et al., 2020). The most recent GLOFs in the region include the 2013 GLOF in Uttarakhand, India; the 2015 Lemthang Tsho flood in western Bhutan; and overflows in several locations in Chitral, Pakistan in 2016. These GLOFs damaged major infrastructure, including hydropower dams and roads. Himalayan GLOFs are particularly damaging, more so than in other regions because they form at extreme altitudes and the rivers extend long distances downstream, affecting large geographic areas (Chen et al., 2007). The GLOF event of August 4, 1985, from Dig Tsho glacial lake in the Mount Everest region was another devastating example. The flow of water increased two to four times the magnitude of maximum monsoon flood levels. This catastrophe destroyed the Namche Small Hydel Electricity Project, which was at its completion phase to supply electricity to Namche Bazar, the gateway community to Mount Everest (Nyaupane & Chhetri, 2009).

Avalanches are also one of the biggest hazards in the Himalaya and are common during the winter months (Vaidya et al., 2019). Between 2000 and 2015, two million people were exposed to, and more than 153,000 people were directly affected by, avalanches in Afghanistan (World Bank, 2017). At least 124 people were killed in Afghanistan's Panjshir Province in 2015 as a result of 40 avalanches triggered by heavy snowfall, and in 2017, 137 people were killed in southern Afghanistan near the Pakistani border (World Bank, 2017). A catastrophic avalanche killed 16 Sherpas in the Everest region on April 18, 2014. On October 14 the same year, another avalanche initiated by an unanticipated blizzard claimed the lives of 43 people, including 21 trekkers in the Annapurna region (Nyaupane, 2015). Among all the countries in the Himalaya, Nepal is the most vulnerable because it is home to the longest and highest stretch of the Himalayan range. Wang et al. (2015) rank Nepal the fourth most climate-vulnerable country in the world.

Other potential climate change impacts in the Himalaya include ecosystem shifts, changes to traditional agriculture and food trade systems, increases in heat-related illnesses and vector-borne diseases, and impacts on trekking and mountaineering (Klein et al., 2004; Nyaupane et al., 2014; Nyaupane & Chhetri, 2009; Xu et al., 2019). Climate change has also led to the introduction of invasive plant species and pests into the ecosystem (Adhikari et al., 2020; Thakur et al., 2020). The disturbance on the ecosystem and loss of biodiversity severely affect communities and the entire economic system, including tourism (UNISDR & UNEP, 2007). Drought is another hazard in the western and northwestern regions of the Himalaya, including the Tibetan Plateau, Afghanistan, northern Pakistan, northwest India, and northwest Nepal (Vaidya et al., 2019). The whole region faces severe water shortages during the dry season.

In addition to these seismic and climate-related natural disasters, the region is facing pressure on its fragile natural resources, such as forests, rivers, and sensitive ecosystems, as a result of deforestation, overharvesting of non-timber forest products (e.g., medicinal plants), overgrazing, overfishing, illegal hunting, unsustainable mining, and unsustainable tourism (Wester et al., 2019). The region is very rich in biodiversity, including megafauna like rhinoceros, tiger, musk deer, pangolin, red panda, snow leopard, and many species of birds and reptiles, which have become threatened or endangered as a result of habitat loss, coupled with illegal poaching to meet the demand of wildlife products on the international market (Rupa Kumar et al., 2006; Wester et al., 2019). Himalayan communities have been using these resources for millennia, but recent commercialization and external demand, including by tourism, have put more pressure on the Himalayan ecosystems and natural resources.

## Making the Himalaya Region More Resilient

As noted earlier, many of the threats and catastrophes in the region occur naturally. However, many are human-caused by poorly planned infrastructure, deforestation, inappropriate land uses, unsustainable management, and mismanagement of resources. There are two approaches to mitigating disasters: addressing foundational issues at the conceptual level and operationalizing these through planning and policy-making. Both of these approaches are discussed in more depth in the following section.

Many Himalayan challenges can be mitigated through better land-use planning, which takes into account the mountain characteristics discussed earlier. For example, because of steep slopes and fragile topography, most of the land is not suitable for intensive agriculture or infrastructure development. Bhutan and Nepal protect more than 40% of the forests, which is encouraging but given the fragile mountain environment, more areas need to be protected. Other countries in the region need to prioritize the protection of the Himalayan ecosystems to reduce their vulnerability to natural disasters. Himalayan hazards and issues transcend national borders, as the consequences of activities upstream can impact the people living downstream in another country. Therefore, understanding Himalayan challenges, capitalizing on opportunities, and mitigating disasters can be better addressed by utilizing ecosystem-based approaches.

The montane fluvial systems play a very important role in water security, livelihoods, and economic development, including agriculture and tourism. The Greater Himalaya are known as Asia's water tower, because they are the main source of water for more than ten major rivers in Asia (Xu et al., 2019). However, many of these rivers and their tributaries flow through multiple countries, and their use and management are inconsistent. Most of the region's river systems are unmanaged or mismanaged. For example, the Brahmaputra, which is approximately 4,000 km long, originates in Tibet (near Mount Kailash) and flows through India and Bangladesh to the Bay of Bengal. Koshi, a major tributary of the Ganges, flows through Tibet, Nepal, and India. Massive flooding is largely an outcome of the mismanagement of rivers. The rivers are exploited at each level from their sources to their deltas. This not only threatens the ecosystem itself but the livelihoods of the people who rely on the rivers for drinking water, sanitation, irrigation, and spiritual purposes. The sustainable management of rivers, therefore, requires transfrontier collaboration for more equitable water use, water quality maintenance, and stronger management policies.

The International Center for Integrated Mountain Development (ICIMOD) is an example of such an organization that aims to enable sustainable and resilient development for improved and equitable livelihoods in the region through knowledge sharing and regional cooperation (ICIMOD, 2020). ICIMOD was established as a regional intergovernmental knowledge center for the eight member countries of the broader mountain region: Afghanistan, Bangladesh, Bhutan, China, India, Myanmar, Nepal, and Pakistan. The organization has recently coordinated the Hindu Kush Himalayan Monitoring and Assessment Programme (HIMAP) as an integrated science and policy initiative for regional cooperation in an effort to translate research into policy. However, ICIMOD is an NGO and lacks strong connections with national governments. Hence, there is a need for an intergovernmental organization, such as SAARC, a supranational alliance of South Asian states, which does not include China or Myanmar, but does include some countries outside of the Himalayan region, such as the Maldives and Sri Lanka.

At the foundational level, there needs to be better integration of Himalayan knowledge systems in policies and practices adopted for the broader region. Mountain communities have millennia of experience living in harsh environments. They are among the most innovative and resilient communities in the world. Through trial and error over generations, the people of the Himalaya, including its Indigenous communities, have developed mechanisms that have helped them adapt and become resilient. However, their knowledge is not adequately disseminated through formal institutional and official channels (Negi et al., 2017). In fact, much of their native knowledge is often ignored by government agencies and outside stakeholders who believe they know what is best for the mountain populations. Local knowledge needs to be integrated while building a resilience infrastructure and sustainable planning strategies (Negi et al., 2017).

At the operational level, the disaster risk reduction framework needs to consider four principle elements (Vaidya et al., 2019). First, building a strong knowledge base, including developing hazard maps for communities, real-time information systems, and danger warning systems can help reduce vulnerability to potential hazards. Data and information are essential for informed decision-making. The data should be shared beyond national boundaries. Although there have been some initiatives to inform the public about crises, such as landslides blocking rivers, and glacial lake outburst floods, the extant information and messaging system to date is very slow and ineffective.

Second, hazard-resilient critical infrastructure, such as water systems, irrigation canals, hotels, airports, and roads, should not be built in hazard-prone areas. In addition, sound planning and development means that these infrastructure elements should not increase an area's vulnerability by clearing forests and destabilizing slopes. Given the biophysical and socioecological complexity of the Himalaya, applications of the ideas developed for other regions of the world should be cautiously applied in this region. More investments need to be made in innovations that are more appropriate for local and regional contexts. For example, cable cars might be a better option instead of roads and bridges in some sensitive, high relief, and steep-sloped areas.

Third, it is important to establish locally and regionally grounded institutions that are resilient through capacity building and local investments rather than relying on outside resources prior to, during, and following disasters. Tackling the effects of calamities after the event is very costly; disaster risk reduction plans can help mitigate the aftereffects of disasters and should be a priority at national and local levels in all developmental and planning processes. Finally, the establishment of funds to provide relief, rehabilitation, and reconstruction through public and private partnerships for disaster insurance is essential for post-disaster mitigation.

A more specific tourism-related disaster mitigation plan is also needed for the region as a whole, and perhaps on a smaller scale, adapted to locality-specific conditions. Tourism is more international and global than many other industries, and is more vulnerable to natural disasters than many other economic sectors. That the scale and intensity of the impacts of disasters cross national borders highlights the need for international collaboration and coordination among governments and other stakeholders to respond effectively to cross-border crises (Kafle, 2017; McKercher & Chon, 2004). Since the challenges faced by the region transcend borders and the Himalaya is one region, more ecosystem-based approaches should be implemented, which requires regional coordination and collaboration among Himalayan countries.

## Conclusion

This chapter aims to provide an understanding of vulnerability and resilience in the Himalayan region. The area's unique biophysical conditions, or mountain characteristics, offer complex constraints and opportunities in the Himalaya. People who live in the mountains are experiencing rapid environmental, social, and economic changes with a higher intensity as a result of climate change, globalization, and geopolitics (Wester et al., 2019).

In the early twentieth century, the West's interest in the Himalaya derived from the region's majestic mountains, its unique cultures, and the myths and stories of Shangri-La and the illusive yeti. Although Himalayan communities are among the most innovative and resilient communities in the world, their adaptiveness and resiliency through their institutions, traditions and practices have been eroded by colonization and replaced by Western-oriented development. The Himalaya have been treated as a laboratory for the West in its experiments with tourism and outsider-imposed development programs.

Without regard for a deeper understanding of the region and often in a neocolonialist fashion, outside interventions have been common since the mid-1900s to ameliorate environmental, social, and economic challenges in the Himalaya. Beginning in the 1960s, the Himalayan people (and their poverty and population growth) were blamed for environmental degradation based on the so-called Theory of Himalayan Degradation (Ives & Messerli, 1989). In response, many development and environmental projects were implemented in the region. These were typically donor-centric and initiated by the West based on assumptions that Himalayan communities' traditions, values, and practices were at fault for environmental degradation and the region's pervasive poverty. During this period, many programs aimed to detach rural people and farmers' livelihoods from natural resources. Several aid agencies worked in the Himalaya because of their geopolitical and recreational interests. Jodha (1991), a prominent Himalayan scholar, argued that Himalayan environmental degradation was not arrested through these development efforts but rather accelerated because the actions and interventions by foreign aid agencies and NGOs failed to recognize mountain characteristics and failed to work within the parameters of local traditions and knowledge.

Later, the new concept of 'Himalayan perceptions' refuted the theory of Himalayan degradation, realizing that political and institutional factors, as well as natural hazards, are the root causes of Himalayan degradation (Ives, 2004). It is worth pointing out that outside forces, including migration, colonization, and the formation of sovereign states in the region and the borders that divide them, have had far greater consequences than internal human characteristics and behaviors. Many Indigenous and village-based knowledge systems have been replaced by so-called 'development' introduced by Western colonialism.

Many scholars from the region warn that the ecological crises in the Himalaya are evolving and worsening. Scientific knowledge about the region has grown significantly in the last few decades, but scientific research has made little impact on the ground in planning and policy-making (Ojha, 2020). The knowledge produced through research is often scattered, discipline-specific, and has not reached decision-makers, including government agencies, local communities, and the private sector (Wester et al., 2019). This is one of the challenges of the epistemology and ontology of knowledge production, dissemination, and application. There are advantages of a more hybrid scientific and traditional knowledge system that are unique to the region to manage resources. Hence, there is a vital need for a more community-engaged, locally-grounded (culturally and scientifically), cross-disciplinary and problem-solving research agenda in the Himalayan region (Ojha, 2020; Wester et al., 2019).

# References

Adger, W.N. (2000). Social and ecological resilience: Are they related? *Progress in Human Geography*, 24(3), 347–364.

Adhikari, S., Dhungana, N., & Upadhaya, S. (2020). Watershed communities' livelihood vulnerability to climate change in the Himalayas. *Climatic Change*, 162(3), 1307–1321.

Alam, M., & Regmi, B.R. (2004). *Adverse Impacts of Climate Change on Development of Nepal: Integrating Adaptation Into Policies and Activities*. Dhaka: Bangladesh Center for Advanced Studies.

Bajracharya, S.R., Maharjan, S.B., Shrestha, F., Sherpa, T.C., Wagle, N., & Shrestha, A.B. (2020). *Inventory of Glacial Lakes and Identification of Potentially Dangerous Glacial Lakes in the Koshi, Gandaki, and Karnali River Basins of Nepal, the Tibet Autonomous Region of China, and India*. Kathmandu: ICIMOD and UNDP.

Bergstrand, K., Mayer, B., Brumback, B., & Zhang, Y. (2015). Assessing the relationship between social vulnerability and community resilience to hazards. *Social Indicators Research*, 122(2), 391–409.

Beirman, D., Upadhayaya, P.K., Pradhananga, P., & Darcy, S. (2018). Nepal tourism in the aftermath of the April/May 2015 earthquake and aftershocks: Repercussions, recovery and the rise of new tourism sectors. *Tourism Recreation Research*, 43(4), 544–554.

Bilham, R. (2004). Earthquakes in India and the Himalaya: Tectonics, geodesy and history. *Annals of Geophysics*, 47(2–3), 839–858.

Bürki, R., Elsasser, H., Abegg, B., & Koenig, U. (2005). Climate change and tourism in the Swiss Alps. In C.M. Hall & J. Higham (Eds.), *Tourism, Recreation and Climate Change* (pp. 155–264). Bristol: Channel View Publications.

Chen, X.Q., Cui, P., Li, Y., Yang, Z., & Qi, Y.Q. (2007). Changes in glacial lakes and glaciers of post-1986 in the Poiqu River basin, Nyalam, Xizang (Tibet). *Geomorphology*, 88(3–4), 298–311.

Cui, P., Wei, F., Chen, X., & He, S. (2008). Geo-hazards in Wenchuan earthquake area and countermeasures for disaster reduction. *Bulletin of the Chinese Academy of Sciences*, 23(4), 317–323.

Espiner, S., & Becken, S. (2014). Tourist towns on the edge: Conceptualising vulnerability and resilience in a protected area tourism system. *Journal of Sustainable Tourism*, 22(4), 646–665.

Faulkner, B. (2001). Towards a framework for tourism disaster management. *Tourism Management*, 22(2), 135–147.

Folke, C. (2006). Resilience: The emergence of a perspective for social—ecological systems analyses. *Global Environmental Change*, 16(3), 253–267.

Folke, C., Hahn, T., Olsson, P., & Norberg, J. (2005). Adaptive governance of social-ecological systems. *Annual Review of Environment and Resources*, 30, 441–473.

Gioli, G., Thapa, G., Khan, F., Dasgupta, P., Nathan, D., Chhetri, N., Adhikari, L., Mohanty, S.K., Aurino, E., & Scott, L.M. (2019). Understanding and tackling poverty and vulnerability in mountain livelihoods in the Hindu Kush Himalaya. In P. Wester, A. Mishra, A. Mukherji, & A. Shrestha (Eds.), *The Hindu Kush Himalaya Assessment: Mountains, Climate Change, Sustainability and People* (pp. 421–455). Cham, Switzerland: Springer.

Goldberg, M.L. (2015). Nepal earthquake facts and figures. *UN Dispatch*, May 19. Retrieved from www.undispatch.com/nepal-earthquake-facts-and-figures/

Gomez Martin, M.B. (2005). Weather, climate and tourism—A geographical perspective. *Annals of Tourism Research*, 32(3), 571–591.

Guo, Y., Zhang, J., Zhang, Y., & Zheng, C. (2018). Examining the relationship between social capital and community residents' perceived resilience in tourism destinations. *Journal of Sustainable Tourism*, 26(6), 973–986.

Gunderson, L. (2000). Ecological resilience—In theory and application. *Annual Review of Ecology and Evolution and Systematics*, 31, 425–439.

Harrison, G.P., & Wallace, A.R. (2005). Sensitivity of wave energy to climate change. *IEEE Transactions on Energy Conversion*, 20(4), 870–877.

Holladay, P J., & Powell, R.B. (2013). Resident perceptions of social-ecological resilience and the sustainability of community-based tourism development in the commonwealth of Dominica. *Journal of Sustainable Tourism*, 21(8), 1188–1211.

ICIMOD (2020). Mission and vision. Retrieved from www.icimod.org/who-we-are/vision-mission/.

IPCC (2012). Summary for policymakers. In C.B. Field, V. Barros, T.F. Stocker, D. Qin, D.J. Dokken, K.L. Ebi, M.D. Mastrandrea, K.J. Mach, G.-K. Plattner, S.K. Allen, M. Tignor, & P.M. Midgley (Eds.), *Managing the Risks of Extreme Events and Disasters to Advance Climate Change Adaptation* (pp. 3–21). Cambridge: Cambridge University Press.

Ives, J.D. (2004). *Himalayan Perceptions: Environmental Change and the Well-Being of Mountain Peoples*. London: Routledge.

Ives, J.D., & Messerli, B. (1989). *The Himalayan Dilemma: Reconciling Development and Conservation*. London: Routledge & United Nations University.

Jodha, N.S. (1991). Sustainable agriculture in fragile resource zones: Technological imperatives. *Economic and Political Weekly*, 26(13), A15–A26.

Jodha, N.S. (2001). *Life on the Edge: Sustaining Agriculture and Community Resources in Fragile Environments*. Oxford: Oxford University Press.

Kafle, S.K. (2017). Disaster risk management systems in South Asia: Natural hazards, vulnerability, disaster risk and legislative and institutional frameworks. *Journal of Geography and Natural Disasters*, 7(3), 1000207.

Kala, C.P. (2014). Deluge, disaster and development in Uttarakhand Himalayan region of India: Challenges and lessons for disaster management. *International Journal of Disaster Risk Reduction*, 8, 143–152.

Klein, J.A., Harte, J., & Zhao, X.-Q. (2004). Experimental warming causes large and rapid species loss, dampened by simulated grazing, on the Tibetan Plateau. *Ecology Letters*, 7(12), 1170–1179.

Körner, C., & Ohsawa, M. (2005). *Ecosystems and Human Well-Being: Current States and Trends*. Washington, DC: Island.

Krishnan, R., Shrestha, A.B., & Ren, G. (2019). Unravelling climate change in the Hindu Kush Himalayas: Rapid warming in the mountain and increasing extremes. In P. Wester, A. Mishra, A. Mukherji, & A.B. Shrestha (Eds.), *The Hindu Kush Himalaya Assessment: Mountains, Climate Change, Sustainability and People* (pp. 57–91). Cham, Switzerland: Springer.

Lew, A.A. (2014). Scale, change and resilience in community tourism planning. *Tourism Geographies*, 16(1), 14–22.

McKercher, B., & Chon, K. (2004). The over-reaction to SARS and the collapse of Asian tourism. *Annals of Tourism Research*, 31(3), 716–719.

Negi, V.S., Maikhuri, R.K., Pharswan, D., Thakur, S., & Dhyani, P.P. (2017). Climate change impact in the Western Himalaya: People's perception and adaptive strategies. *Journal of Mountain Science*, 14(2), 403–416.

Nepal, S.K., & Chipeniuk, R. (2005). Mountain tourism: Towards a conceptual framework. *Tourism Geographies*, 7(3), 313–333.

Nyaupane, G.P. (2015). Mountaineering on Mount Everest: Evolution, economy, ecology, and ethics. In G. Musa, J. Higham, & A. Thompson-Carr (Eds.), *Mountaineering Tourism* (pp. 265–271). London: Routledge.

Nyaupane, G.P., & Chhetri, N. (2009). Vulnerability to climate change of nature-based tourism in the Nepalese Himalayas. *Tourism Geographies*, 11(1), 95–119.

Nyaupane, G.P., Lew, A.A., & Tatsugawa, K. (2014). Perceptions of trekking tourism and social and environmental change in Nepal's Himalayas. *Tourism Geographies*, 16(3), 415–437.

Nyaupane, G.P., Prayag, G., Godwyll, J., & White, D. (2020). Toward a resilient organization: Analysis of employee skills and organization adaptive traits. *Journal of Sustainable Tourism*, 29(4), 658–677.

Ojha, H.R. (2020). Building an engaged Himalayan sustainability science. *One Earth*, 3(5), 534–538.

Pandey, R., Maithani, N., Aretano, R., Zurlini, G., Archie, K.M., Gupta, A.K., & Pandey, V.P. (2016). Empirical assessment of adaptation to climate change impacts of mountain households: Development and application of an adaptation capability index. *Journal of Mountain Science*, 13(8), 1503–1514.

Rai, S.C., & Gurung, T. (2005). *An Overview of Glaciers, Glacier Retreat, and Subsequent Impacts in Nepal, India and China*. Kathmandu: WWF Nepal.

Ritchie, B.W., & Jiang, Y. (2019). Tourism, globalization, and natural disasters. In D.J. Timothy (Eds.), *Handbook of Globalisation and Tourism* (pp. 188–197). Cheltenham: Edward Elgar.

Rupa Kumar, K., Sahai., A.K., Krishna Kumar, K., Patwardhan, S.K., Mishra, P.K., Revadekar, J.V., Kamala, K., & Pant, G.B. (2006). High-resolution climate change scenario for India for the 21st century. *Current Science*, 90, 334–345.

Sella, G.F., Dixon, T.H., & Mao, A. (2002). REVEL: A model for recent plate velocities from space geodesy. *Journal of Geophysical Research: Solid Earth*, 107(B4), 1–31.

Smit, B., Burton, I., Klein, R.J., & Wandel, J. (2000). An anatomy of adaptation to climate change and variability. *Climatic Change*, 45(1), 223–251.

Smit, B., & Wandel, J. (2006). Adaptation, adaptive capacity and vulnerability. *Global Environmental Change*, 16(3), 282–292.

Thakur, S., Negi, V.S., Pathak, R., Dhyani, R., Durgapal, K., & Rawal, R.S. (2020). Indicator based integrated vulnerability assessment of community forests in Indian west Himalaya. *Forest Ecology and Management*, 457, 117674.

UNISDR & UNEP (2007). *Environment and Vulnerability: Emerging Perspectives*. Nairobi: UNEP.

Vaidya, R.A., Shrestha, M.S., Nasab, N., Gurung, D.R., Kozo, N., Pradhan, N.S., & Wasson, R.J. (2019). Disaster risk reduction and building resilience in the Hindu Kush Himalaya. In P. Wester, A. Mishra, A. Mukherji, & A.B. Shrestha (Eds.), *The Hindu Kush Himalaya Assessment: Mountains, Climate Change, Sustainability and People* (pp. 389–419). Cham, Switzerland: Springer.

Wang, S.Y., Fosu, B., Gillies, R.R., & Singh, P.M. (2015). The deadly Himalayan snowstorm of October 2014: Synoptic conditions and associated trends. *Bulletin of the American Meteorological Society*, 96, S89–S94.

Wang, S.Y., Wu, N., Kunze, C., Long, R., & Perlik, M. (2019). Drivers of change to mountain sustainability in the Hindu Kush Himalaya. In P. Wester, A. Mishra, A. Mukherji, &

A.B. Shrestha (Eds.), *The Hindu Kush Himalaya Assessment: Mountains, Climate Change, Sustainability and People* (pp. 17–56). Cham, Switzerland: Springer.

Wester, P., Mishra, A., Mukherji, A., & Shrestha, A.B. (Eds.). (2019). *The Hindu Kush Himalaya Assessment: Mountains, Climate Change, Sustainability and People.* Cham, Switzerland: Springer.

Wilson, G.A. (2013). Community resilience: Path dependency, lock-in effects and transitional ruptures. *Journal of Environmental Planning and Management*, 57(1), 1–26.

World Bank (2017). *Disaster Risk Profile: Afghanistan.* Washington, DC: The World Bank.

World Economic Forum (2014). Competitiveness dataset. Retrieved from http://www3. weforum.org/docs/GCR2014-15/GCI_Dataset_2006-07-

Xu, J., Badola, R., & Chhetri, N. (2019). Sustaining biodiversity and ecosystem services in the Hindu Kush Himalaya. In P. Wester, A. Mishra, A. Mukherji, & A.B. Shrestha (Eds.), *The Hindu Kush Himalaya Assessment: Mountains, Climate Change, Sustainability and People* (pp. 127–156). Cham, Switzerland: Springer.

Xu, J., Grumbine, R.E., Shrestha, A., Eriksson, M., Yang, X., Wang, Y., & Wilkes, A. (2009). The melting Himalayas: Cascading effects of climate change on water, biodiversity, and livelihoods. *Conservation Biology*, 23, 520–530.

# 4 Conservation and Restoration of Alpine Ecosystems in the Himalaya

## New Challenges for Adventure Tourism in the 21st Century

*Alton C. Byers and Milan Shrestha*

## Introduction

"Perhaps the time is not so far distant when travel agencies will include tours to the highest mountain in the world in their itineraries."

(Erwin Schneider, 1963, p. 194)

As evidenced by climber-cartographer Erwin Schneider's statement above, scientists and climbers in the 1950s and 1960s seem to have had a difficult time imagining the prospect of people actually wanting to visit the Mt. Everest (8,848 meters) region for pleasure. That novices would someday routinely pay USD $80,000.00 for a guided climb of Everest, as well as other peaks higher than 8,000 meters, would have been unthinkable. However, even in the 1950s, the Himalaya were already on their way to becoming high mountain tourist destinations. Beginning with Annapurna in 1950 (Herzog, 1952), the region's highest peaks were 'conquered', one after the other, by international climbing expeditions in what had been an historically off-limits, or at least rarely visited, mountain region of the world. Most expeditions were followed by the publication of books and lecture tours, usually necessary to fund the next expedition (e.g., Fiennes, 2000), for a public with a seemingly insatiable appetite for adventures above the 'death zone' (altitudes above 8,000 meters where life cannot be sustained for long) (Denise, 2014). Descriptions of amputated toes being swept out of railroad cars (Herzog, 1952), *yeti* tracks in the snow (Alter, 2020; MacInnes, 1979; Taylor, 2017), and the low odds (one in three) of surviving any high-altitude expedition in the Himalaya (Florino, 2015) only added to the intrigue.

By the early 1950s, technological advances, and more complete understandings of human physiology at high altitudes, played key roles in the ability of humans to climb to, and survive in, some of the harshest and most unforgiving environments of the world. Although expedition leader John Hunt preferred to believe that it was the 'human spirit' alone that propelled Edmund Hillary and Tenzing Norgay to the summit of Mt. Everest on May 28, 1953 (Hunt, 1953), others argue that newly developed fabrics, materials, boot design, ice climbing tools and methods,

DOI: 10.4324/9781003030126-6

bottled oxygen, and an emphasis on proper hydration at high elevations played an equally, if not a more, important role (Tuckey, 2014).

But the excitement and romance of that first ascent was soon eclipsed by ascents without oxygen (Messner, 1999), scaling all 'Seven Summits' (i.e., the highest peaks of each of the world's seven continents) (Bass et al., 1998), the guiding of amateur and wealthy climbers to the summit of Everest and other 8,000 meters+ peaks (Krakauer, 1997), or being the first of a particular group (e.g., youngest/oldest/legless/blind/having high blood pressure) to climb Everest (Grills, 2004; Tabei, 2017; Weihenmayer, 2002).

If one could not, or had no desire to, climb the highest mountains, trekking[1] offered an opportunity to experience firsthand the diverse cultures and environments of the Himalaya, including visits to the basecamps of the world's most famous mountains (Bezruchka, 1972; Nepal, 2000, 2007). For a fee (an average of $250.00 for up to four climbers in 2021),[2] tourists can also climb dozens of so-called 'trekking peaks' of less than 6,500 meters, experiencing all of the joys, and frequently the hazards, of high mountain expeditions (Kuniyal, 2002; O'Connor, 1989). Although very few westerners had visited Nepal, Bhutan, or Tibet prior to the 1950s, British civil servants, plant hunters, climbers, and explorers had been using the *dak* bungalows (rest houses) of northern India (Dhaku, 2016) since first established in the 1840s as part of the Raj mail service system. When compared to other Himalayan countries, the peaks and valleys of northern India were well explored by westerners beginning in the 19th century. It would be another 100 years before Nepal, Bhutan, and Tibet opened their borders to westerners and, eventually, to tourism.

High-altitude climbers and lower-altitude trekkers have three things in common. For one, they must acclimate, often over a period of many days or weeks, to the thinner air and decreased atmospheric pressures of the higher elevations, or otherwise suffer the consequences of acute mountain sickness (Houston, 1998; see also Pun, Bhandari, and Basnyat chapter on high-altitude sickness in this volume). For another, the approach climb onto the mountain will take them through multiple vegetation and climate zones, ranging from subtropical forest, to broadleaf evergreen forest, to the subalpine formations of fir, birch, and rhododendron and beyond, unless, of course, the journey is expedited by helicopter or airplane (Hadley et al., 2013). Third, they must spend at least some time crossing through, or staying in, the alpine zone between 4,000 and 5,500 meters, also referred to as the 'land above the trees' (Hadley et al., 2013; Körner, 1999). Here, the forests, fields, and villages trekkers have been walking through for days give way to the open shrub-grasslands and cushion plants of the higher, colder alpine ecosystems, the last before entering the nival zone of perpetual snow and ice above.

Until the 1970s, the alpine zone in the Himalaya was inhabited only seasonally by livestock herders and their livestock (primarily yak, or sheep and goats), tending their herds in the monsoon when forage was most abundant and lush. As more climbing expeditions and trekkers arrived, however, local people sought to take advantage of the unprecedented money-making opportunities these tourists presented by building new lodges and other tourist facilities (Figures 4.1 and 4.2).

*Figure 4.1* Namche Bazaar, a Sherpa village in the Mt. Everest region of Nepal, in 1973 (Photo A. Byers).

*Figure 4.2* The same scene as in Figure 4.1 in 2018, where the once-traditional mountain town has now been replaced with more than 60 three- or four-story lodges (photo A. Byers).

By the early 2000s, many accommodations featured yak dung- and wood-heated dining rooms, hot showers, imported alcoholic and nonalcoholic beverages, TV, *Pringles, Snickers*, and a range of western comfort foods manufactured outside the region.

From the 1950s onward, the numbers of trekkers and climbers visiting the Himalaya grew exponentially. For example, while only a few dozen visited the Everest region of Nepal in the late 1960s (Byers, 2005), over 60,000 arrived during the 2019 season, a number which doubles when support staff (e.g., porters and guides) are added (Byers et al., 2021). Many Everest climbers, both guided and unguided, became motivational speakers (Everest Climbers Speakers, 2021). Every spring, the preferred climbing season in the Himalaya, the news media has bemoaned the heaps of garbage and human waste left behind in basecamps (Hickok, 2018), now complicated by climate change and the appearance of dead bodies and historic refuse at the higher altitudes (Khadga, 2019; Nyaupane & Chhetri, 2009).

However, in spite of the media's obsession with climbing deaths, garbage, and the ethics of guiding amateurs up dangerous mountains, other Himalayan occurrences were apparently not as interesting nor newsworthy, perhaps, even though they were resulting in serious impacts. Among others, this included the widespread degradation of the fragile alpine ecosystems through the pressures imposed by new lodges, unsustainable land use practices, and growing human and livestock presence.

This chapter focuses on the little known and poorly studied, yet vitally important, alpine ecosystems of the Himalaya. In it, we examine the geoecology of alpine ecosystems, why these high-altitude environments are so vitally important, and how contemporary adventure tourism has resulted in negative impacts related to resource use and solid waste management. We will then examine how current trends can be reversed, or at least minimized, through a combination of scientific research, the involvement of local people, and ultimately the involvement of all stakeholders.

## The Himalayan Alpine Ecosystem

Alpine ecosystems in the Himalaya are generally found between 4,000 and 5,500 meters in elevation. Approximately 830 alpine plants have been documented in the Western and Central Himalaya (Kashmir to Nepal) (Körner, 1999; Polunin & Stainton, 1984), and 284 in Sagarmatha National Park (E. Byers, 2020). Although the total number of alpine plants within the entire Himalayan chain has yet to be determined, an informed guess would suggest something in the range of 1200–1600 alpine species, which appear to be increasing in number, frequency, and diversity as a result of climate change (Salick et al., 2014).[3]

Alpine plants are well adapted to cold and harsh environments by their perennial nature, which emphasizes greater growth and energy efficiency during the short growing season, their low growth forms, large underground root systems, and protective leaf and stem coverings (Hadley et al., 2013). Monsoonal climate patterns dominate, with peak rainfall occurring generally between May and September. Rainfall amounts decrease with altitude and from the eastern to western regions of the range. Conditions are cool, moist, and usually cloud covered (Meihe et al., 2015), although the upper limit of the alpine zone (approximately 5,500 meters) is frequently above the monsoonal clouds (Cronin, 1979).[4]

Alpine ecosystems in the Himalaya are of particular importance not only for local people, but also as sources of freshwater supplies for millions of people living downstream (Bandyopadhyay et al., 1997; Viviroli & Weingartner, 2008). These environments have been used for transhumance livestock grazing for centuries (Byers, 1996; Meihe et al., 2015; Nagy & Grabherr, 2009; Stevens, 1993) and are often the locations of sacred sites, such as *beyuls* (hidden valleys of refuge established by Guru Rinpoche in the 8th century (Reinhard, 1978; Skog, 2010). Fully half of the plant species in the Himalayan alpine zone are used by local people for medicinal and aromatic purposes (Buntaine et al., 2007; Byers et al., 2020a; Olsen & Larsen, 2003; Salick et al., 2014) related to the high concentrations of alkaloids, glycosides, polyphenols, and terpenes found in both leaves and rootstocks.

Especially when alpine hillsides are covered with resplendent wildflowers, it is tempting to imagine that these ecosystems have been relatively untouched by humans, their cattle, and other forms of extractive use. Yet despite appearances, most alpine ecosystems have been altered by grazing, fire, plant collection, and other forms of human disturbance for centuries (Meihe et al., 2015), if not for millennia (Byers, 2005). Telltale signs include a lack of rhododendron shrub cover (long since cut and burned); an abundance of shrubby cinquefoil (*Potentilla spp.)* and primrose (*Primula spp.)* within the new pastures, as both are unpalatable to cattle; and regular horizontal deposits of charcoal within soil profiles (Byers et al., 2020b). The impacts of these disturbances seem to be less severe in the wet alpine ecosystems of the Eastern Himalaya, such as in the Kanchenjunga and Makalu-Barun regions of Nepal, which tend to heal quickly because of a wetter climate, more developed soils, and more abundant vegetation. In drier climates, however, such as Khumbu, due west of Makalu-Barun, the removal of such soil-binding shrubs such as juniper and dwarf rhododendron can lead to accelerated soil loss as well as a bare, degraded look within a relatively short period of time (Byers, 2005; Byers et al., 2020a).

Alpine environments in general are characterized by thin soils, cold and harsh climates, and slow-growing vegetation that is highly susceptible to disturbances as a result of turf mining, overgrazing, and the harvesting of shrubby vegetation for fuel (Hillary, 2000; Byers, 2005; Nagy & Grabherr, 2009). In most cases, once the 'geomorphic glue' of alpine shrubs and protective alpine turf is removed from the thin alpine soils, mass wasting and soil loss processes accelerate and can take decades, if not centuries, to heal (Byers, 2005, 2013; Körner, 1999). Additionally, alpine plants have extremely slow growth rates. One sample of shrub juniper harvested near the seasonal village of Chukkung, for example, showed a diameter of 5.5 cm (2.1 inches) and age of 157 years (Byers, 2005). Even in the more forgiving and wetter environments of the eastern Himalaya, other non-sustainable practices related to tourism, such as poorly managed human and solid waste disposal, can lead to even more serious health and environmental issues.

## Impacts of Tourism

As mentioned previously, mountain climbing, trekking, eco- and adventure tourism began to increase in popularity in the West in the early 1970s. This was largely related to the trends and circumstances of the time, such as the establishment of Earth Day, the younger generation's greater emphasis on nature-based activities and pursuits, counterculture movement in the West, a greater abundance of free time, more disposable income, and access to the Himalaya, including overland road access from Europe, expanding airline services and cheap international flights and travel. Nepal and India were already famous as the destination of young 'overlanders' in the 1960s and 1970s, consisting primarily of 'hippies', who took public transportation from Europe, across the Middle East, and into South Asia. The legality of hashish, a perceived and desired spirituality of the East, and delicious cakes and pies at the journey's end in Nepal and India were major attractions as well. Some, however, began to venture beyond the urban centers of Nepal and India to explore the relatively unvisited routes to mountaineering basecamps (e.g., Everest and Annapurna) that ultimately resulted in the first trekking guidebooks (e.g., Bezruchka, 1972; McCue, 1999) and the trekking tourism that followed. This trend encouraged a whole new generation of eco- and adventure tourists to explore the high mountains of Nepal, Bhutan, Tibet, India, and Pakistan in the following years.

A number of popular books, such as *Magic and Mystery in Tibet* (David-Néel, 2014 (original 1929)), *Lost Horizon* (Hilton, 1933), and *The Snow Leopard* (Matthiessen, 2008 (original 1978)) were assuredly catalysts for the hundreds of thousands of young westerners who journeyed eastward in the 1980s and 1990s in search of similar adventure, romance, and experiences (Liechty, 2017, 2018). Tibet was not opened to foreigners until the early 1980s, but thereafter enjoyed several decades of backpacker tourism (Chen & Weiler, 2014), with the primary destinations being Lhasa and areas within the recently established Qomolangma Nature Preserve (Taylor-Ide et al., 1992). Bhutan, with its highly publicized 'gross national happiness', was a latecomer to tourism and more restrictive in its admission of foreigners, preferring the high-end, short-stay tourist to the low budget backpacker (Nyaupane & Timothy, 2010; Tshering, 2021).

While the 1970s and 1980s saw a modest increase in the number of trekkers and climbing expeditions to the Himalaya, both began to increase dramatically by the mid-1990s. Villages that had remained essentially unchanged for centuries began to transform into dozens of two- and three-story lodges of hand-hewn stone, usually designed and built by craftsmen from other regions. Lodges were capped by roofs of brightly painted corrugated metal, featuring rooms with flush toilets, electric blankets, TV, and Wi-Fi (see Figure 4.2).

Communal dining rooms offered pizza, spring rolls, 'yak steaks', beer, whiskey, and dozens of imported foods arriving in plastic, metal, and paper packaging.

Strangely, applications to climb Everest *increased* dramatically after the devastating loss of life during the spring season of 1996 (Krakauer, 1997), adding additional pressures in the form of more expeditions, porters, pack animals, and demands for food, lodging, and cattle feed. On some popular trekking routes in Nepal, mules replaced the traditional *dzopkio* (yak-cattle crossbreeds) used for carrying supplies because of their greater tolerance of low altitudes, but they also brought with them increased trail damage because of their metal shoes, dung-covered trails, and bottlenecks at bridges (Byers, 2019).

Lodge development in the alpine zones, usually along the main trail to a world-famous high mountain base camp (e.g., Kanchenjunga, Everest, Annapurna, or K2) lagged somewhat behind that of the lower elevations, largely because of remoteness, altitude (>4,000 meters), and distance from material supply centers. While basic accommodations in the form of tea houses existed in the late 1990s, the 2000s saw the development of more luxurious lodges in the alpine with Western-style ceramic bathrooms, Western foods, imported (by porter back) ply-wood-walled rooms, and dining centers heated each night by wood stoves burning subalpine fuelwood carried from lower elevations, alpine shrubs, cushion plants, and yak dung. The differences between subalpine and alpine lodge impacts, however, only became obvious after a number of years, and were directly related to the comparative resilience and fragility of each ecosystem.

### *Lodge Construction*

In any discussion of tourism's impacts upon Himalayan alpine ecosystems, the construction of modern lodges to accommodate visitors should be front and center. The transformation of Lobuche, a small village in the Everest region, as seen in Figures 4.3 and 4.5 is an excellent example. Figure 4.3 shows how in the 1950s, the site was a one-building *goth* (shelter for yak herders) but by the late 2000s had become a 'tourist village', home to approximately 15 tourist lodges. In between, the evolution of tourist accommodations in the Himalaya generally follows a pattern where (a) there are no lodges, where tourist numbers are limited and individuals stay in people's homes or camp in their own tents, (b) tea houses are developed, providing basic meals and rustic, usually communal (bunk house) accommodations, (c) one or two pioneer lodges are established within a village, with traditional income-generating activities and architecture still dominating, and (d) dozens of lodges are built that dominate the cultural and economic landscapes. Because of the fragility of alpine ecosystems as discussed previously, these new 'tourist colonies' create a range of new problems that impact the land, water, air, and people's health. Several are discussed briefly below.

### *Fuelwood*

Most villages located in the subalpine zones have access to forests, usually designated and regulated for community use by local governments or protected area authorities, where dead wood can be collected seasonally for fuelwood for

*Figure 4.3* Lobuche (4,940 meters) seasonal settlement in 1956 (Photo F. Müller). The groups of people are most likely porters associated with the 1956 Swiss Everest expedition.

cooking, water heating, and for heating communal dining areas. As forests do not exist in the alpine, and because importing fuelwood from further down valley can be prohibitively expensive, the early lodges needed to find alternative sources of fuel that were locally available and, preferably, free. Shrub juniper (*J. indica*) thus began to be harvested in earnest to fulfil these new demands. Cushion plants were also dried and used, including the football-sized, extremely slow-growing *Arenaria* sp. and other cushion plants that dotted the hillslopes (Figure 4.5).

These problems were exacerbated by trekking agencies that utilized tent camping for their clients, since the cooks needed a reliable source of fuel for preparing at least two hot meals a day. Porters, until recently, were left to take care of themselves at the end of the day, and kept warm at night by building fires fueled

*Figure 4.4* Lobuche in 2020. At least a dozen lodges were constructed in the interim, mostly in the early 2000s and beyond (Photo L.S. Sherpa).

with shrub juniper dwarf rhododendron, and other local woody matter (see Sanjay K. Nepal's chapter 'Moral geographies, porters and social exclusion in trekking tourism in the Himalayan region' in this volume). Finally, climbing expeditions not only carried up shrub juniper to their basecamps for fuel, they burned tons of it annually in pre-climb *pujas*—religious ceremonies, where the mountain gods are asked for the blessing of a safe climb.

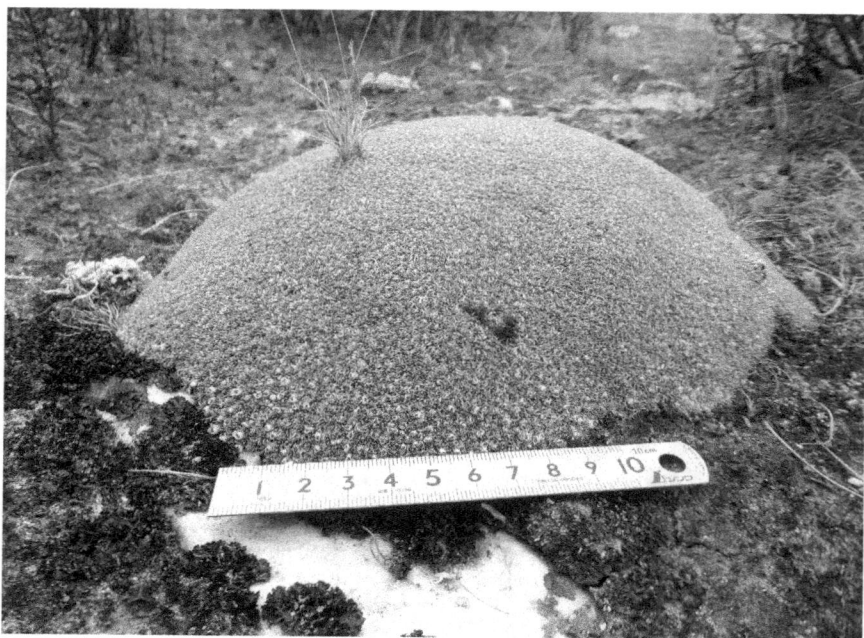

*Figure 4.5  Arenaria polytrichoides*, a cushion plant that takes hundreds of years to reach a diameter of 10 cm, is sometimes harvested and dried as another source of fuel unless protected by local or protected area authorities (Photo E. Byers).

In terms of total amounts of shrub juniper burned annually, in 2004, one alpine community in the Sagarmatha National Park reported harvesting 2,000 *doko* (baskets) of shrub juniper per tourist season for the 15 lodges that existed at that time (Byers, 2005). At 35 kilograms per basket, this equates to 7,000 kilograms, or 7.7 US tons per season, or 14,000 kilograms (15.4 US tons) per year for one village only. As three more tourist settlements of a similar size existed along the remaining route to the Everest basecamp, more than 42,000 kilograms (46.2 US tons) of shrub juniper were being harvested each year within the upper Imja Khola watershed, a situation that was clearly unsustainable.

### Alpine Turf

Shallow alpine soils typically form a dense and protective layer of turf. This top turf consists of fine soil, roots, and detritus capped by a continuous cover of grasses and forbes. Traditionally, the turf has been used to build walls and corral buildings for yak herds, alternating layers of turf with layers of local stone. When mined in traditionally limited quantities, similar to shrub juniper harvesting, the impacts are minimal and can remain so for hundreds of years. However, when dozens of new lodges begin mining the turf for wall building, or as floors for courtyards and patios, major damage can ensue with results similar to those described earlier related to shrub juniper removal.

## Pack Animals

Throughout the Himalaya, yaks and *zopkios* (yak/cattle crossbreeds) have been used for millennia to transport goods from one village to another on a limited and seasonal basis for trade between countries (e.g., the salt trade between Tibet and Nepal), and for crossing high-altitude passes from one valley to the next (von Fürer-Haimendorf, 1975). Trekking and climbing tourism dramatically increased demand for pack animals along basecamp and other popular trails, such that what had formerly been several trips per season became hundreds. The increased use of pack animals had a number of interesting economic impacts besides more income for the owner. These included growing and selling hay in lower altitude villages, which became far more profitable than growing and selling potatoes and other foods (A.R. Sherpa, personal communication, 2015). On the negative side, increased trail disturbance and erosion resulted from the increased number of pack animals, creating a new 'impact corridor' from wandering animals on either side of the main trail. As noted previously, at lower altitudes (<3,500 meters), mules began replacing both porters and *dzopkios* in the 2010s in Nepal. However, they rapidly fell into disfavor among local communities, who complained of their damage to trails with their shoe-clad hooves, dangerous disregard for humans, and excessive manure (Byers, 2019).

## Human Waste

With nearly all lodges now offering Western flush toilets, septic systems are often incapable of controlling the volumes of human waste produced annually, and/or were poorly built in the first place and now leak (Byers et al., 2021). A second source of contamination are the outhouses situated either near or directly over water courses or stream beds.

Popular climbing basecamps produce prodigious amounts of human waste per year, the 12,000 pounds (5,455 kilograms) from each season at the Everest basecamp (Nepal) receiving the most press (Gurubacharya, 2019; Hickok, 2018). Thus far, Himalayan waste management has consisted of transporting the effluent in plastic-lined barrels to dumping sites located several kilometers away from and below the basecamp regions, which in turn may contaminate both downstream and groundwater supplies if deposited in the vicinity of seasonal watercourses.

## Solid Waste

Increases in remote area trekking and climbing have resulted in increased demand for imported food, alcoholic beverages, and other consumer goods. Differing from traditional systems where all food came from the farm or local market, most international-standard food comes packaged in plastic, glass, metal, or paper, which in areas where recycling facilities are not available ultimately ends up in landfills (Byers et al., 2021). Approximately 1,000 US tons (907,185 kilograms) of solid waste is deposited in landfills each year in Nepal's Everest region, most of which is burned regularly and then buried once the garbage pit is full. During the burning process, toxic poisons are released into the air, and the buried garbage can contaminate groundwater supplies for decades. In popular basecamps, semi-annual and internationally publicized clean-up expeditions are held to address the problem cosmetically, but the challenge of accumulating human effluent, solid waste, and human bodies higher up in the nival persists.

## Discussion

Rhoades (1997) maintains that many of the socio-economic, cultural, and demographic changes sweeping across the Himalaya in the last several decades were influenced by forces exogenous to the region, as the wider market economy began to penetrate and supplant the stable and self-sufficient subsistence economy. The introduction of adventure tourism also meant an increasing reliance on a cash economy and the specialization of an unstable single enterprise—tourism—rather than the more stable agropastoralist livelihoods.

This trend was also enabled by internal or endogenous forces, which were largely connected with the socio- economic, cultural, and political changes occurring at the household level. These included higher literacy rates, better public health support, a growing middle class, and off-farm income opportunities. The interactions between exogenous and endogenous forces at different geographic scales produce a more complex nature of forces that impact the Himalayan region (Nyaupane et al., 2014). All in all, the remote areas popular for adventure tourism embraced the governmental concept of *bikas* (development), or modernization trends, which shuns remoteness as backward and a sign of feudalism. At the same time, the concept has often resulted in unintended consequences (Pigg, 1993) such as those described earlier. Although human impacts upon Himalayan alpine environments predate the advent of tourism, these impacts have most certainly been exacerbated by tourism development and created more volatile hospitality-dependent livelihoods.

The impacts seen in the alpine environments are a result of the convergence of both the natural and human factors associated with tourism in the region (Figure 4.6). As tourism started to replace trade and agropastoralism as a key livelihood source for mountain inhabitants, the Himalayan landscapes and their alpine environments also witnessed unprecedented changes as noted previously.

In 2021, alpine ecosystems throughout the Himalaya remained among the world's most neglected landscapes, even within most national parks and protected areas. This is probably partially a result of their remoteness, high altitudes, and difficult terrain encountered by management authorities compared to the more comfortable and easier lower elevations. A history of negative relations between park and forest authorities and local people can also stymie management progress,

EXOGENOUS FORCES

Globalized World
↓
Land system        Exotic Destinations        Governance
↓

Alpine turf        **TOURISM**        Lodge building
↓
Natural factors ←→ **ALPINE ENVIRONMENTS** ←→ Human factors
↑
Pack animals        **LIVELIHOOD**        Fuelwood
↑

Solid waste        Increased Accessibility        Human waste
↑
Remoteness

EXOGENOUS FORCES

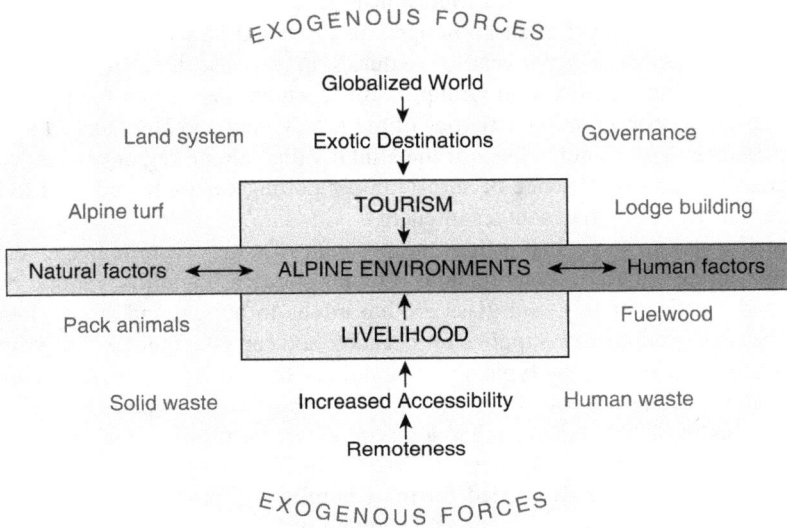

*Figure 4.6* Coupled natural and human systems of Alpine tourism and environmental impacts.

especially when conservation mandates (e.g., restrictions on harvesting shrub juniper or medicinal/aromatic plants) are rejected by local people as just another government imposition upon their lives and livelihoods.

Fortunately, a number of success stories exist, mostly within the realm of community-based conservation and restoration of alpine ecosystems across the Himalaya. For example, one study (Byers, 2005) concluded that the presence of disturbed and denuded hillslopes and accelerated soil loss in the Mt. Everest region of Nepal was primarily the result of recent shrub juniper harvesting for fuelwood, primarily by recently built lodges. This information was presented to local lodge owner committees in a series of workshops that followed the research, where stakeholders were quick to grasp the connections between resource use (or misuse), positive visitor experiences, and incomes. In response, lodge owners formed the Khumbu Alpine Conservation Committee (KACC), which immediately discontinued the harvesting of all shrub juniper for fuelwood by lodges and expedition groups in the Everest basecamp; replaced fuelwood with kerosene, propane, and solar batteries, with plans for the eventual development of a small hydropower facility; restored a porter's shelter in Lobuche that offered food and lodging at prices below those charged by local lodges; and constructed a KACC office and community center where visitor educational programs in altitude sickness and alpine conservation were regularly conducted.

A similar research process and community-based project was established in the Hinku valley to the east (Byers, 2013). There, the popularity of Mera Peak (6,400 meters), Nepal's highest 'trekking peak', resulted in a similar process of rapid lodge construction following the Maoist insurgency of 1996–2006, with a concurrent and heavy removal of shrub juniper for fuelwood in a valley that had only known a few yak herders per year for centuries. In response, lodge owners formed the Mera Alpine Conservation Group in 2008, which then began the process of replicating the conservation activities of the KACC that had commenced several years before. A 2017 field check of the Hinku valley alpine regions documented its rapid restoration, absence of shrub juniper cutting for fuelwood, and lodges actively using alternative sources of energy.

Next to alpine shrub harvesting, the problem of solid waste management in Himalayan alpine ecosystems remains a major concern. Landfills, routine burning, and plastics, metals, and glass are hazardous to human and animal health, contaminate groundwater supplies for decades, and can take hundreds of years to decompose because of the typically cold and dry conditions of elevations above 4,000 meters. The pattern of tourism, lodge establishment, imported foods, and landfills as the local 'management solution' exists in all remote alpine regions where tourism is present (Byers et al., 2020b). In 2019, a sustainable solid waste management plan was developed for the Khumbu region of Nepal based upon the principles of source segregation (separating plastics, glass, metals, etc.), pre-processing (shredding, compacting), and transporting to local airstrips to return to Kathmandu for recycling (Byers et al., 2020b). The plan was scheduled for testing by the local government in 2019–2020 until the global COVID-19 pandemic broke out in February 2020, which resulted in the cancellation of most tourism to the region. Once implementation can resume, however, the plan might provide a model for other high mountain parks and their alpine ecosystems throughout the montane world.

Managing human waste in the high and cold environments of the alpine will also remain a challenge for the foreseeable future. As mentioned previously, the thousands of kilograms of human waste generated each season in the Everest basecamp have been removed and dumped in landfills further south for a number of years. Because of the associated health hazards, however, this is not a sustainable long-term solution, and more locally appropriate and innovative technologies are needed. For example, solar composting of human effluent may be a promising and cost-effective solution that has yet to be adequately researched or tested.

Within the international adventure tourism sector, increased awareness of the fragility of the Himalayan alpine, as well as the need for tourism's involvement in supporting and funding local and national park alpine conservation activities, is urgently needed. While there have been some promising beginnings (see Byers, 2005), more work is needed to create a more meaningful stewardship among local people for the sustainability of these programs and reduce the dependency on external resources. At the same time, the involvement of international climbing clubs, the outdoor retail industry, trekking and climbing outfitters, and all those entities that are directly and indirectly benefitting from the services provided by Himalayan alpine ecosystems should be involved to reduce their impacts.

Finally, non-tourism items (but tourism-related) that could threaten alpine ecosystems include climate change, the COVID-19 pandemic, current trends in unregulated road building, and continued demand for the caterpillar fungus *yarsagumba* (*Ophiocordyceps sinensis*). With warming global temperatures, more aggressive plants from lower altitudes are migrating upward to higher elevations, pushing alpine plants increasingly further upward to the finite resource of available land. Native alpine plants are even starting to colonize the bare areas left by receding glaciers (E. Byers, 2019).

COVID-19 destroyed the 2020 trekking and climbing season across the Himalayan chain. However, as with the impacts of earthquakes, palace massacres, and civil wars upon tourist numbers in the past, arrivals are expected to rebound at some point in the near future. The extent of the long-term impacts of the tourism hiatus, both positive (e.g., landscape recovery and resting periods) and negative (e.g., increased poaching of animals and rare plants) (Bahukhandi et al., in press) remains to be seen.

During the past decade there has been an unprecedented increase in road building in Nepal and other Himalayan countries, which is providing new and easier access to remote regions that formerly took many days to reach by foot. The large-scale extraction of alpine juniper, dwarf rhododendron, and other rare medicinal and aromatic plants has been reported, which in the long run could do far more damage by tourism, and on a far greater scale, than that documented to date.

Finally, the collection of the caterpillar fungus *yarsagumba* throughout much of the Himalayan alpine has been of tremendous economic benefit to hundreds of thousands of villagers, but this has happened sometimes at the cost of various social (e.g., violence, conspicuous consumerism) and environmental ills (Byers et al., 2020b). The environmental problems reflect those of unregulated adventure tourism discussed in this chapter which, with the exception of wildlife poaching common in the *yarsagumba* collection camps, include unmanaged garbage accumulations, sanitation and water contamination, and excessive shrub juniper harvesting for fuelwood.

## Conclusion

For hundreds of years, alpine environments in the Himalaya remained quiet, high-altitude backwaters, inhabited only seasonally by a handful of livestock herders and their yaks, *dzopkios*, sheep, and goats. Natural alpine landscapes were nevertheless modified, slowly but surely by cutting and burning shrubs to increase pasture size, but for the most part these transformations were hardly recognizable to outsiders. The growth and popularity of trekking, mountaineering, pilgrimages, and other high-altitude tourism pursuits eventually caused major and, for the most part, positive changes in local incomes, but often at the cost the natural environment in the form of unregulated fuelwood harvesting, accelerated erosion, unsightly and unhealthy solid waste accumulations, and contaminated air and water. A few isolated, community-based projects have been successful in reversing some of these negative land use trends, and some progress has recently been made in addressing the chronic problem of solid waste management in the high mountain environment. Still, the field of human impacts upon Himalayan alpine ecosystems is relatively new, and much more work, research, and partici-patory projects with local communities will be needed before proven solutions to contemporary problems can be found.

## Notes

1  'Trekking' is primarily a British term that, while the purist would argue refers to multi-day trips through remote regions, is still synonymous with hiking, hill walking, or back-packing. 'Glamping', in the mountain sense of the word, is a more modern term that refers to hiking and camping under luxurious conditions.
2  See: www.nepalmountaineering.org/article-Fee%20Structure for a list of current fees for Nepal's trekking peaks.
3  Globally, alpine ecosystems cover 3 percent of the earth's land surface and contain over 10,000 species of plants which, according to Körner (1999), makes them one of the most biodiverse ecosystems per unit area in the world.
4  In his book *The Arun*, Ted Cronin refers to the alpine ecosystems of the remote Barun valley, east Nepal, as 'one step from the moon' in terms of their climate, lack of air, and overall inhospitality to life.

## References

Alter, S. (2020). *Wild Himalaya*. New Delhi: Aleph Book Company.
Bahukhandi, K., Agarwal, S., & Singhal, S. (in press). Impact of lockdown COVID-19 pandemic on Himalayan environment. *International Journal of Environmental Analytical Chemistry.* https://doi.org/10.1080/03067319.2020.1857751
Bandyopadhay, J., Rodda, J.C., Kattelmann, R., Kundzewicz, Z.W., & Kraemer, D. (1997). Highland waters—a resource of global significance. In B. Messerli & J.D. Ives (Eds.), *Mountains of the World: A Global Priority* (pp. 131–155). New York: The Parthenon Publishing Group.
Bass, D., Wells, F., & Ridgeway, R. (1998). *Seven Summits*. New York: Grand Central Publishing.
Bezruchka, S. (1972). *Trekking in Nepal: A Traveler's Guide*. Kathmandu: Sahayogi Press.
Buntaine, M.T., Mullen, R.B., & Lassoie, J.P. (2007). Human use and conservation planning in alpine areas of northwestern Yunnan, China. *Environment, Development and Sustainability*, 9, 305–324.
Byers, A.C. (1996). Historical and contemporary human disturbance in the upper Barun valley, Makalu-Barun national park and conservation area, east Nepal. *Mountain Research and Development*, 16(3), 235–247.
Byers, A.C. (2005). Contemporary human impacts on alpine landscapes in the Sagarmatha (Mt. Everest) National Park, Khumbu, Nepal. *Annals of the Association of American Geographers*, 95(1), 112–140.
Byers, A.C. (2013). Contemporary human impacts on subalpine and alpine ecosystems of the Hinku valley, Makalu-Barun national park and buffer zone, Nepal. *Himalaya: The Journal of the Association for Nepal and Himalayan Studies*, 33(1 & 2), 25–41.
Byers, A.C. (2019). Too many mules on the Everest trail. *Nepali Times*, October 31. Retrieved from www.nepalitimes.com/banner/too-many-mules-on-the-everest-trail/
Byers, A.C., Byers, E., Shrestha, M., Thapa, D., & Sharma, B. (2020a). Impacts of *yartsa gunbu* harvesting on alpine ecosystems in the Barun Valley, Makalu-Barun national park, Nepal. *Himalaya*, 39(2), 44–59.

Byers, A.C., Gustafsson, T., Shrestha, M., & Chhetri, N. (2020b). A sustainable solid waste management for the Sagarmatha (Mt. Everest) national park and buffer zone, Nepal. *Mountain Research and Development*, 40(3), A1–A9.

Byers, E. (2019). Wildflowers of the melting glaciers. *ECS-Nepal*, April. Kathmandu: ECS-Nepal. Retrieved from http://ecs.com.np/features/wildflowers-of-the-melting-glaciers

Chen, H., & Weiler, B. (2014). Chinese donkey friends in Tibet—evidence from the cyberspace community. *Journal of China Tourism Research*, 10(4), 475–492.

Cronin, E. (1979). *The Arun: A Natural History of the World's Deepest Valley*. Boston: Houghton Mifflin Company.

David-Néel, A. (2014). *Magic and Mystery in Tibet*. Mansfield Centre, CT: Martino Publishing.

Denise, T. (2014). *The Death Zone: Murder on Mt. Everest*. Raven Press.

Dhaku, N.A. (2016). The forgotten dak bungalows from the British era. *Dawn*, December 11. Retrieved from www.dawn.com/news/1301710

Everest Climbers Speakers (2021). Retrieved March 28, 2021, from https://speaking.com/category/everest-climbers/

Fiennes, R. (2000). *Beyond the Limits: The Lessons Learned From a Lifetime's Adventures*. London: Little, Brown & Company.

Florino, J. (2015). Climbing Everest: Who makes it to the top? *Climbing*, December. Retrieved from www.climbing.com/news/climbing-everest-who-makes-it-to-the-top/#:~:text=Four%20died%20trying%20and%20only,a%202.2%25%20risk%20of%20death

Fürer-Haimendorf, C. von (1975). *Himalayan Traders*. London: John Murray.

Grills, B. (2004). *The Kid Who Climbed Everest*. Guilford, CT: Lyons Press.

Gurubacharya, B. (2019). 8,000 kilograms of human poop estimated left on Mt Everest this year. *The Associated Press/Global News*, June 23. Retrieved from https://globalnews.ca/news/5423926/mount-everest-trash/

Hadley, K., Price, L., & Grabherr, G. (2013). Mountain vegetation. In M. Price, A.C. Byers, D.A. Friend, T. Kohler, & L.W. Price (Eds.), *Mountain Geography: Physical and Human Dimensions* (pp. 183–220). Berkeley: University of California Press.

Herzog, M. (1952). *Annapurna*. New York: E.P. Dutton & Company, Inc.

Hickok, K. (2018). How much trash is on Mount Everest? *LiveScience*, July 15. Retrieved from www.livescience.com/63061-how-much-trash-mount-everest.html

Hillary, E. (2000). *View from the Summit*. Reading: Cox & Wyman, Ltd.

Hilton, J. (1933). *Lost Horizon*. New York: Harper Perennial.

Houston, C. (1998). *Going Higher*. Seattle: The Mountaineers.

Hunt, J. (1953). *The Ascent of Everest*. London: Hodder & Stoughton.

Khadga, K.S. (2019). Melting glaciers expose dead bodies. *BBC News*, March 21. Retrieved from www.bbc.com/news/science-environment-47638436

Körner, C. (1999). *Alpine Plant Life*. Berlin: Springer.

Krakauer, J. (1997). *Into Thin Air*. New York: Anchor Books.

Kuniyal, J.C. (2002). Mountain expeditions: Minimising the impact. *Environmental Impact Assessment Review*, 22(6), 561–581.

Liechty, M. (2017). *Far out: Countercultural Seekers and the Tourist Encounter in Nepal*. Chicago: University of Chicago Press.

Liechty, M. (2018). Himalayas in the western mind's eye: "Incredibly spiritual and marvellous". *Economic and Political Weekly*, 53(19), 32–36.

MacInnes, H. (1979). *Look Behind the Ranges*. Middlesex: Penguin Books.

Matthiessen, P. (2008). *The Snow Leopard*. New York: Penguin Books.

McCue, G. (1999). *Trekking in Tibet: A Traveler's Guide*. Seattle: The Mountaineers.

Meihe, G., Pendry, C., & Chaudhary, R. (2015). *Nepal: An Introduction to the Natural History, Ecology and Human Environment of the Himalayas*. Edinburgh: Royal Botanical Garden.

Messner, R. (1999). *All Fourteen 8000ers*. Seattle: The Mountaineers.

Nagy, L., & Grabherr, G. (2009). *The Biology of Alpine Habitats*. Oxford: Oxford University Press.

Nepal, S.K. (2000). Tourism in protected areas: The Nepalese Himalaya. *Annals of Tourism Research*, 27(3), 661–681.

Nepal, S.K. (2007). Tourism and rural settlements in Nepal's Annapurna region. *Annals of Tourism Research*, 34(4), 855–875.

Nyaupane, G.P., & Chhetri, N. (2009). Vulnerability to climate change of nature-based tourism in the Nepalese Himalayas. *Tourism Geographies*, 11(1), 95–119.

Nyaupane, G.P., Lew, A.A., & Tatsugawa, K. (2014). Perceptions of trekking tourism and social and environmental change in Nepal's Himalayas. *Tourism Geographies*, 16(3), 415–437.

Nyaupane, G.P., & Timothy, D.J. (2010). Power, regionalism and tourism policy in Bhutan. *Annals of Tourism Research*, 37(4), 969–988.

O'Connor, B. (1989). *The Trekking Peaks of Nepal*. Seattle: Cloudcap Press.

Olsen, C.S., & Larsen, H.O. (2003). Alpine medicinal plant trade and Himalayan mountain livelihood strategies. *Geographical Journal*, 159(3), 243–254.

Pigg, S.L. (1993). Unintended consequences: The ideological impact of development in Nepal. *South Asia Bulletin*, 13(1&2), 45–58.

Polunin, O., & Stainton, A. (1984). *Flowers of the Himalaya*. Delhi: Oxford University Press.

Reinhard, J. (1978). Khembalung: The Hidden Valley. *Kailash, A Journal of Himalayan Studies*, 6(1), 5–35.

Rhoades, R.E. (1997). *Pathways Towards a Sustainable Mountain Agriculture for the 21st Century: The Hindu Kush-Himalayan Experience*. Lalitpur, Nepal: International Centre for Integrated Mountain Development.

Salick, J., Ghimire, S., Fang, Z., Dama, S., & Konchar, K. (2014). Himalayan alpine vegetation, climate change and mitigation. *Journal of Ethnobiology*, 34(3), 276–293.

Schneider, E. (1963). Foreword to the map of the Mount Everest region. In T. Hagen, G. Dyhrenfurth, C. von Fürer-Haimendorf, & E. Schneider (Eds.), *Mount Everest: Formation, Population and Exploration of the Everest Region* (pp. 182–195). London: Oxford University Press.

Sherpa, A.R. (2015). Director. The partners Nepal, September 2015, Khumbu, Nepal regarding the increasing values of hay due to the increased demand from climbing expeditions.

Skog, L.A. (2010). *Beyul Khumbu: Sherpa Constructions of a Sacred Landscape*. Unpublished thesis, Portland State University.

Stevens, S. (1993). *Claiming the High Ground: Sherpas, Subsistence, and Environmental Change in the Highest Himalayas*. Berkeley: University of California Press.

Tabei, J. (2017). *Honouring High Places: The Mountain Life of Junko Tabei*. Calgary, AB: Rocky Mountain Books.

Taylor, D. (2017). *Yeti: The Ecology of a Mystery*. New Delhi: Oxford University Press.

Taylor-Ide, D., Byers, A., & Campbell, J.G. (1992). Mountains, nations, parks and conservation: A case study of the Mt. Everest area. *GeoJournal*, 27(1), 105–112.

Tshering, T. (2021). High value low volume. *KuenselOnline*, August 21, 2015. Retrieved from https://kuenselonline.com/high-value-low-volume/

Tuckey, E. (2014). *Everest, The First Ascent: The Untold Story of Griffith Pugh, the Man Who Made It Possible*. London: Ebury Publishing.

Viviroli, D., & Weingartner, R. (2008). "Water towers"—A global view of the hydrological importance of mountains. In E. Wiegandt (Ed.), *Mountains: Sources of Water, Sources of Knowledge. Advances in Global Change Research* (pp. 15–20). Dordrecht: Springer.

Weihenmayer, E. (2002). *Touch the Top of the World: A Blind Man's Journey to Climb Farther Than the Eye Can See*. New York: Plume.

# 5 High-Altitude Medical Conditions Among Travelers in the Himalaya Mountains

*Matiram Pun, Sanjeeb S. Bhandari
and Buddha Basnyat*

## Introduction

Mountains occupy one-fifth of the earth's surface and provide one of the best natural laboratories for research into the effects of low oxygen on the human body, climate change, astronomical observations, altitude training, and tourism. The wealth of information collected from mountain environments in the past has contributed significant scientific knowledge about hypoxia (the lack of oxygen in the blood) and altitude illnesses, and the genetic adaptation of different groups of highlanders around the world.

With increased altitude, the barometric pressure falls and the density of air decreases, which means a person's blood contains less oxygen. The human body naturally tries to adjust to this low oxygen environment within minutes or hours. Human physiology responds to low oxygen levels (hypoxia) immediately with faster and deeper breathing and an increasing heart rate. Fully adjusting to a new altitude and lower oxygen level may take hours or days; this depends on the individual, for each person's physiology has its own timeframe for acclimatizing (West, 2004). Hence, it is crucial to allow one's body to acclimatize gradually by not going too high too fast, because the effects of hypoxia can be quick and dramatic during mountain ascents. Other subtle and deleterious effects of staying at a high altitude can appear over days, weeks, and months. For example, mountains are home to more than 140 million people who permanently live at high altitudes (Penaloza & Arias-Stella, 2007). Even certain high altitude residents (e.g. inhabitants of the Andes in South America) are not spared the deleterious effects of low oxygen and as a result suffer chronic mountain sickness (Beall, 2014; Monge, 1942).

Mountains have become increasingly popular tourist destinations and now attract large numbers of visitors from lowland regions (Godde et al., 2000; Richins & Hull, 2016). With the development of transportation and access, remote montane regions are becoming increasingly accessible to the public (Reisman et al., 2017; West, 2008). As a result, more people are visiting high altitude terrains for tourism activities, such as trekking, climbing, trail running, pilgrimages, cultural heritage visits, scientific fieldwork, and even conferences and business travel. Improved transportation infrastructure, including roads, cable cars, and

DOI: 10.4324/9781003030126-7

airstrips, have enabled people to ascend to higher elevations. Increased accessibility and the resultant growth in mountain tourism have encouraged nontraditional mountain tourist segments, such as children, the aged, individuals with comorbidities, and people with transplants, to visit (Luks, 2016; Nafstad et al., 2016; Parati et al., 2018). This expanding tourist market brings with it a range of new medical challenges that were not a significant concern in the past when only trekkers (usually well prepared and fit) and trained climbers visited these austere destinations.

This chapter describes and examines the physiological effects of high altitude travel on the human body and the medical complications that accompany it. It offers insight into the causes, medical diagnoses, and treatments of altitude sicknesses, as well as the importance of awareness and good preparation for travelers at high elevations. Although this chapter draws upon research in many highland areas of the world, its focus is the Himalaya. The chapter first explains the bodily effects of high altitude exposure, including hypoxia-induced illnesses and the exacerbation of existing comorbidities. It then turns to a discussion of health-related patterns and trends in the Himalaya and concludes with a statement of the importance of education and preparation for high-altitude travel experiences.

## High-Altitude Exposure

High altitude, by definition, lies above 2,500 meters. As noted in the introduction, high altitude spaces are becoming increasingly popular tourist destinations, especially in famous mountain regions such as Mount Kilimanjaro, the Andes, the Rocky Mountains, and the European Alps, yet no other mountains in the world have as many extreme elevations as the Himalaya. Every year, millions of tourists ascend to extreme elevations and suffer the consequences of high altitude sicknesses, the exacerbation of existing comorbidities, physical injuries, or other maladies that they would not otherwise experience in lowland destinations (Basnyat & Murdoch, 2003; Lawrence & Reid, 2016; Nafstad et al., 2016; Yang et al., 2020).

The Himalaya are the highest mountains on Earth and have become increasingly popular among trekkers and other types of tourists. Each year, many Himalayan tourists fall ill from exposure to extreme elevations, whether by hiking or ascending by airplane, motor vehicle, or pack animal. Some tourists recover quickly, while others may experience permanent injuries or succumb to their altitude-induced illnesses and die. As such, from a Himalayan tourism perspective, a detailed discussion about altitude sickness is warranted, particularly for destination managers who have to deal with mountain rescues, providing health care, and repatriating the ill or deceased.

Earth's atmosphere is comprised of critical gases, including nitrogen (78.08%), oxygen (20.95%), argon (0.93%), carbon dioxide (0.04%), and others such as hydrogen, helium, neon, methane, and krypton in trace amounts. At higher elevations, the barometric pressure (i.e., the pressure exerted by the atmosphere's various gases) falls. For example, the barometric pressure at sea level is 760 mmHg.

It drops to 400 mmHg at the Everest Base Camp and declines even further to 253 mm Hg at the top of Mount Everest (8848 meters). At these elevations, the percentage of gaseous components in the air remains constant, but because of the decline in barometric pressure, oxygen levels decrease, resulting in hypoxia which, as noted earlier, occurs with a lack of oxygen for corporeal functions.

Lowlanders are individuals or populations born and raised in low-altitude regions, below 1500 meters (Table 5.1). Individuals who permanently dwell below around 500 meters are considered either sea level or near sea level residents. These people, when quickly exposed to high altitudes are at risk of suffering from altitude illnesses. The term 'highlanders' is specific to a population that has resided at high altitudes (over ~2,500 meters) for generations or centuries. These

*Table 5.1* Altitude classification, physiological response, and altitude effects

| Classification of altitude/populations | Physiological changes | Altitude problems |
|---|---|---|
| Near sea level: 0–488 meters (0–1,600 ft) | No altitude effects on physiological processes. | There are no altitude related effects. |
| Low altitude (low landers) 488–1,520 meters (1,500–5,000 ft) | No altitude effects on physiological processes in an otherwise known healthy person. | There are no altitude effects. However, there may be impairments in elite athlete competitions to an athlete who is born and raised at sea level. |
| Intermediate: 1,520–2,440 meters (5,000–8,000 ft) | Breathing and heart rate increase slightly. Exercise performance will decrease. Blood oxygen saturation should typically be above 90%. | It is unlikely one will suffer from altitude illness. However, an extremely susceptible individual might get AMS symptoms if they ascend rapidly and perform vigorous physical activity. With proper acclimatization and training schedule, those athletes can perform well to this altitude. |
| High altitude (High Landers): 2,440–4,270 meters (8,000–14,000 ft) | The physiological responses become more pronounced. There will be increased depth and frequency of breathing, elevated heart rate, and frequent urination. Oxygen saturation is usually <90%. $SpO_2$ will decrease upon physical exertion and during sleep. | AMS and other acute altitude illnesses usually occur at this altitude range (above 2,500 meters) among newly arriving low landers. Most of the hiking trails, teahouses, ski resorts, and mining locations are located in this altitude range. Hence, many low landers who opt to sojourn at altitude for recreation or other purposes happen to hang around this range. Therefore, more altitude illness cases are observed in this range such as AMS, HAPE, and HACE. |

| Classification of altitude/populations | Physiological changes | Altitude problems |
|---|---|---|
| Very high altitude 4,270–5,490 meters (14,000–18,000 ft) | Breathing, heart rate, frequency of urination will increase significantly. $SpO_2$ will decrease markedly. Physical activity will be challenging. Sleep will be poor and people awaken more frequently. Appetite may decrease. | Rapid ascent with vehicle, plane or horse/yak ride to this altitude is dangerous. Slow and graded ascent with acclimatization suggested at this altitude to avoid severe altitude illnesses. The likelihood of having severe acute altitude illnesses such as severe AMS, HAPE, and HACE is high. Individuals who have a prior history of severe AMS, HACE, and HACE are strongly advised to be careful and be on preventive medications for those illnesses. |
| Extreme altitude: >5,490 meters (>18,000 ft) | Physiological responses become extreme and optimal to defend extremely low oxygen (hypoxia). Individuals will be fatigued even with minimal exertion, i.e., struggle to maintain regular activities. Every activity at this altitude becomes a massive effort. Sleep is disturbed with frequent awakening and periodic breathing. Usually, climbers are the only ones exposed to this altitude for longer term in the mountains. | The level of oxygen in blood drops profoundly low. Body attempts to adapt with extreme frequency of breathing (deeper and faster). There is an increase in frequency and amount of urination. Sleep becomes disrupted and people often awake due to periodic breathing multiple times at night. Human body struggles to acclimatize due to profound levels of hypoxia. Longer terms of stays at this altitude leads to progressive physiologic deterioration such as weight loss, hallucination, impaired neurocognitive functions. The altitude is, apparently, the ceiling for long-term human adaptation and habitation. |

Note: The altitude classification is based upon Bartsch and Saltin (2008) and Davis and Hackett (2017).

groups have adapted to the effects of high altitude through favorable changes in their genetic structure and bodily functions. Several groups have been well studied, including the Sherpas of Nepal, the Tibetans, the Andeans in South America, and the Ethiopian highlanders (Beall, 2014; Simonson, 2015). Although they are all highlanders, they differ significantly in phenotypes from one another, but by using different coping mechanisms, they have evolved to survive at high altitudes.

Above around 2,500 meters, the human body begins to experience physiological responses, and the probability of having altitude sickness increases. Physiological reactions to high altitude exposure are normal responses in the human body. High altitude exposure means arriving at an elevation above what a person is used to. For example, someone born and raised or currently living at 500 meters being exposed to 2,500 meters will experience physiological changes (Table 5.1), but for those living at 2,500 meters, the elevation has fewer effects.

The stress of reduced oxygen triggers a number of normal physiological responses. These responses typically begin within a minute and are a continuous process involving multiple systems in the body. However, the full physiological response to a new altitude takes time. The higher a person goes, the heavier the stress and the longer it takes to become accustomed to physiological changes. The physical effects of increased altitude endure as people continue to ascend. If altitude is gained slowly, the human body can keep pace better with the speed of ascent. However, if the ascent is rapid, especially by motor vehicle, airplane, horse, or yak, the human body is usually unable to acclimatize quickly enough to adapt, which increases the risk of altitude sickness. Normal physiological responses include deeper and faster breathing, increased heart rate, increased urine production and tiring quickly with physical activity or exercise. These are normal and non-dangerous responses and represent the acclimatization process. The higher the altitude, the more robust the corporeal responses will be. The process of acclimatization is gradual and differs from one individual to the other.

When people stay at high altitudes for weeks or months, certain long-term physiological changes begin to occur, such as an increase in red blood cells, which enables the blood to carry higher amounts of oxygen. This acclimatization process is known as 'subacute high altitude exposure', which happens when acute exposure has been adjusted by physiological responses (acclimatization) and some long-term effects start to manifest themselves. These effects may be good or bad and have been known to occur among all types of tourists.

Chronic high altitude exposure occurs when someone is exposed to high altitudes from many months, years, or generations (Arias-Stella et al., 1973). This is common among many people who work in the tourism sector. There is a growing trend of lowlanders migrating to high altitudes for employment, including in tourism. Many Han Chinese have migrated to the Tibetan Plateau. Many lowlander and mid-hill Nepalese are moving to the High Himalaya to work in teahouses and lodges, and to serve climbers and trekkers as porters, cooks, and guides. They stay at high altitudes for many months (season to season) or years. Some of them migrate permanently. This is a prime example of chronic high altitude exposure, which has deleterious effects depending upon the altitude they come from and the altitude to which they migrate. On the other hand, highland populations such as those mentioned earlier, have lived at high altitudes for many generations, which enables a genetic adaptation that tolerates lower pressure and lower oxygen levels. However, recent migrants and visitors are not genetically adapted to high altitude conditions.

Chronic mountain sickness happens among long-term residents of high altitudes with the gradual development of progressive polycythemia, pulmonary hypertension, and hypoxemia (Leon-Velarde & Reeves, 1999). Symptoms include fatigue, weakness, somnolence, and slowed mentation, which may progress to complete mental incapacitation (Leon-Velarde et al., 2003). Symptoms can be relieved with a descent to lower altitudes. This is an extremely uncommon condition among Sherpa and Tibetans but more common among Andeans. The condition is also known as Monge's disease, named after Dr. Carlos Monge, who first described it in the Andes (Monge, 1942).

### Types of Acute Altitude Illness, Prevention, and Treatment Options

One of the most common immediate effects of elevation-caused hypoxia is the high altitude headache (HAH) (Broome et al., 1994; Burtscher, 1999), which can manifest in frontal, lateral, or occipital forms. Initially, headaches can be of mild to moderate severity, but with increasing altitude, they often develop into severe and even incapacitating forms. These headaches are aggravated by exertion, bending, straining, and dehydration. Individuals who are prone to migraines will have an increased likelihood of HAH. The HAH is the primary criterion for acute mountain sickness (AMS) in the Lake Louise Scoring System (Table 5.2). Approximately, 80% of high altitude travelers experience HAHs (International Headache Society, 2004). This is a benign condition but should be treated with rest, warmth, and rehydration (Broome et al., 1994) (see Table 5.3 for treatment option).

*Table 5.2* 2018 Lake Louise acute mountain sickness score

| AMS symptoms | AMS self-report scoring | |
|---|---|---|
| | *AMS symptoms severity* | *Score* |
| Headache | None at all | 0 |
| | A mild headache | 1 |
| | Moderate headache | 2 |
| | Severe headache, incapacitating | 3 |
| Gastrointestinal symptoms | Good appetite | 0 |
| | Poor appetite or nausea | 1 |
| | Moderate nausea or vomiting | 2 |
| | Severe nausea and vomiting, incapacitating | 3 |
| Fatigue and/or weakness | Not tired or weak | 0 |
| | Mild fatigue/weakness | 1 |
| | Moderate fatigue/weakness | 2 |
| | Severe fatigue/weakness, incapacitating | 3 |
| Dizziness/light-headedness | No dizziness/light-headedness | 0 |
| | Mild dizziness/light-headedness | 1 |
| | Moderate dizziness/light-headedness | 2 |
| | Severe dizziness/light-headedness, incapacitating | 3 |

**Total acute mountain sickness score**
**Score interpretation:** To have a positive definition of AMS, one must have headache ($\geq 1$) and the total score of at least three ($\geq 3$).

*(Continued)*

*Table 5.2* (Continued)

| AMS symptoms | AMS self-report scoring | |
|---|---|---|
| | *AMS symptoms severity* | *Score* |
| **AMS clinical functional score** | **Clinical symptoms** | **Score** |
| Overall, if you had AMS symptoms, how did they affect your activities? | Not at all | **0** |
| | Symptoms present, but did not force any change, inactivity, or itinerary | **1** |
| | My symptoms forced me to stop the ascent or to go down on my own power | **2** |
| | Had to be evacuated to a lower altitude | **3** |
| The self-report questionnaire should be followed by the clinical functional assessment. This question can be asked by the medical personnel or examiner. | | |

Note: The consensus score (LLS) was published in 1993 (see Roach et al., 1993) and modified in 2018 (see Roach et al., 2018).

*Table 5.3* Acute altitude illnesses—prevention and treatment

| Condition | Sign and symptoms | Prevention | Treatment |
|---|---|---|---|
| High altitude headache (HAH) | Headache of any type at high altitude | Graded ascent | Paracetamol Ibuprofen |
| Acute mountain sickness (AMS) | Headache and at least one of these: fatigue or weakness; dizziness or lightheadedness; gastrointestinal symptoms (nausea or vomiting, anorexia) | Graded ascent Acetazolamide 125 mg twice daily (250 mg is robust) | Descent Acetazolamide 250 mg twice daily Dexamethasone for severe AMS Supplemental oxygen |
| High-altitude cerebral edema (HACE) | Ataxic, disoriented with or without AMS Severe AMS often progresses to HACE if untreated or continuing ascent | Dexamethasone 8 mg daily in divided doses | Dexamethasone 8 mg stat, then 4 mg every 6 hours |
| High-altitude pulmonary edema (HAPE) | Dyspnea at rest Cough Decreased exercise tolerance Chest tightness Crepitations or wheeze in the back of the patient (on careful listening or by doctor) Central cyanosis (blue lips) Very high respiratory and heart rates | Appropriate ascent profile Nifedipine 60 mg modified release in two divided doses per day | Descent Supplemental oxygen Portable hyperbaric chamber Nifedipine 60 mg in two divided doses per day (slow release) |

Source: Based upon information in Wilderness Medical Society (2019) and Luks et al. (2014).

Acute mountain sickness (AMS) is a syndrome identified by a collection of symptoms induced by high altitude hypoxia (Basnyat & Murdoch, 2003). Symptoms include headache, poor appetite, nausea or vomiting, fatigue or weakness, and dizziness/lightheadedness (Roach et al., 2018). AMS is often likened to the symptoms of a hangover and is divided into mild, moderate, and severe. The incidence of AMS depends upon the altitude of exposure (absolute altitude gain), rate of ascent, previous history of altitude illness, and prior acclimatization. It is estimated that about a third (in some years much more than a third) of the population suffers from AMS during high altitude exposure (Figure 5.1, Panel C). The number of tourists (1998–2016) visiting the Everest region of Nepal (also month-wise distribution) and incidence of AMS from the Nepali Himalaya are illustrated in Figure 5.1.

Severe AMS often leads to brain swelling, or 'high altitude cerebral edema' (HACE) (Table 5.3). HACE is a malignant, neurological condition in which brain swelling (the accumulation of fluid in the brain) occurs. The higher the altitude the more likely a person is to experience HACE. It is a fatal condition and often progresses from severe AMS if not treated, or if the person ignores his/her AMS and continues to ascend. This is clinically diagnosed with severe headache, ataxia, altered sensorium, a seizure in some cases, and ultimately a coma (Basnyat & Murdoch, 2003). When lowlanders are exposed to very high altitudes, HACE is expected to occur approximately 0.5–1.0% of the time (Hackett et al., 1976). This condition requires urgent medical attention, supplemental oxygen, and evacuation.

High altitude pulmonary edema (HAPE) often develops one or two days after arriving at a new altitude. Most HAPE cases materialize within the first five days of high altitude exposure. This condition is an accumulation of fluid/water in the lungs and is fatal if the patient is not evacuated and treated immediately (Basnyat & Murdoch, 2003). The person suffering from HAPE exhibits a cough, extreme tiredness even at rest, chest tightness, and might expend a great deal of energy just to breathe, even while resting. In a severe case of HAPE, a person may have blue lips and fingers, a high heart and breathing rate, a cough with pink frothy sputum, and even a mild fever. They often cannot continue trekking or climbing on their own. Depending upon how high they are, their genetic predisposition, and susceptibility, the incidence of HAPE may vary from 0.2% to 5.0% (Luks et al., 2017; Maggiorini et al., 1990). HAPE is a malignant and fatal condition that requires urgent medical attention.

### Preexisting Medical Conditions and High Altitudes

Numerous other medical problems besides acute altitude illnesses sometimes manifest during high altitude exposure or are exacerbated by hypoxia, reduced barometric pressure, cold, dryness, and radiation (Basnyat et al., 2000). Among people with cardiovascular problems, acute exposure and hypoxia will increase blood pressure and can exacerbate hypertension, especially during the first few days of travel. However, this condition usually plateaus or stabilizes after some time. Prolonged residency at a high altitude is not associated with increased

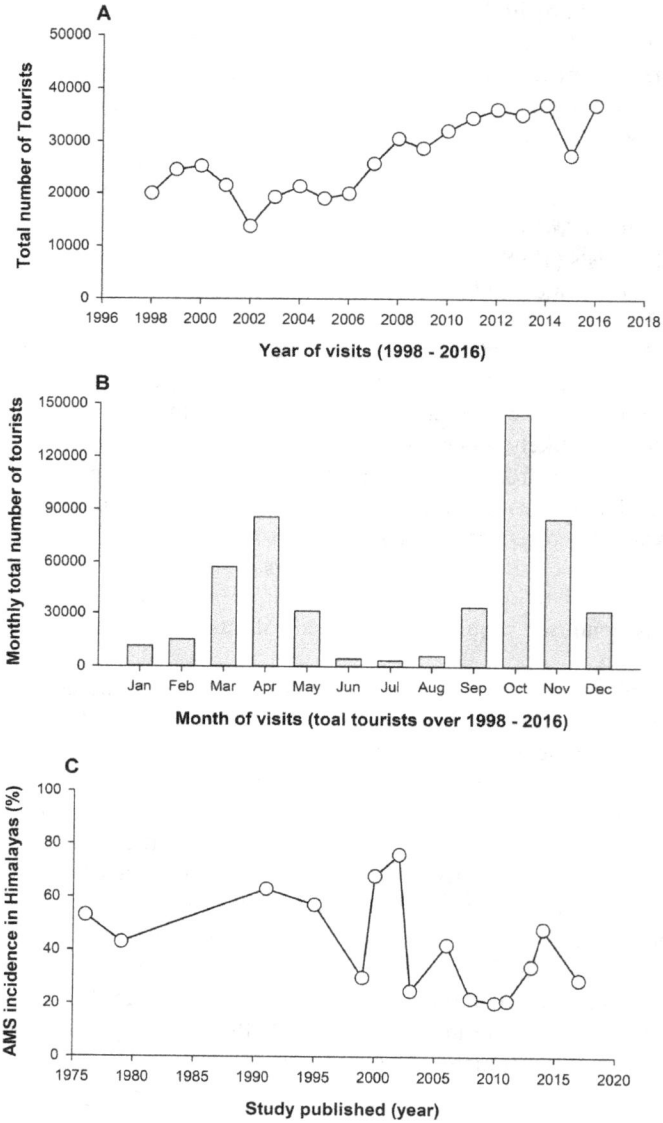

*Figure 5.1* Trends of trekkers in the Everest region and the incidence of AMS in Nepal Himalaya (Panel A. Trends of tourists over the years (1998–2016), Panel B. Trends of tourists over months in a year, Panel C. AMS incidence trends over the years from Nepal Himalaya based on various publications)

Note: Data source for **Panel A** and **B** is the Nepal Tourism Board. **Panel C** is based on several litera-ture sources: (Basnyat et al., 1999, 2000, 2003, 2008, 2011; Cumbo et al., 2002; Hackett et al., 1976; Hackett & Rennie, 1979; Kayser, 1991; MacInnis et al., 2013; McDevitt et al., 2014; Murdoch, 1995; Newcomb et al., 2011; Shah et al., 2006; Zafren et al., 2017).

blood pressure, but patients with symptomatic coronary artery disease or previous ischemic heart disease (a heart attack) may be adversely affected by a rapid ascent and extreme altitudes. People with heart disease seem to do well at moderate elevations (Hofstetter et al., 2017; Parati et al., 2018), but vigorous physical exercise may be dangerous. Usually, after acclimatization, high altitude does not increase stress to the heart. Individuals who have septal defects between chambers of the heart may be at an increased risk of altitude sickness, especially high altitude pulmonary edema. Some of these individuals might be undiagnosed at sea level, with the condition appearing only at upper elevations.

People with pulmonary (lung) problems are at increased risk at high elevations, especially pulmonary embolisms as discussed previously. Upper respiratory tract infections with runny nose, colds, and coughs are very common at high altitudes. By themselves, these are rather benign but might signal a predisposition to other altitude illnesses or pneumonia. During extreme altitude exposure, many climbers experience a very dry hacking and recurrent cough. It is common to see climbers with this condition at the Everest Base Camp. This cough is known as 'Khumbu Cough' after the Khumbu Valley in the Everest Region and is so severe that climbers have been known to break their ribs coughing (Mason, 2013). Chronic Obstructive Pulmonary Disease (COPD) of moderate intensity can tolerate moderate altitudes and air travel, but patients with severe forms of COPD may experience worsening symptoms even at moderate altitudes. Patients with asthma may benefit from the clean environment of high altitudes as long as they do not suffer from cold air-induced asthma (Luks & Swenson, 2007). People also may experience irritating hacking coughs after being at high altitudes for a couple of weeks. This may be due to irritation of the airway by cold, dry air. This is usually a dry cough with no other symptoms, which should be carefully assessed or observed since it could be a symptom of HAPE.

Neurological conditions are sometimes aggravated at high elevations. If someone has a brain tumor and is asymptomatic (even undiagnosed) at sea level, they might experience neurological symptoms, such as seizures or focal neurological deficits, at higher elevations due to hypoxic brain swelling. Brain swelling from hypoxia may lead to tumors compressing adjacent brain areas, leading to neurological symptoms. High altitude hypoxia may decrease people's seizure threshold and make them more susceptible to seizures, especially if they are already prone to seizures (Basnyat, 1997; Dekker et al., 2019). The science on these conditions is not definitive, only anecdotal, but the consensus is that trekkers taking anti-seizure medicines with good seizure control can visit high altitudes if they continue to take their medication. There have also been reports of strokes and transient ischemic attacks during high altitude exposure (Dekker et al., 2019), especially among elderly and individuals with comorbidities. Syncope, subarachnoid hemorrhage, and transient global amnesia have also been reported.

Diabetes is not a contraindication to high altitude travel, but insulin may work differently in these conditions. There is a significant difference in energy expenditure compared to daily work in low altitudes (Richards & Hillebrandt, 2013); thus, diabetics may need to adjust their insulin intake. Likewise, insulin is prone to freezing if not cared for properly. Many diabetics have successfully trekked to very high altitudes.

Pregnancy in itself is physiologically stressful and the fetus is already dependent upon the mother's environment. Thus, pregnant women's exposure to austere environments and hypoxic stress may be a cause of concern (Jean & Moore, 2012). Such conditions may affect the health of the mother and the child.

In terms of age, there is no limit for visiting high altitudes. Older adults often do well at moderately high elevations, but there are risks associated with age-related changes, such as impaired vision, diminished coordination, memory lapses, and the ability to endure stress related to distance traveled in a day or the amount of provisions and possessions carried. There also tend to be other comorbid conditions associated with the progression of age, such as diabetes and hypertension (Parati et al., 2015).

Although there is insufficient research on the effects of high altitude on the well-being of organ transplant recipients, the limited research does suggest that it is safe for solid organ recipients to travel up to 6,200 meters. The acclimatization mechanisms are similar to those of people with no organ transplants (Luks, 2016; van Adrichem et al., 2015).

Hypoxia facilitates tumor and cancer growth by inducing various pathways that help in the process. Lowlanders with cancer, especially certain types of tumors, are at increased risk at very high altitudes. The incidence and prevalence of cancer at high elevations vary, and there may be an increased risk of certain types of cancer among high altitude residents (Pun, 2019; Thiersch & Swenson, 2018).

High elevations may also exacerbate eye problems, especially among those who have had corneal corrective surgeries (e.g. radial keratotomy—LASIK, laser-assisted in situ keratomileusis). Corneal edema can occur with altitude exposure and lead to refractive errors (Mader & Tabin, 2003). Very high altitude exposure commonly leads to retinal hemorrhaging, although these cases are usually resolved upon descent. Other problems related to eyes include lateral rectus palsy, cortical blindness, and amaurosis fugax, as well as snow blindness (photokeratitis).

Various other genetic or metabolic disorders may also be amplified at high elevations. Prothrombotic conditions such as protein C deficiency and antiphospholipid syndrome may precipitate venous thrombosis in the brain. This sometimes leads to the neurological symptoms of a stroke, even when these conditions may otherwise be asymptomatic at sea level. Therefore, any condition that increases the chances of a blood clot, either metabolically or drug induced, may cause problems during high altitude travel (Basnyat et al., 2001).

### Other Health Conditions at High Elevations

As newly arriving travelers breathe more and heavier, urinate more and exert themselves more as they carry their backpacks or luggage, they lose a lot of fluid, thus running the risk of dehydration. Likewise, gastrointestinal problems, such as diarrhea, dysentery, or even enteric fever, are common among Western travelers in mountainous areas of developing countries. Changes in food habits and hygiene contribute to gastrointestinal stress. Trekkers usually experience excess gas and flatulence, which is known as High Altitude Flatulence Expulsion (HAFE) (Auerbach & Miller, 1981). Diarrhea contributes to dehydration as well.

Sleep disturbance is a common phenomenon at high elevations due to frequent awakenings, periodic breathing, difficulty breathing, and bizarre dreams (Bloch et al., 2015). Periodic breathing refers to a few cycles of deep breathing followed by a period of cessation of breathing for a few seconds. This is especially endemic at very high altitudes. Sleep is disturbed in many ways: the new environment, cold temperatures, and uncomfortable beds. Inadequately prepared travelers may experience numerous practical problems besides high altitude illness. One of the most common problems relates to cold temperatures (Procter et al., 2018). Prolonged exposure to below freezing temperatures may cause frostbite at the extremities or exposed body parts, commonly the nose and ears, or hypothermia, which can be life threatening. Once the body is very cold and has experienced hypoxic exposure, cognitive functions and dexterity will have already been severely compromised, which can be fatal.

Mountain adventure tourism has some relatively unique medical issues to deal with beyond those described so far. For example, some trekkers enjoy alcohol and drugs during their montane journeys. Increasing numbers of Himalayan teahouses and lodges are serving alcohol. This alone can have serious health implications, as alcohol can hinder acclimatization, and using these substances can also result in people getting injured or lost.

High altitude travel often involves adventure activities such as hiking, trekking, climbing, skiing, and snowboarding (McIntosh et al., 2012). The risks of falling and injury, sprains, or avalanche burial are very real. Other wilderness-related injuries, such as animal attacks or bug bites, are also very common. Unfortunately, there are many reports of carbon monoxide poisoning due to a faulty stove or heating system in tents or encampments. Moreover, higher elevations and snow-capped mountains reflect brighter light and also have more radiation, which frequently results in sunburns and snow blindness.

## Patterns and Trends to Consider in Himalayan Tourism

Everyone who visits the Himalaya as a tourist, regardless of their activities and travel purposes, is at risk of altitude illness. Likewise, new migrants to high altitude localities, such as tourism service employees, are prone to chronic high altitude problems. Absolute altitude gain, the rate of ascent, a previous history of altitude illnesses, a genetic predisposition, and concurrent medical conditions will increase one's chances of experiencing some level of altitude illness. Presently, there is no medical test or examination that can predict who gets altitude illness.

Contrary to popular belief, even physical fitness and good training do not protect travelers against altitude sickness. Some mountain travelers try to hide their symptoms, especially if they are part of an organized trekking group. There is peer pressure to keep up, and people tend to avoid showing weaknesses. They do not want to disturb the team's travel schedule, and often the team leader does not want to compromise the trek. This has often led to the conclusion that organized high altitude treks have higher risk factors than individual ascents or lower elevation group adventures. Some individuals isolate themselves in their rooms after suffering from altitude illness, which can be a sign of ominous things to come. Such behavior is indicative of severe altitude illness and strongly indicates the need for urgent assessment and care. There have been numerous cases in the past in which trekkers exhibiting this type of behavior have died.

Altitude sickness is a specialized area of medical practice and research. Not every physician has adequate knowledge about attitude illnesses. Much of the work in this area derives from the Himalaya where, given their extreme elevations, trekkers and other tourists commonly face many health-related challenges. Research shows that preparing for a high altitude holiday experience in terms of knowledge about related illness is essential, as is traveling with informed companions who are prepared to recognize the signs of physical problems or diminished psychological capacities.

In the Himalaya, there are many myths, folkloric traditions, and misconceptions about altitude illness. For example, one belief is that the sickness derives from the scent of certain high-elevation flowers. This is a common belief among the mountain pilgrims who visit Gosainkunda (4,350 meters), a sacred lake in Langtang, Nepal. Several traditional cultural practices are still followed, such as chewing raw garlic, ginger, or coca leaves, in hopes of preventing altitude illness (Basnyat, 2014). There is no evidence that these practices help prevent sicknesses.

Table 5.4 provides practical recommendations for pilgrims and other tourists in the Himalaya. High altitude pilgrims are one of the most affected travel groups in the Tibetan, Nepali, and Indian Himalaya (Basnyat, 2002). They often fast in preparation for their journeys, and they tend to be elderly and ill-prepared for high elevations (Basnyat, 2006). They are strong devotees, many of whom believe they will go to heaven if they die on the way to their highland destination, but their deaths are preventable with proper preparation and treatment (Yadav, 2009).

*Table 5.4* Practical recommendations for reducing altitude-related problems among pilgrims and general tourists

---

*Practical recommendations for pilgrims, general tourists, and other stakeholders*

---

***Improved awareness of altitude illness and other problems in pilgrims***
1. If proper acclimatization is not an option, advise using acetazolamide (in those with no serious sulfa allergy), 125 mg bid, starting a day before the trip and continuing for about 3 days into the trip.
2. To make the trip less strenuous, advise simple aerobic fitness program starting some months prior to the trip.
3. Pre-acclimatization: advise spending one week between 2,000 and 3,000 meters as close to the time of pilgrimage as possible.
4. Recommend avoiding dehydration by drinking 2–3 L of fluid per day and proper footwear for slippery trails and warm clothes to avoid hypothermia.

Although there is little evidence to suggest that traditional herbal remedies help prevent altitude sickness, there are several drugs that have proven to be effective. Over-the-counter painkillers, such as acetaminophen and ibuprofen, are effective in preventing AMS (Kanaan et al., 2017; Lipman et al., 2012), particularly the headaches. Likewise, Acetazolamide (Diamox) prevents AMS and helps trekkers sleep better. Dexamethasone prevents and treats HACE, and Dexamethasone and Nifedipine prevent and treat HAPE.

## Education and Preparation

Although medical treatments are crucial in alleviating the symptoms and problems associated with high-altitude travel, it is particularly important for travelers to educate themselves about the dangers and challenges of visiting higher elevations and prepare accordingly. Some of the problems discussed so far in this chapter can be partly mitigated through proper self-education and preparation, even if, as noted before, everyone undergoes certain physiological changes when they ascend to high elevations regardless of their level of preparation. Knowledge and preparation are key in managing the human corporeal responses to changes in elevation.

Consulting with a physician before undertaking pilgrimages, trekking holidays, or other types of activities in the highlands is recommended, particularly for those with existing medical conditions. Learning to recognize the symptoms of AMS and a willingness to slow the ascent and seek treatment are critical to minimizing the effects of high altitude. Likewise, tourists and migrant workers knowing proper treatments that they can administer themselves, as well as which treatments must be administered by medical professionals is important.

Carrying proper medications, medical kits, and perhaps a battery-operated pulse oximeter if available (Luks & Swenson, 2011), as well as plentiful water, suitable travel gear, sturdy hiking boots, and warm clothes is necessary (Table 5.5). Beyond carrying proper equipment and medications, understanding new environments, appropriate behaviors, and the crucial principle of acclimatization and slow ascent are extremely important for high altitude travelers to understand. Vigorous physical exertion, and alcohol and drug use, for example, should be avoided. Likewise, the general rule for safe ascents among trekkers is to avoid climbing more than 500 meters per day above the 2,500 meters mark and to rest for a full day after three to four days of hiking. Travelers also need to understand the repercussions of limited communication, language barriers, the constraints of traveling in developing countries, and how to cope with unfamiliar food and hygiene. These are among the most salient means of ensuring a safe and enjoyable high-elevation holiday or pilgrimage experience.

*Table 5.5* General checklist for high altitude trekking

| Items | Amount | Comments |
| --- | --- | --- |
| Acetaminophen (500 mg) | 20 tablets | For HAH and AMS (possibly) prevention and treatment |
| Ibuprofen (400 mg) | 20 tablets | For HAH and AMS prevention and treatment |
| Throat lozenges | 20 | Altitude-induced dry cough and sore throat |
| Sunglasses | 1 | Prevention of snow blindness (ultraviolet radiation blocking) |
| Sunscreen (SPF 40–50) | 1 | Prevention of sunburn |
| Lip protection (e.g. lip balm) | 1 | Prevention and treatment of lip cracks or healing of fissured lips |
| Water disinfection (e.g. iodine tablets or chlorine solution) | 1 | Safe drinking water is quite important, have powerful water disinfectant. |
| Blister care (moleskin and adhesive foam) | 2 | Hiking for many days, boots can result in blisters. |
| Wound care package (e.g. handyplast/gauge) | 10 | Trauma/injuries are often part of wilderness adventure. |
| Hand sanitizer and antiseptic wipes | 20 | For the prevention of contagious diseases |
| Insect repellents | 3 | It is good to have but may not be necessary at high altitudes. |
| Rehydration sachets | 10 | For common travelers' diarrhea |
| Condoms | Few packets | Contraception |
| Loperamide (2 mg) | 5 tablets | For diarrhea (noninfective or treated) |
| Antiprotozoal/antibacterial | | Travelers' diarrhea |
| vaccination | | Overall preventive measures of infectious diseases |
| Current medications for medical treatment | | Continue taking medications as prescribed. |

## Conclusion

Mountains, including the Himalaya, are attractive tourist destinations for low-landers, which poses a risk of altitude illnesses, also known in medical terms as hypoxia. The underlying cause of high altitude sicknesses is a lack of oxygen, which continues to thin as increased elevation is achieved. There are many hypoxic conditions, ranging from headaches to brain swelling and liquid in the lungs. Some hypoxic illnesses are mild and can be treated rather easily, but in some cases they are extreme and fatal. Proper medical care is required and pre-trip learning and preparation are key in terms of treatment and prevention. Risk of high altitude sicknesses is higher among tourists, pilgrims, and their support staff, such as porters and cooks, who are born and raised in low-elevation regions. The more unprepared people travel to high mountains quickly for a short period of time for leisure or work, the greater the issue of high altitude will be. Although the chapter outlined causes, symptoms, medical diagnoses and treatment options, more emphasis must be placed on prevention. Most tourist destinations in the Himalaya are in remote locations that lack adequate health care facilities. This warrants all stakeholders involved in tourism and mountaineering to take immediate steps to create awareness about the causes of mountain sickness, its symptoms, preventions, and treatment options. High altitude travelers should always be prepared and listen to their own bodies. Their bodies will inform them when problems arise. For severe illnesses, there should be effective mountain rescue operations in place initiated by the government and tourism associations. As tourism in the Himalaya continues to grow, these health imperatives will become increasingly important to know and act upon to ensure positive visitor experiences and to maintain the image of these Asian highlands as a desirable destination in the countries that share them.

## Acknowledgment

We would like to thank Mrs. Arta Johnson for reading the manuscript from a general reader's perspective. Arta's comments and feedback helped us avoid too much technical jargon.

## References

Arias-Stella, J., Kruger, H., & Recavarren, S. (1973). On the pathology of chronic mountain sickness. *Pathologia et Microbiologia*, 39(3), 283–286.

Auerbach, P., & Miller, Y.E. (1981). High altitude flatus expulsion (HAFE). *Western Journal of Medicine*, 134(2), 173–174.

Bartsch, P., & Saltin, B. (2008). General introduction to altitude adaptation and mountain sickness. *Scandinavian Journal of Medicine & Science in Sports*, 18, 1–10. doi:10.1111/j.1600-0838.2008.00827.x

Basnyat, B. (1997). Seizure and hemiparesis at high-altitude outside the setting of acute mountain sickness. *Wilderness & Environmental Medicine*, 8(4), 221–222.

Basnyat, B. (2002). Pilgrimage medicine. *British Medical Journal*, 324, 745.

Basnyat, B. (2006). The pilgrim at high altitude. *High Altitude Medicine & Biology*, 7(3), 183–184.

Basnyat, B. (2014). High altitude pilgrimage medicine. *High Altitude Medicine & Biology*, 15(4), 434–439.

Basnyat, B., Cumbo, T.A., & Edelman, R. (2000). Acute medical problems in the Himalayas outside the setting of altitude sickness. *High Altitude Medicine & Biology*, 1(3), 167–174.

Basnyat, B., Gertsch, J.H., Johnson, E.W., Castro-Marin, F., Inoue, Y., & Yeh, C. (2003). Efficacy of low-dose acetazolamide (125 mg BID) for the prophylaxis of acute mountain sickness: A prospective, double-blind, randomized, placebo-controlled trial. *High Altitude Medicine & Biology*, 4(1), 45–52.

Basnyat, B., Graham, L., Lee, S.D., & Lim, Y. (2001). A language barrier, abdominal pain, and double vision. *The Lancet*, 357, 2022.

Basnyat, B., Hargrove, J., Holck, P.S., Srivastav, S., Alekh, K., Ghimire, L.V., . . . Swenson, E.R. (2008). Acetazolamide fails to decrease pulmonary artery pressure at high altitude in partially acclimatized humans. *High Altitude Medicine & Biology*, 9(3), 209–216.

Basnyat, B., Holck, P.S., Pun, M., Halverson, S., Szawarski, P., Gertsch, J., . . . Farrar, J. (2011). Spironolactone does not prevent acute mountain sickness: A prospective, double-blind, randomized, placebo-controlled trial by SPACE Trial Group (spironolactone and acetazolamide trial in the prevention of acute mountain sickness group). *Wilderness & Environmental Medicine*, 22(1), 15–22.

Basnyat, B., Lemaster, J., & Litch, J.A. (1999). Everest or bust: A cross sectional, epidemiological study of acute mountain sickness at 4243 meters in the Himalayas. *Aviation, Space, and Environmental Medicine*, 70(9), 867–873.

Basnyat, B., & Murdoch, D.R. (2003). High-altitude illness. *The Lancet*, 361, 1967–1974.

Basnyat, B., Subedi, D., Sleggs, J., Lemaster, J., Bhasyal, G., Aryal, B., & Subedi, N. (2000). Disoriented and ataxic pilgrims: An epidemiological study of acute mountain sickness and high-altitude cerebral edema at a sacred lake at 4300 m in the Nepal Himalayas. *Wilderness & Environmental Medicine*, 11(2), 89–93.

Beall, C.M. (2014). Adaptation to high altitude: Phenotypes and genotypes. *Annual Review of Anthropology*, 43, 251–272.

Bloch, K.E., Buenzli, J.C., Latshang, T.D., & Ulrich, S. (2015). Sleep at high altitude: Guesses and facts. *Journal of Applied Physiology*, 119(12), 1466–1480.

Broome, J.R., Stoneham, M.D., Beeley, J.M., Milledge, J.S., & Hughes, A.S. (1994). High altitude headache: Treatment with ibuprofen. *Aviation, Space, and Environmental Medicine*, 65(1), 19–20.

Burtscher, M. (1999). High altitude headache: Epidemiology, pathophysiology, therapy and prophylaxis. *Wiener klinische Wochenschrift*, 111(20), 830–836.

Cumbo, T.A., Basnyat, B., Graham, J., Lescano, A.G., & Gambert, S. (2002). Acute mountain sickness, dehydration, and bicarbonate clearance: Preliminary field data from the Nepal Himalaya. *Aviation, Space, and Environmental Medicine*, 73(9), 898–901.

Davis, C., & Hackett, P. (2017). Advances in the prevention and treatment of high altitude illness. *Emergency Medicine Clinics*, 35(2), 241–260.

Dekker, M.C.J., Wilson, M.H., & Howlett, W.P. (2019). Mountain neurology. *Practical Neurology*, 19(5), 404–411.

Godde, P.M., Price, M.F., & Zimmermann, F.M. (Eds.). (2000). *Tourism and Development in Mountain Regions*. Wallingford: CABI.

Hackett, P.H., & Rennie, D. (1979). Rales, peripheral edema, retinal hemorrhage and acute mountain sickness. *American Journal of Medicine*, 67(2), 214–218.

Hackett, P.H., Rennie, D., & Levine, H.D. (1976). The incidence, importance, and prophylaxis of acute mountain sickness. *The Lancet*, 2(7996), 1149–1155.

Hofstetter, L., Scherrer, U., & Rimoldi, S.F. (2017). Going to high altitude with heart disease. *Cardiovascular Medicine*, 20(4), 87–95.

International Headache Society (2004). The International Classification of Headache Disorders: 2nd edition. *Cephaialgia*, 24(1), 9–160.

Jean, D., & Moore, L.G. (2012). Travel to high altitude during pregnancy: Frequently asked questions and recommendations for clinicians. *High Altitude Medicine & Biology*, 13(2), 73–81.

Kanaan, N.C., Peterson, A.L., Pun, M., Holck, P.S., Starling, J., Basyal, B., . . . Basnyat, B. (2017). Prophylactic Acetaminophen or Ibuprofen results in equivalent acute mountain sickness incidence at high altitude: A prospective randomized trial. *Wilderness & Environmental Medicine*, 28(2), 72–78.

Kayser, B. (1991). Acute mountain sickness in Western tourists around the Thorong Pass (5400 m) in Nepal. *Journal of Wilderness Medicine*, 2(2), 110–117.

Lawrence, J.S., & Reid, S.A. (2016). Risk determinants of acute mountain sickness and summit success on a 6-day ascent of Mount Kilimanjaro (5895 m). *Wilderness & Environmental Medicine*, 27(1), 78–84.

Leon-Velarde, F., McCullough, R.G., McCullough, R.E., & Reeves, J.T. (2003). Proposal for scoring severity in chronic mountain sickness (CMS). *Advances in Experimental Medicine and Biology*, 543, 339–354.

Leon-Velarde, F., & Reeves, J.T. (1999). International consensus group on chronic mountain sickness. *Advances in Experimental Medicine and Biology*, 474, 351–353.

Lipman, G.S., Kanaan, N.C., Holck, P.S., Constance, B.B., & Gertsch, J.H. (2012). Ibuprofen prevents altitude illness: A randomized controlled trial for prevention of altitude illness with nonsteroidal anti-inflammatories. *Annals of Emergency Medicine*, 59(6), 484–490.

Luks, A.M. (2016). Travel to high altitude following solid organ transplantation. *High Altitude Medicine & Biology*, 17(3), 147–156.

Luks, A.M., McIntosh, S.E., Grissom, C.K., Auerbach, P.S., Rodway, G.W., Schoene, R.B., . . . Hackett, P.H. (2014). Wilderness Medical Society practice guidelines for the prevention and treatment of acute altitude illness: 2014 update. *Wilderness & Environmental Medicine*, 25(4), S4–S14.

Luks, A.M., & Swenson, E.R. (2007). Travel to high altitude with pre-existing lung disease. *European Respiratory Journal*, 29(4), 770–792.

Luks, A.M., & Swenson, E.R. (2011). Pulse oximetry at high altitude. *High Altitude Medicine & Biology*, 12(2), 109–119.

Luks, A.M., Swenson, E.R., & Bärtsch, P. (2017). Acute high-altitude sickness. *European Respiratory Review*, 26(143), 160096.

MacInnis, M.J., Carter, E.A., Freeman, M.G., Pandit, B.P., Siwakoti, A., Subedi, A., . . . Rupert, J.L. (2013). A prospective epidemiological study of acute mountain sickness in Nepalese pilgrims ascending to high altitude (4380 m). *PLoS One*, 8(10), e75644.

Mader, T.H., & Tabin, G. (2003). Going to high altitude with preexisting ocular conditions. *High Altitude Medicine & Biology*, 4(4), 419–430.

Maggiorini, M., Buhler, B., Walter, M., & Oelz, O. (1990). Prevalence of acute mountain sickness in the Swiss Alps. *British Medical Journal*, 301(6756), 853–855.

Mason, N.P. (2013). Altitude-related cough. *Cough*, 9(1), 23.

McDevitt, M., McIntosh, S.E., Rodway, G., Peelay, J., Adams, D.L., & Kayser, B. (2014). Risk determinants of acute mountain sickness in trekkers in the Nepali Himalaya: A 24-year follow-up. *Wilderness & Environmental Medicine*, 25(2), 152–159.

McIntosh, S.E., Campbell, A., Weber, D., Dow, J., Joy, E., & Grissom, C.K. (2012). Mountaineering medical events and trauma on Denali, 1992–2011. *High Altitude Medicine & Biology*, 13(4), 275–280.

Monge, C. (1942). Life in the Andes and chronic mountain sickness. *Science*, 95(2456), 79–84.

Murdoch, D.R. (1995). Symptoms of infection and altitude illness among hikers in the Mount Everest region of Nepal. *Aviation, Space, and Environmental Medicine*, 66(2), 148–151.

Nafstad, P., Stigum, H., Wu, T., Haldorsen, Ø.D., Ommundsen, K., & Bjertness, E. (2016). Acute mountain sickness among tourists visiting the high-altitude city of Lhasa at 3658 m above sea level: A cross-sectional study. *Archives of Public Health*, 74(1), 1–7.

Newcomb, L., Sherpa, C., Nickol, A., & Windsor, J. (2011). A comparison of the incidence and understanding of altitude illness between porters and trekkers in the Solu Khumbu region of Nepal. *Wilderness & Environmental Medicine*, 22(3), 197–201.

Parati, G., Agostoni, P., Basnyat, B., Bilo, G., Brugger, H., Coca, A., . . . Torlasco, C. (2018). Clinical recommendations for high altitude exposure of individuals with pre-existing cardiovascular conditions: A joint statement by the European society of cardiology, the council on hypertension of the European society of cardiology, the European society of hypertension, the international society of mountain medicine, the Italian society of hypertension and the Italian society of mountain medicine. *European Heart Journal*, 39(17), 1546–1554.

Parati, G., Ochoa, J.E., Torlasco, C., Salvi, P., Lombardi, C., & Bilo, G. (2015). Aging, high altitude, and blood pressure: A complex relationship. *High Altitude Medicine & Biology*, 16(2), 97–109.

Penaloza, D., & Arias-Stella, J. (2007). The heart and pulmonary circulation at high altitudes: Healthy highlanders and chronic mountain sickness. *Circulation*, 115(9), 1132–1146.

Procter, E., Brugger, H., & Burtscher, M. (2018). Accidental hypothermia in recreational activities in the mountains: A narrative review. *Scandinavian Journal of Medicine & Science in Sports*, 28(12), 2464–2472.

Pun, M. (2019). Re: "High altitude and cancer mortality" by Thiersch and Swenson (High Alt Med Biol 2018;19:116–123). *High Altitude Medicine & Biology*, 20(1), 101.

Reisman, J., Deonarain, D., & Basnyat, B. (2017). Impact of a newly constructed motor vehicle road on altitude illness in the Nepal Himalayas. *Wilderness & Environmental Medicine*, 28(4), 332–338.

Richards, P., & Hillebrandt, D. (2013). The practical aspects of insulin at high altitude. *High Altitude Medicine & Biology*, 14(3), 197–204.

Richins, H., & Hull, J.S. (Eds.). (2016). *Mountain Tourism: Experiences, Communities, Environments and Sustainable Futures*. Wallingford: CABI.

Roach, R.C., Bartsch, P., Oeltz, O., & Hackett, P.H. (1993). Lake Louise AMS scoring consensus committee: The Lake Louise acute mountain sickness scoring system. In J.R. Sutton, C.S. Houston, & G. Coates (Eds.), *Hypoxia and Molecular Medicine* (pp. 272–274). Burlington, VT: Queen City Printers.

Roach, R.C., Hackett, P.H., Oelz, O., Bartsch, P., Luks, A.M., MacInnis, M.J., & Baillie, J.K. (2018). The 2018 Lake Louise acute mountain sickness score. *High Altitude Medicine & Biology*, 19(1), 4–6.

Shah, M.B., Braude, D., Crandall, C.S., Kwack, H., Rabinowitz, L., Cumbo, T.A., . . . Bhasyal, G. (2006). Changes in metabolic and hematologic laboratory values with ascent to altitude and the development of acute mountain sickness in Nepalese pilgrims. *Wilderness & Environmental Medicine*, 17(3), 171–177.

Simonson, T.S. (2015). Altitude adaptation: A glimpse through various lenses. *High Altitude Medicine & Biology*, 16(2), 125–137.

Thiersch, M., & Swenson, E.R. (2018). High altitude and cancer mortality. *High Altitude Medicine & Biology*, 19(2), 116–123.

van Adrichem, E.J., Siebelink, M.J., Rottier, B.L., Dilling, J.M., Kuiken, G., van der Schans, C.P., & Verschuuren, E.A. (2015). Tolerance of organ transplant recipients to physical activity during a high-altitude expedition: Climbing Mount Kilimanjaro. *PLoS One*, 10(12), e0145566,

West, J.B. (2004). The physiologic basis of high-altitude diseases. *Annals of Internal Medicine*, 141(10), 789–800.

West, J.B. (2008). A new approach to very-high-altitude land travel: The train to Lhasa, Tibet. *Annals of Internal Medicine*, 149(12), 898–900.

Wilderness Medical Society (2019). Wilderness medicine clinical practice guidelines. Retrieved January 3, 2020, from https://wms.org/research/guidelines

Yadav, S. (2009). Urgent action is needed to prevent deaths among Himalayan pilgrims, says climbing federation. *British Medical Journal*, 338, b2628.

Yang, S.L., Ibrahim, N.A.F., Jenarun, G., & Liew, H.B. (2020). Incidence and determinants of acute mountain sickness in Mount Kinabalu, Malaysia. *High Altitude Medicine & Biology*, 21(3), 265–272.

Zafren, K., Pun, M., Regmi, N., Bashyal, G., Acharya, B., Gautam, S., . . . Basnyat, B. (2017). High altitude illness in pilgrims after rapid ascent to 4380 M. *Travel Medicine and Infectious Disease*, 16, 31–34.

# 6 Geopolitics in the Himalaya

## Territorial Conflict, Political Instability, Uprisings, and Tourism

*Dallen J. Timothy*

## Introduction

The world is ablaze with geopolitical conflicts with many roots, including ethno-religious dissonance, religious extremism, overlapping territorial claims, and control of natural resources (Al-Kanany, 2020; Metreveli & Timothy, 2010; Pak, 2016; Timothy, 2013). Some struggles are short-lived, while others have lasted for centuries with no end in sight. The history and heritage of many places revolve around political discord, wars, and struggles for recognition. While these events and their commemorative places often serve as important heritage tourism attractions, active conflicts typically deter tourism's growth and development in many different ways, including the destruction of natural resources and cultural heritage assets, impeding access and funding, and ruining a place's image.

The Himalayan region is beset with many of the types of conflicts described earlier. Some of the world's most dangerous areas and deepest conflicts are found in the Himalaya in areas that have great tourism potential, yet relatively few scholars have addressed the interactions between conflict and tourism in this region. Therefore, this chapter examines some of the geopolitical conflicts rooted in the Himalaya and how these affect resource management and tourism development. The chapter first describes tourism, conflict, and geopolitics from a general perspective and then hones in on how these manifest in the Himalayan region, most notably in the India-Pakistan border conflict, the Sino-Indian border dispute, the China-Bhutan border conflict, Nepal's Maoist insurgency, and Bhutan's refugee crisis—all of which have salient and unique implications for tourism.

## Geopolitics and Conflict in the Himalaya

Since the beginnings of the Anthropocene, humans have been embroiled in conflict and contentious relationships with one another. Until the establishment of the concept of Westphalian sovereignty in the 1648 Peace of Westphalia treaties, states and countries were in many cases nebulous entities loosely bounded by unclear frontier areas where the limits of sovereign control were fuzzy and often overlapping. As such, many kingdoms, principalities, and other political-territorial entities were in a constant state of disagreement and armed conflict over

DOI: 10.4324/9781003030126-8

their territorial extents. With the 1648 definition of the modern state in international law and geopolitics, conditions improved dramatically as national borders became clearer and functioned better as lines that demarcate the limits of national control. Nevertheless, territorial conflicts continue on every continent.

The most common international or intranational conflicts in history have had to do with territorial disputes, resource disagreements, inter-ethnic conflict, and religious disputations. There are hundreds of ongoing border and territorial disputes, many of which may never be resolved (Brunet-Jailly, 2015). Border disputes manifest in diplomatic démarches and protests on the milder end of the spectrum to outright wars and armed insurgencies on the extreme end (Timothy, 2001, 2013). The locations of international borders are directly related to natural resource claims (e.g., oil, water, natural gas, fishing rights, and coastal access), which lies at the core of most border disputes. Other conflicts arise when a state is populated by more than one nation, or when a dominant ethnic group superimposes power over minority groups within the same state.

Although some conflicts have become heritage tourist attractions as historic events, ongoing conflict is also seen in its own right as a potential attraction for certain types of extreme tourism, including tourists who are willing to take severe risks to venture into active war zones or areas of terrorist and rebel control (Mahrouse, 2016; Pelton, 2007; Timothy, 2013; Timothy, Prideaux & Kim, 2004). For the most part, however, the effects of conflict on tourism are rather predictable. Geopolitical instability causes immediate downturns in tourist arrivals, heavy layoffs and unemployment in the destination, canceled flights and tour packages, infrastructure and attraction closures, and frequent destruction of cultural and natural tourism assets (Al-Kanany, 2020; Butler & Suntikul, 2013; Fyall et al., 2006; Hall et al., 2003; Mansfeld & Pizam, 2006).

The Himalaya are a hotbed of contested territories, national boundaries, natural resources, and ethnic communities (Murton, 2019; Murton & Lord, 2020; Yeh, 2012). There are dozens of major and minor border and territorial disputes in Afghanistan, Pakistan, China, India, Bhutan, and Nepal. Most of them involve China with its claims and counterclaims. Table 6.1 illustrates some of the extant

*Table 6.1* A selection of current border and territorial disputes in the Himalaya

| Conflict | Belligerents | Currently Controlled by |
|---|---|---|
| Doklam Valley and Western Haa region | Bhutan and China | Bhutan |
| Arunachal Pradesh | China and India | India |
| Kalapani, Lipulekh, and Limpiyadhura | India and Nepal | India |
| Aksai Chin | China and India | China |
| Jammu and Kashmir | India and Pakistan | Divided between India and Pakistan |
| Siachen Glacier | India and Pakistan | India |
| Shaksgam Valley | China and India | China |

(*Continued*)

*Table 6.1* (Continued)

| Conflict | Belligerents | Currently Controlled by |
|---|---|---|
| Kula Kangri and mountainous area | Bhutan and China | China |
| Demchok | China and India | Divided between India and China |
| Nelang | China and India | India |
| Sang | China and India | India |
| Pulam Sunda | China and India | India |
| Chumur area | China and India | India |
| Kaurik | China and India | India |
| Bara Hoti | China and India | India |
| Shipki La | China and India | India |
| Susta village | India and Nepal | India |
| 'The Finger' of Sikkim | Indian and China | India |

Source: compiled by the author from various sources

border and territorial disputes in the Himalayan region. All of the area's conflicts have important implications for tourism, most notably degrowth and interrupted tourism development (Hannam, 2013; Mushtaq & Zaman, 2014). Several of these are examined in the sections that follow.

### The India-Pakistan Border Impasse

In 1947, the United Kingdom granted independence to British India, and the territory was partitioned into a predominantly Muslim state (Pakistan) and a chiefly Hindu state (India). Many South Asian scholars believe that the partition was an intentional effort by the British to sow discord in the region. Despite the intentions of the framers of independence and partition to create Muslim and Hindu majority states, the new international border divided ethnic and religious groups, locking many Hindus and Sikhs inside Pakistan and scores of Muslims inside the boundaries of India. The new border divided the Bengali people between India and Bangladesh, which was originally East Pakistan until its independence in 1971, as well as the Sikh homeland, Punjab, on the western border separating Sikhs from one another, and truncating access to many of their sacred sites on the Pakistani side of the border. The fear among all groups was so intense that the division prompted severe inter-communal violence in Punjab and elsewhere, characterized by intimidation, fighting, and massacres that some observers have characterized as genocide, enacted as a means of ethnically cleansing each side of the border of those who did not belong (Khan, 2017; White-Spunner, 2017). Setting the new boundary and the resulting violence led millions of people to flee to the side where they felt safest and most socially and religiously connected. This resulted in large waves of mass migration and millions of refugees on both sides of the border.

During independence, the princely states of British India could decide if they wanted to join Pakistan, India, or remain independent. A Hindu maharaja ruled the princely state of Jammu and Kashmir, even though the majority of its inhabitants were Muslims. During the 1947 mayhem, many of the principality's Muslim citizens were massacred by radical Hindus and Sikhs, ostensibly with the support of the prince. To quell some of the discord among his Muslim citizenry, the ruler decided to remain independent rather than join India (Paul, 2005), much to the dismay of the Pakistanis, who thought Jammu and Kashmir should be theirs owing to its majority Muslim population. In an effort to win the territory, Pakistan sent armed insurgents to take the region or divide it along religious lines. In retaliation, the maharaja rescinded the state's independence and decided to unite Jammu and Kashmir with India—a move which provoked a war that lasted until the end of 1948 with a brokered ceasefire (Schofield, 2000). In the Karachi Agreement, a ceasefire line was established and would be overseen by the United Nations. This new de facto boundary divided the region of Jammu and Kashmir between Indian and Pakistani control.

Since the division of British India, four official wars between the belligerent neighbors have taken place (in 1947, 1965 1971, and 1999), but many additional armed clashes have ensued at other times. In 1972, the 1947 ceasefire line became the 'line of control' (LoC) between the Pakistani and Indian controlled portions of Jammu and Kashmir. The section currently under Indian rule includes the regions of Jammu and Kashmir and Ladakh. The areas under Pakistani administration are comprised of Gilgit-Baltistan and Azad Kashmir (Timothy, 2019).

The violence associated with the partition of British India lies at the heart of the current hostile relationship between the two neighbors. After the 1947 partition, relations between Pakistan and India deteriorated drastically, with the core of the problem today continuing to be the disputed region of Jammu and Kashmir, with Pakistan and India both claiming sovereignty rights over the entire former princely state. Three quarters of a century later, the LoC and its adjacent territories are still in conflict, with frequent armed incursions, general political instability, and a constant threat of militant strikes (Timothy, 2019).

Jammu and Kashmir has significant tourism potential, but naturally the ongoing conflict has reduced the industry's ability to flourish. At present, the entire Pakistan-India border is all but entirely sealed against cross-border travel for Pakistanis and Indians, with only a few exceptions (Chhabra, 2018; Timothy, 2003, 2019; Timothy & Olsen, forthcoming). The two governments have conceded in recent years for tourism purposes and other human mobility needs. For example, with appropriate visas, a limited number of lorry drivers are permitted to cross the LoC, and a cross-border bus service was established some time ago to enable local Kashmiri 'tourists' to cross over for the purposes of small-scale trade and visiting relatives. Similarly, after years of intensive negotiations, the Kartarpur Corridor was opened in 2019 between the Indian village of Dera Baba Nanak, India, and the Pakistani village of Kartarpur. The gurdwara at Kartarpur is one of Sikhism's holiest shrines; it was built on the spot where Guru Nanak, the founder of Sikhism, died in 1539 and was one of holiest sites to have been left on the Pakistani side of the partition line, only 4 km from the border of India. Bilateral agreements allowed Indian Sikhs to visit this sacred site beginning November 9, 2019, on visa-free same-day foot journeys (Timothy & Olsen, forthcoming).

Although the COVID-19 pandemic curtailed pilgrims' use of this border opening after March 2020, it is expected to resume operations once the border reopens.

In both of these cases, the LoC local travel and the Kartarpur Corridor, the application process for border-crossing permits is laborious and lengthy, deterring many people from crossing (Malik, 2013; Singh, 2020). Foreign tourists and nonlocal Indians and Pakistanis are not allowed to cross the LoC or get close to it at any location (Timothy, 2019), and only Indian citizens can cross the border at the Kartarpur Corridor (Timothy & Olsen, forthcoming). The Wagah crossing near Lahore and Amritsar is the only place on the entire 4,063 km Indo-Pakistani border (including the LoC) where foreigners are permitted to cross. An international Delhi-Lahore coach service was initiated in 1999, which can be used by foreigners, as well as Indians and Pakistanis who have the appropriate visas (Malik, 2013), although the visa acquisition process for Pakistanis and Indians is extremely burdensome, time-consuming, and expensive. Although agreements between South Asian states through the South Asian Association for Regional Cooperation (SAARC) insist on freedom of travel between member states for SAARC citizens, its treaties are unheeded by the two hostile neighbors (Timothy, 2003). In addition to the cumbersome nature of the visas, Indian visas for Pakistanis and Pakistani visas for Indians restrict where one can travel and which ports of entry they are allowed to use. These and other restrictions are salient challenges to cross-border tourism (Irfan & Ali, 2021; Rasul & Manandhar, 2009).

Other concerns include the environmental costs of the conflict and a lack of outside investments in tourism. Even though the Himalayan environment in Jammu and Kashmir is one of the most biodiverse regions on the earth, the ongoing war directly damages natural and cultural heritage and prevents the involvement of foreign environmental and development NGOs from operating to help protect nature and address sustainable development projects (Ahmad, 2019). Likewise, a legitimate fear of violence translates into few outside financiers being willing to invest in a region that is plagued by constant armed conflict. This, together with the direct impacts of the conflict, continues to keep people in conditions of poverty and desperation (Ahmad, 2019).

Tourism in the disputed area has a great deal of potential, but most of its tourism development has been relegated to Ladakh and the vicinity of Srinagar. Many parts of the area are too geopolitically sensitive and dangerous for extensive tourism development. Violence along the LoC has prevented the growth and success of tourism in many otherwise desirable areas (Bhat, 2013; Dar, 2014). Jammu and Kashmir is one of the world's most dangerous war zones, which naturally taints the region's tourism image and directly impacts the natural and cultural assets in the region (Bhat et al., 2020; Islam, 2014; Vogel & Field, 2020).

### The Sino-India Border Disputes

There are several ongoing territorial disputes along the Chinese-Indian border, some of which involve large territories (see Table 6.1). In the Western Himalaya, several disputes have soured relations between China and India since the 1950s, the most notable of these being the region known as Aksai Chin which, although disputed between China and India, is currently controlled by China. There are several other areas along Tibet's border with Ladakh, Himachal Pradesh, and Uttarakhand that are contested between the two countries, with India currently controlling the majority of them. In the east, the northernmost tip of Sikkim, nick-named 'the Finger', has been a hotspot between the two superpowers since the 1960s, and India's northeastern state, Arunachal Pradesh, is the subject of Chinese claims and Indian counterclaims.

Much of the Sino-Indian territorial dispute derives from agreements India made with Tibet before the latter's takeover by China. The McMahon Line forms the border between India and China at Arunachal Pradesh and Tibet. It was agreed upon by Tibet and British India as part of the 1914 Simla Convention, but the Chinese maintain that Tibet was not a sovereign state at that time (it was under Chinese suzerainty) and was therefore ineligible to be party to the international treaty (Wagner, 2020). Arunachal Pradesh is an integral part of India and functions as such, but China continues to maintain its claims over the territory. In 1950, China took full control over Tibet and integrated it into the Chinese state. The area of Aksai Chin had until that time been a functional part of India's Jammu and Kashmir region, but a border between China and India in and around Aksai Chin had never been fully established because of the area's isolation, lack of population and natural resources, and the nebulous sovereignty of the states and proto-states that surrounded it. Nonetheless, India's cartographic version of the border was located further toward Tibet and China's view of the border placed it further into Ladakh. For China, the area had strategic importance because it connected the autonomous areas of Tibet and Xinjiang. For the Indians, it was a functional part of Jammu and Kashmir.

In 1962, the month-long Sino-Indian War broke out when India discovered and filed a complaint that China had built a road in Aksai Chin, and China invaded several Indian border stations in Aksai Chin, Sikkim, and Arunachal Pradesh. At the end of the 1962 war, the Chinese maintained what is now the Line of Actual Control (LAC), the de facto border between the two countries which set the current de facto boundary of Aksai Chin, and it continues to assert its claim over Arunachal Pradesh (Guo, 2007). Since the 1962 confrontation, the Sino-Indian border has experienced many armed skirmishes and standoffs in the area of Aksai Chin, Sikkim, and Arunachal Pradesh. At the time of writing, the most recent standoff between India and China occurred in 2020 (Wagner, 2020).

On May 5–6, 2020, Indian and Chinese soldiers positioned at the Chinese LAC in Ladakh/Aksai Chin exchanged heated arguments, pelted each other with rocks, and became involved in physical altercations. On May 10, the same thing happened further east on Sikkim's border with Tibet. A month later, the skirmishes turned deadly, with 20 Indian border patrol officers and an unnamed number of Chinese soldiers being killed at the Aksai Chin LAC. These latest clashes derived from India's efforts to improve its roads on its side of the LAC and increase its military presence inside the disputed border area. These efforts, as well as India's decision in 2019 to initiate direct national governance over Ladakh, upset the Chinese (Wu & Myers, 2020). Similar flareups occurred on the border of Arunachal Pradesh and Tibet in September 2020, with a heavier buildup of Chinese military near the border in the Eastern Himalaya. India responded with an increase in troops in Arunachal Pradesh. China still claims Arunachal Pradesh as its territory, calling it South Tibet, and does not recognize Indian sovereignty over the area. There is a constant state of tension between the two countries, with the September 2020 incidents, being among the latest escalations in hostility between the neighboring nuclear powers. In the words of one Chinese official, "We have never recognised the so-called Arunachal Pradesh illegally established on the Chinese territory" (quoted in Priyanka, 2020, n.p.). Although the territorial contestation between China and India has largely been about the placement of the international border, it has been fueled by India's growth as an Asian economic powerhouse, which China sees as its biggest competitor (Goswami, 2011). Amid the strife in the Eastern Himalaya, concerns have also been voiced that tensions may escalate in the same region in the future due to water insecurity, with China's water scarcity and its water diversion plans on the Brahmaputra River, which originates in Tibet and runs through disputed Arunachal Pradesh (Davis et al., 2021; Pak, 2016).

These continuing border conflicts have a number of important implications for tourism. Interestingly, Chinese authorities have denied visas to Indians from Arunachal Pradesh in recent years under the claim that the Indian state is part of China and its 1.2 million inhabitants are therefore citizens of China and thus do not require visas to enter the country (Goswami, 2011). Likewise, China has also attempted to impede Asian Development Bank funding for development projects, including tourism, in the state (Goswami, 2011; Rahman, 2014). The militarization of the border areas by India has also been accompanied by increased tourism development efforts, with the improved transportation infrastructure serving both purposes and the rising military presence lending credence to India's ability to control the frontier areas more securely for national interests, including tourism (Davis et al., 2021).

China's continuous claims over parts of India's Himalayan states translates into extra travel requirements. Visiting India's northeastern states and parts of Ladakh is more heavily regulated than visiting other parts of the country. Although the internal permit system was established during the British colonial era to limit trade between Indians and the tribes of the northeastern territories, ostensibly to protect the Indigenous people from outside interference and exploitation (Gill, 2020; Shackley, 1995) and to protect the interests of the British (Mahanta, 2016), the system remains in place but is now partially aimed at preventing tourists' involvement in civil unrest among tribal communities and sheltering tourists in border areas that are contested by China (Cook, 2019). Citizens of China and Pakistan, both in Himalayan territorial disputes with India, are not allowed to apply for travel permits to the border areas without special permission from the Ministry of Home Affairs.

Citizens of India are required to obtain Inner Line Permits (ILPs) to visit certain states and regions. Foreign tourists must acquire Protected Area Permits (PAPs) or Restricted Area Permits (RAPs) to visit certain states, although there are a few exceptions. PAPs require tourists to travel in groups of at least two, have their journeys organized by official travel agencies, enter the affected regions only at approved checkpoints, and have strict time limitations. RAPs can be issued for individuals traveling on their own.

Because of their sensitive border situations with China, permits are required to visit the Himalayan states of Sikkim and Arunachal Pradesh. To visit Arunachal Pradesh, Indian citizens need an ILP from the state government. Foreigners must obtain a PAP and travel in groups of at least two people, although individual travelers may visit a limited number of areas in the state. Although Indian citizens generally do not require an ILP to visit most parts of Sikkim, access to certain areas requires the permit. Foreign tourists must acquire a PAP to visit Sikkim, and in certain remote parts of the state where trekking is popular they require an additional RAP (Cook, 2019). Similarly, permits are required for travel in many parts of Ladakh outside the towns of Leh and Kargil, and tourists are not allowed to enter areas adjacent to the territories controlled by China or Pakistan (Vogel & Field, 2020).

According to Rizal and Asokan (2013), the border conflicts between China and India are one of the biggest impediments to tourism development in parts of India. Mitra and Lama (2013) argue that the cumbersome and time-consuming task of seeking permits is one of the most significant barriers to tourism development in India's Eastern Himalayan states. The states that impose these permit restrictions see them as an opportunity to monitor and control tourism, as well as to help protect Indigenous cultures (Mahanta, 2016).

### China-Bhutan Border Rows

On the northernmost edge of Bhutan is a large mountainous area that is currently contested between Bhutan and China. Presently, it is under Chinese control, but Bhutan claims ownership of the area. Physiologically, the region is flanked by the Tibetan Plateau and extremely high mountains, including Kangphu Kang I (7,204 meters) and Kula Kangri (7,538 meters), which Bhutan claims as its highest elevation. The landscape is replete with glaciers and glacial lakes, and hiking is permitted by the Chinese. Although the area is unpopulated, China has built limited infrastructure (i.e., unpaved mountain roads for military access) and placed military compounds in the area (Smith, 2013). Although Bhutan claims the area as its national territory, it is unable to utilize it for transhumance or tourism purposes. Because of its natural beauty, the contested area has significant tourism potential, but at present only China benefits from the limited number of trekkers who approach it from the Tibetan side.

A second conflicted area is the Doklam Valley and the Western Haa region near the tripoint of Bhutan, China, and India. It is home to agricultural villages, and tourism began there in 2002 with a focus on natural scenery and mountain hiking, as well as cultural landscapes. There is a strong Indian military presence to help defend Bhutan's territorial claims against the Chinese counterclaims. On

June 16, 2017, the Chinese military began extending a road onto the disputed territory. Bhutan notified China of its displeasure with this encroachment on territory it controlled and considered its sovereign territory. Indian and Chinese troops appeared to fortify their positions on opposite sides of the border. The tensions were eventually eased a couple months later, with China reinserting its claim to Doklam and Bhutan rejecting the Chinese claims (Safi, 2017).

### Nepal's Maoist Rebellion

The Nepali Civil War (also known as the Maoist Insurgency) took place between 1996 and 2006 between the government of Nepal and the country's Maoist Communist Party. During the ten-year conflict, the communist rebels carried out many criminal actions, including kidnappings, massacres, theft, executions, and bombings of bridges, telecommunications installations, power plants, schools, hotels, and police outposts in urban and rural areas in an attempt to devastate the government, overthrow the monarchy, and establish a communist people's republic (Lawoti, 2010; Thapa, 2004; Zharkevich, 2019). More than 17,000 police officials, military personnel, guerillas, and civilians were killed, and tens of thousands of people were displaced (Adhikari & Samford, 2013). During the conflict, the rebels ruled much of the country's rural territory, while most cities remained under government control. The armed conflict eventually ended with a peace accord in 2006. The monarchy was dissolved in May 2008 and a constitutional republic was established, integrating the Maoist Communist Party into conventional politics (Nyaupane & Budruk, 2009).

Before the Maoist revolution, Nepal had become a successful tourist destination, largely based on the image of a pristine, albeit mysterious and exotic, Himalayan 'Shangri La'—an image perpetuated by early explorers, climbers, hippies, and travel writers from the 1950s to the 1990s because of the country's natural beauty, pristine cultures, and relative isolation (Bhattarai et al., 2005). It was also regarded as a safe, tranquil and peaceful country (Thapa, 2004).

The terror attacks in the United States on September 11, 2001, intensified the world's concerns over armed rebellions, including that of the Maoists in Nepal, leading in part to the rebels being labeled terrorists in 2002 by the US government, the Nepali government, and various international organizations. Right away, Nepal was featured on the travel warning lists of several national governments, and travel agencies and tour companies began directing their clients to safer destinations (Bhattarai et al., 2005). Even though tourists themselves were not targeted (although some tourism infrastructure was), the rebel leaders warned tourists not to come to Nepal because their safety could not be guaranteed (Thapa, 2004). These incidents, of course, resulted in a measurable decline in international arrivals. The insurgency slowed tourism growth in the country, particularly in 2001 and 2002, although arrivals and growth ratios remained low until 2006. Tourist arrivals exceeded 1999 numbers only in 2007, once the war had ended (Nepal, 2010).

Despite this notable decline, tourists did not avoid Nepal during the late 1990s and early 2000s at the scale they normally do in conflict situations (Nepal, 2010).

In fact, reports suggest that tourism coexisted simultaneously with the insurgency and that most foreign visitors enjoyed their holidays in Nepal during the war (Hepburn, 2010); the country continued to be seen as a relatively safe destination even during the armed conflict (Upadhayaya et al., 2011). This was the case for a number of reasons. First, the Nepali civil war was overshadowed by the September 11 terror attacks and the newly-minted wars in Afghanistan and Iraq, resulting in the Nepali conflict receiving less condemnation and recognition by the global media (Pettigrew, 2013). It was thus a lesser-known conflict in the major tourism markets. Second, the Nepal Tourism Board intensified its efforts to counter-market against the conflict by launching its 'Destination Nepal' campaign in 2002 to stimulate domestic travel and to enact an international repositioning strategy to portray Nepal as a safe, attractive, and reliable destination (Thapa, 2004). Lastly, the insurgency maintained a dual policy of harm (to elite-owned hotels and government agencies) and no harm (to tourists) and therefore did not target individual tourists or tourist groups directly. Despite the warfare and attacks on tourism infrastructure, Upadhayaya et al. (2011) maintain that there are no records of foreign tourist fatalities from the conflict.

Besides the obvious decline in international arrivals, the insurgency had other negative effects on tourism and its resources (Nepal, 2010; Upadhayaya et al., 2011). Beginning in 1996, Nepal's positive tourism image was tainted, as it began to be seen on the global stage as a war-prone destination. Secondly, the Maoists usurped control over many conservation areas, which increased cases of poaching and illicit wildlife trade (Baral & Heinen, 2005). Their control over nature preserves also led to coerced 'donations' from tourists to help protect nature sanctuaries, such as the Annapurna Conservation Area. Likewise, trekking tourists were extorted by the revolutionaries, who forcefully collected 'access taxes' from independent travelers on popular trekking trails.

Armed attacks also took place. Several rural resorts, visitor centers, and wildlife camps were bombed, resulting in heavy destruction and extended closures (e.g., Begnas Lake Resort and Gaida Wildlife Camp in Chitwan National Park). Similarly, urban lodges and hotels were bombed in Kathmandu, Gorkha, and Pokhara, leading to the closures of a dozen multinational corporation-owned accommodation establishments in various parts of the country. An armed attack on Lukla Airport, the main gateway to the Everest region, closed the airport for a long time, weakening travel options to one of the country's busiest trekking regions (Upadhayaya et al., 2011).

Declining visitor arrivals resulted in intense competition between hotels, which meant they had to lower their prices substantially. This, together with communist labor unions forcing the closure of some hotels through their lack of cooperation with the government regarding wage equity, part-time versus full-time employment, and other work conditions, led a number of hotels to close their doors permanently. Some were transformed into department stores or other facilities (Upadhayaya et al., 2011). Likewise, many shops and retail businesses were closed throughout the country; owners were reluctant to remain open for fear of reprisal or extortion from the guerillas (Pettigrew, 2013). The conflict also spelled the near collapse of Nepal Airlines in 2007 due to insolvency; its name was associated with the ongoing war and foreign carriers were able to offer ostensibly more reliable air services (Nepal, 2010).

Within just a few years of the end of the Maoist rebellion, tourism had showed a remarkable recovery (Upadhayaya, 2013), and today, with the exception of the interruption caused by the COVID-19 pandemic, it continues to grow as one of Asia's most popular destinations, having received 1.2 million foreign tourists in 2019—a 100% increase over 2010. One unexpected outcome of the war is the recent development of the 'Guerrilla Trail'—a unique trekking route that traverses rural areas in western Nepal, retracing the footsteps of the Maoist insurgents. The hiking trail blends Indigenous culture, natural scenery, and the heritage of the 'people's war', and aims to reconcile the conflict and utilize the country's dark heritage to develop harmony and peaceful coexistence between Nepal's various ethnic groups (Rana & Bhandari, 2018).

### *Bhutan's Nepali Minority and the Refugee Crisis*

The population of Bhutan is mixed with various Indigenous people, and a large portion of people of Tibetan and South Asian descent. The majority of the Bhutanese population, the Ngalop, speaks Dzongkha and adheres to Buddhism. Approximately one-quarter of the population are of Nepali origin (known as Lhotshampa (southerners)), speak Nepali, and mostly practice Hinduism (Pulla, 2016). Although the earliest migrants of Nepali origin arrived in Bhutan in the seventeenth century, most Lhotshampa immigrated to Bhutan from Nepal and northern India in the nineteenth and early twentieth centuries to farm the country's lowlands. They populated the southernmost part of Bhutan and produced much of the country's food. By law, they were restricted on where they could live, being relegated to the southernmost parts of the kingdom (Evans, 2010; Tez, 2020).

Concerns over the democratic movement in Nepal and its spillover effects and threats to the Bhutanese monarchy in the late 1980s, and a growing Bhutanese nationalist narrative saw the enactment of laws and policies that excluded much of the country's Nepali origin population and favored the northern Bhutanese cultures (Pulla, 2016). Bhutanese government actions in the 1980s and 1990s included cultural suppression by incentivizing intermarriage and cultural integration, required wearing of traditional Bhutanese attire, required new buildings to be built in traditional Bhutanese style, reaffirmed Dzongkha as the national language of the country, ended teaching in Nepalese, and allowed citizenship only for immigrants who could prove they had arrived at least ten years before 1958. This resulted in a large number of Lhotshampa being labeled illegal immigrants and being expelled using armed forces to Nepal through India (Evans, 2010).

The Bhutanese of Nepali origin struggled against these policies, which they deemed to be oppressive and targeting their culture. In 1990, skirmishes, protests, and anti-government violence ensued in the south with demands for equitable human rights and respect for Nepali culture. Many Lhotshampa were arrested and increasing numbers were deported (Hutt, 1996; Tez, 2020). India sided with the Bhutanese government, offering to stand guard on the kingdom's southern border.

Nepal sided with the Lhotshampa, dubbing them 'freedom fighters' and accepting many refugees into camps inside Nepali territory, increasing tension between the two countries. Out of over 130,000 Bhutanese refugees who were in refugee camps in Nepal, more than 113,000 have been resettled through UNHCR, a UN refugee agency, in eight Western countries. Over 96,000 refugees have been resettled in the US alone (Mørch, 2016; Sinha, 2016; UNOCHA, 2019).

Tourism began in Bhutan in 1974, when the government opened the country to outsiders as a means of earning foreign exchange. By the end of the 1980s, tourism had grown substantially and had become an important source of income, although compared to other countries of South Asia, it was still an untouched and relatively isolated destination state (Ritchie, 2008), which is both a function of its mystical ambiance, location in the Himalaya, and its tourism policies that aim to protect its culture and environment (Nyaupane & Timothy, 2010). Although it is likely that the conflict in southern Bhutan in the 1980s and 1990s affected tourism growth and tourist arrivals somewhat, the majority of tourism in Bhutan occurs in the central and western parts of the country, and most international tourists arrive by air in Paro, with relatively few arriving overland from India in the south or spending much time in the region populated by the Lhotshampas. In addition, many observers have suggested that the refugee crisis in Bhutan is an 'unknown' catastrophe or one which the world has forgotten and has received much less global attention than other refugee crises (Ghosh, 2010; Rizal, 2004), which likely translates into fewer people knowing about the situation and it not affecting tourism as much as refugee crises have done in other parts of the world (e.g., Myanmar and southern Europe) (Tsartas et al., 2020).

Likewise, some of the nationalist policies put in place in the 1980s, which oppressed the Nepali minority, have simultaneously benefited tourism. For example, the Driglam Namzha requirement for Bhutanese citizens to wear the national dress of the Ngalop majority, practice Bhutanese etiquette, speak Dzongkha, and construct all buildings according to traditional Ngalop architectural style has helped maintain elements of traditional culture that have become an important part of the Bhutanese tourism experience (Nyaupane & Timothy, 2010).

## Conclusion

Geopolitical contention and armed disputes occur in all corners of the globe. In the Himalaya, these contestations are a result of border disagreements and territorial claims, natural resource ownership, military and tourism access in remote borderlands, ethnic strife, and struggles for socio-economic equality. Every country in the region is affected by these issues in one form or another, and conflict and dissonance are at the forefront of public policy, international relations, SAARC supranational agreements, and the development of tourism.

So many borderland areas and rural regions, where the majority of these conflicts occur, have substantial tourism potential, yet they remain some of the least developed parts of Afghanistan, India, Pakistan, China, Bhutan, and Nepal. Several conflicts have diminished in recent years (e.g., the Nepali insurgency and the Bhutanese refugee predicament), but these are the ones that have focused on socioreligious, cultural, and ethnic differences. These problems have not, however, been eradicated, as the continued existence of Lhotshampa refugee camps in Nepal and India illustrates. Yet, the loudest geopolitical conflicts remain those associated with overlapping territorial claims between India and China, and India and Pakistan. Increased militarization and sporadic armed skirmishes characterize these latter troubles, making these areas some of the most volatile in the world and hindering any effective cross-border cooperation in tourism or the growth of tourism in the borderlands on either side.

Some countries such as Nepal have shown considerable resilience and have been able to adopt post-conflict recovery techniques that have enabled its recovery from the downward effects of its civil war. Even though Nepal remains on the US government's travel advisory list (as of May 1, 2021) for potential political violence, the country's tourism sector had continued to perform well following the end of guerilla warfare in 2006, until the COVID-19 pandemic shattered tourism there as in all the Himalayan countries. The ethnic discord in Bhutan has not affected tourism as much as similar situations have in other countries (e.g., Ethiopia, Ukraine, Azerbaijan-Armenia, and Bosnia and Herzegovina), largely because of the global media's silence on the matter and an overemphasis on the kingdom's unique Gross Happiness Index. Although China and India have thriving tourism sectors, their common border continues to be hotly contested, affecting not only the people who live there but depriving them of a livelihood that has been shown to succeed in other parts of the Himalayan and South Asian region. Finally, the official Indian-Pakistani border has shown some signs of change, such as the Kartarpur Corridor, but the LoC in Jammu and Kashmir continues to be extremely volatile and an improbable tourism region.

## References

Adhikari, P., & Samford, S. (2013). The Nepali state and the dynamics of the Maoist insurgency. *Studies in Comparative International Development*, 48(4), 457–481.

Ahmad, O. (2019). The costs of heightened conflict in the Himalayas. *Inter Press Service*, September 10. Retrieved July 22, 2020, from www.ipsnews.net/2019/09/costs-heightened-conflict-himalayas/

Al-Kanany, M.M.R. (2020). Extremist iconoclasm versus real Islamic values: Implications for heritage-based tourism development in Iraq. *Journal of Heritage Tourism*, 15(4), 472–478.

Baral, N., & Heinen, J.T. (2005). The Maoist people's war and conservation in Nepal. *Politics and the Life Sciences*, 24(1–2), 2–11.

Bhat, A.A., Majumdar, K., & Mishra, R.K. (2020). Local support for tourism development and its determinants: An empirical study of Kashmir region. *Asia Pacific Journal of Tourism Research*, 25(11), 1232–1249.

Bhat, Z.A. (2013). Tourism industry and pilgrimage tourism in Jammu and Kashmir: Prospects and challenges. *International Journal of Research in Management & Technology*, 2, 105–113.

Bhattarai, K., Conway, D., & Shrestha, N. (2005). Tourism, terrorism and turmoil in Nepal. *Annals of Tourism Research*, 32(3), 669–688.

Brunet-Jailly, E. (Ed.). (2015). *Border Disputes: A Global Encyclopedia*. Santa Barbara, CA: ABC-CLIO.

Butler, R.W., & Suntikul, W. (Eds.). (2013). *Tourism and War*. London: Routledge.

Chhabra, D. (2018). Soft power analysis in alienated borderline tourism. *Journal of Heritage Tourism*, 13(4), 289–304.

Cook, E. (2019). Permits for North East India and what you need to know. Retrieved February 22, 2021, from www.tripsavvy.com/permits-for-north-east-india-1539554.

Dar, H. (2014). The potential of tourism in border destinations: A study of Jammu and Kashmir. *African Journal of Hospitality, Tourism and Leisure*, 4(2), 1–12.

Davis, A.E., Gamble, R., Roche, G., & Gawne, L. (2021). International relations and the Himalaya: Connecting ecologies, cultures and geopolitics. *Australian Journal of International Affairs*, 75(1), 15–35.

Evans, R. (2010). The perils of being a borderland people: On the Lhotshampas of Bhutan. *Contemporary South Asia*, 18(1), 25–42.

Fyall, A., Prideaux, B., & Timothy, D.J. (2006). War and tourism: An introduction. *International Journal of Tourism Research*, 8(3), 153–155.

Ghosh, P. (2010). *Bhutanese Refugees: A Forgotten Saga*. London: Minerva.

Gill, S. (2020). Planning to visit the Northeast? You may require an ILP in some of the states. Don't know what it is and how to get it? Here's a guide. Retrieved March 30, 2021, from www.outlookindia.com/outlooktraveller/explore/story/70169/permits-required-to-travel-to-northeast-india

Goswami, N. (2011). China's 'aggressive' territorial claim on India's Arunachal Pradesh: A response to changing power dynamics in Asia. *Strategic Analysis*, 35(5), 781–792.

Guo, R. (2007). *Territorial Disputes and Resource Management: A Global Handbook*. New York: Nova Science.

Hall, C.M., Timothy, D.J., & Duval, D.T. (Eds.). (2003). *Safety and Security in Tourism: Relationships, Management and Marketing*. New York: Haworth.

Hannam, K. (2013). "Shangri-La" and the new "great game": Exploring tourism geopolitics between China and India. *Tourism Planning & Development*, 10(2), 178–186.

Hepburn, S. (2010). Shades of darkness: Silence, risk and fear among tourists and Nepalis during Nepal's civil war. *Journeys*, 11(1), 133–155.

Hutt, M. (1996). Ethnic nationalism, refugees and Bhutan. *Journal of Refugee Studies*, 9(4), 397–420.

Irfan, E., & Ali, Y. (2021). Religious tourism in Pakistan: Scope, obstacles and strategies. *Journal of Convention & Event Tourism*, 22(2), 134–154.

Islam, A.U. (2014). Impact of armed conflict on economy and tourism: A study of the state of Jammu and Kashmir. *IOSR Journal of Economics and Finance*, 4(6), 55–60.

Khan, Y. (2017). *The Great Partition: The Making of India and Pakistan* (2nd ed.). New Haven: Yale University Press.

Lawoti, M. (2010). Evolution and growth of the Maoist insurgency in Nepal. In M. Lawoti & A.K. Pahari (Eds.), *The Maoist Insurgency in Nepal: Revolution in the Twenty-First Century* (pp. 3–31). London: Routledge.

Mahanta, R. (2016). Inner Line Permit as an instrument of protecting identity: Benefits, costs and effectiveness. In M.P. Bezbaruah (Ed.), *Identity Aspirations, Developmental Backlogs and Governance Issues in Northeast India* (pp. 55–65). Guwahati, India: Department of Economics, Gauhati University.

Mahrouse, G. (2016). War-zone tourism: Thinking beyond voyeurism and danger. *ACME: An International Journal for Critical Geographies*, 15(2), 330–345.

Malik, S.A. (2013). Development of difficult region through travel, trade and tourism: A case study of twin border districts Rajouri and Poonch. *International Journal of Marketing, Financial Services and Management Research*, 2(2), 56–65.

Mansfeld, Y., & Pizam, A. (Eds.). (2006). *Tourism, Security and Safety: From Theory to Practice*. Amsterdam: Elsevier.

Metreveli, M., & Timothy, D.J. (2010). Effects of the August 2008 war in Georgia on tourism and its resources. In O. Moufakkir & I. Kelly (Eds.), *Tourism, Progress and Peace* (pp. 134–147). Wallingford: CABI.

Mitra, A., & Lama, M. (2013). Tourism development in Arunachal Pradesh: Opportunities and challenges. *Productivity*, 53(4), 371–378.

Mørch, M. (2016). Bhutan's dark secret: The Lhotshampa expulsion. *The Diplomat*, September 21. Retrieved February 3, 2021, from https://thediplomat.com/2016/09/bhutans-dark-secret-the-lhotshampa-expulsion/

Murton, G. (2019). Facing the fence: The production and performance of a Himalayan border in global contexts. *Political Geography*, 72, 31–42.

Murton, G., & Lord, A. (2020). Trans-Himalayan power corridors: Infrastructural politics and China's belt and road initiative in Nepal. *Political Geography*, 77, 102100.

Mushtaq, A., & Zaman, K. (2014). The relationship between political instability, terrorism and tourism in SAARC region. *Journal of Economic Info*, 1(1), 23–40.

Nepal, S. (2010). Tourism and political change in Nepal. In R. Butler & W. Suntikul (Eds.), *Tourism and Political Change* (pp. 147–159). Oxford: Goodfellow.

Nyaupane, G.P., & Budruk, M. (2009). South Asian heritage tourism: Conflict, colonialism, and cooperation. In D.J. Timothy & G.P. Nyaupane (Eds.), *Cultural Heritage and Tourism in the Developing World: A Regional Perspective* (pp. 127–145). London: Routledge.

Nyaupane, G.P., & Timothy, D.J. (2010). Power, regionalism and tourism policy in Bhutan. *Annals of Tourism Research*, 37(4), 969–988.

Pak, J.H. (2016). China, India, and war over water. *Parameters*, 46(2), 53–67.

Paul, T.V. (2005). *The India-Pakistan Conflict: An Enduring Rivalry*. Cambridge: Cambridge University Press.

Pelton, R.Y. (2007). *The World's Most Dangerous Places: Professional Strength*. New York: Harper Collins.

Pettigrew, J. (2013). *Maoists at the Hearth: Everyday Life in Nepal's Civil War*. Philadelphia: University of Pennsylvania Press.

Priyanka (2020). China increases military presence near Arunachal Pradesh border, India moves troops. *India.com*, September 16. Retrieved January 3, 2021, from www.india.com/news/india/china-increases-military-presence-near-arunachal-pradesh-border-india-moves-troops-4142628/

Pulla, V. (2016). *The Lhotsampa People of Bhutan: Resilience and Survival*. New York: Palgrave.

Rahman, M.Z. (2014). Territory, tribes, turbines: Local community perceptions and responses to infrastructure development along the Sino-Indian Border in Arunachal Pradesh. *ICS Occasional Papers*, 7, 1–35.

Rana, S., & Bhandari, C. (2018). Guerrilla Trail: A tourism product in the Maoist rebel's footsteps. *Journal of Tourism*, 19(1), 33.

Rasul, G., & Manandhar, P. (2009). Prospects and problems in promoting tourism in South Asia: A regional perspective. *South Asia Economic Journal*, 10(1), 187–207.

Ritchie, M. (2008). Tourism in the kingdom of Bhutan: A unique approach. In J. Cochrane (Ed.), *Asian Tourism: Growth and Change* (pp. 273–284). Amsterdam: Elsevier.

Rizal, D. (2004). The unknown refugee crisis: Expulsion of the ethnic Lhotsampa from Bhutan. *Asian Ethnicity*, 5(2), 151–177.

Rizal, P., & Asokan, R. (2013). A comparative study of tourism industry in north-eastern states of India. *IOSR Journal of Business and Management*, 12(4), 56–62.

Safi, M. (2017). Chinese and Indian troops face off in Bhutan border dispute. *The Guardian*, July 5. Retrieved December 3, 2020, from www.theguardian.com/world/2017/jul/06/china-india-bhutan-standoff-disputed-territory

Schofield, V. (2000). *Kashmir in Conflict: India, Pakistan and the Unending War*. London: IB Tauris.

Shackley, M. (1995). Just started and now finished: Tourism development in Arunachal Pradesh. *Tourism Management*, 16(8), 623–625.

Singh, S. (2020). Kartarpur Corridor: Over 59,000 visited kartarpur gurdwara since Nov last year. *Hindustan Times*, March 8. Retrieved February 22, 2021, from www.hindustantimes.com/cities/kartarpur-corridor-over-59-000-visited-kartarpur-gurdwara-since-nov-last-year/story-msJpPO55kLJWtmBqy21BGL.html

Sinha, A.C. (2016). Evicted from home, nowhere to go: The case of Lhotshampas from Bhutan. In T.B. Subba & A.C. Sinha (Eds.), *Nepali Diaspora in a Globalised Era* (pp. 230–245). London: Routledge.

Smith, P.J. (2013). Bhutan-China border disputes and their geopolitical implications. In B.A. Elleman, C.H. Schofield, & S. Kotkin (Eds.), *Beijing's Power and China's Borders: Twenty Neighbors in Asia* (pp. 23–35). London: M.E. Sharpe.

Tez, B.R.S. (2020). Ethnic issues of the Lhotshampas in Bhutan. *World Affairs*, 24(4), 136–151.

Thapa, B. (2004). Tourism in Nepal: Shangri-La's troubled times. In C.M. Hall, D.J. Timothy, & D.T. Duval (Eds.), *Safety and Security in Tourism* (pp. 117–138). New York: Haworth.

Timothy, D.J. (2001). *Tourism and Political Boundaries*. London: Routledge.

Timothy, D.J. (2003). Supranationalist alliances and tourism: Insights from ASEAN and SAARC. *Current Issues in Tourism*, 6(3), 250–266.

Timothy, D.J. (2013). Tourism, war, and political instability: Territorial and religious perspectives. In R. Butler & W. Suntikul (Eds.), *Tourism and War* (pp. 12–25). London: Routledge.

Timothy, D.J. (2019). Tourism, border disputes and claims to territorial sovereignty. In R.K. Isaac, E. Çakmak, & R. Butler (Eds.), *Tourism and Hospitality in Conflict-Ridden Destinations* (pp. 25–38). London: Routledge.

Timothy, D.J., & Olsen, D.H. (forthcoming). Sacred spaces and the 'other': Social distance, functional distance, and two pilgrimage sites in Asia. In R.N. Progano, J.M. Cheer, K. Kato, & X.M. Santos (Eds.), *Host Communities and Pilgrimage Tourism: Asia and Beyond*. Singapore: Springer.

Timothy, D.J., Prideaux, B., & Kim, S.S. (2004). Tourism at borders of conflict and (de) militarized zones. In T.V. Singh (Ed.), *New Horizons in Tourism: Strange Experiences and Stranger Practices* (pp. 83–94). Wallingford: CABI.

Tsartas, P., Kyriakaki, A., Stavrinoudis, T., Despotaki, G., Doumi, M., Sarantakou, E., & Tsilimpokos, K. (2020). Refugees and tourism: A case study from the islands of Chios and Lesvos, Greece. *Current Issues in Tourism*, 23(11), 1311–1327.

UNOCHA (2019). *Press Statement: US Ambassador Randy Berry Visits Bhutanese Refugee Settlements*. United Nations Office for the Coordination of Humanitarian Relief. Retrieved

July 10, 2021, from https://reliefweb.int/report/nepal/press-statement-us-ambassador-randy-berry-visits-bhutanese-refugee-settlements

Upadhayaya, P.K. (2013). Peace through tourism: A critical look at Nepalese tourism. *Nepal Tourism and Development Review*, 1(1), 15–40.

Upadhayaya, P.K., Müller-Böker, U., & Sharma, S.R. (2011). Tourism amidst armed conflict: Consequences, copings, and creativity for peace-building through tourism in Nepal. *The Journal of Tourism and Peace Research*, 1(2), 22–40.

Vogel, B., & Field, J. (2020). (Re)constructing borders through the governance of tourism and trade in Ladakh, India. *Political Geography*, 82, 102226.

Wagner, C. (2020). The Indian-Chinese confrontation in the Himalayas: A stress test for India's strategic autonomy. *Stiftung Wissenschaft und Politik*, 39, 1–4.

White-Spunner, B. (2017). *Partition: The Story of Indian Independence and the Creation of Pakistan in 1947*. London: Simon & Schuster.

Wu, J., & Myers, S.L. (2020). Battle in the Himalayas. *New York Times*, July 18. Retrieved August 3, 2020, from www.nytimes.com/interactive/2020/07/18/world/asia/china-india-border-conflict.html

Yeh, E.T. (2012). Transnational environmentalism and entanglements of sovereignty: The tiger campaign across the Himalayas. *Political Geography*, 31(7), 408–418.

Zharkevich, I. (2019). *Maoist People's War and the Revolution of Everyday Life in Nepal*. Cambridge: Cambridge University Press.

# 7 Moral Geographies, Porters, and Social Exclusion in Trekking Tourism in the Himalayan Region

*Sanjay K. Nepal*

## Introduction

Tourism spaces in the Himalayan region are contested grounds due to conflicts between stakeholders engaged in planning, developing, and providing services to tourists (Butz, 2002; Nepal, 2000a, 2000b). This raises many questions about moral geographies of tourism space. For instance, what are the moral discourses used in constructing spaces of trekking and mountaineering tourism? How do hosts and guests engaged in trekking and mountaineering in the Himalaya learn "to do tourism" ethically and morally in that context? How does tourism accentuate or minimize socially and culturally-embedded practices of injustice and inequality prevalent among the poorer and weaker sections of society that are involved in tourism?

Relevant to the questions raised earlier, this chapter seeks to answer three primary questions: 1) what is the role of porters in Himalayan mountaineering and trekking tourism? 2) how do porters perceive themselves (vis-a-vis others) and their roles in the tourism industry? and 3) what performative practices (spatial and temporal) contribute to the social exclusion and discrimination of porters? These questions are relevant to moral geographies as they "focus on power to prescribe who may do what, where and with whom: moral geographies operate not only through a dominative power of control and exclusion but also through performative powers of spatial practice" (Matless, 1994, p. 396, as cited in Smith, 2000b). The analysis presented here is primarily focused on the Nepalese Himalaya, which is the primary research location for tourism studies conducted by this author. However, whenever possible, references are made to the wider Himalayan region.

The remainder of this chapter is organized under three sections. The second section briefly contextualizes moral geographies of tourism. The third section forms the main body of this chapter, as it focuses on porters' role in trekking tourism in the Himalayan region, porters' views of themselves and others, and the spaces of exclusion within which they interact with the others. This section is based in part on data collected in the Mt. Everest area of Nepal and is supported by other studies conducted in the broader Himalayan region. Spaces of social exclusion and practices within trekking tourism that contribute to social exclusion

DOI: 10.4324/9781003030126-9

of porters and undermine their overall importance to the industry are discussed in detail. The final section offers some key conclusions and suggestions for policy, practice, and research.

## Moral geographies of tourism

Broadly speaking, moral geography refers to "a particular style of geographical investigations with a moral dimension" (Smith, 2004, p. 197). Based on Harvey's (1993, 1996) views of social justice, and Smith's (1997) work on ethics, this body of knowledge is relatively recent, i.e., post-1990, and is framed arguably as prompting the moral turn in geography (Smith, 1997). A critical synthesis on this topic can be found in Smith's (2000a) *Moral Geographies: Ethics in a World of Difference.* From a social perspective, Cresswell (1996) suggests the place of an act, whether the act is right or wrong, plays a role in our understanding of what is good, just, and appropriate. All interpretations of place have moral content, reflecting what is considered good (or bad) in a geographical and historical context (Smith, 2000a). Adopting a moral perspective shows how values are implicated in the places people create (Smith, 2000b). Indeed, places and their interpretations are intrinsically moral projects: "imposing order or locating economic or social purpose or by imbuing particular places with value according to what they represent" (Smith, 2000a, p. 45). Moral geographies work around the way individuals act for themselves as ethical subjects in relation to a moral code defined by a social group (Holloway, 1998; Matless, 1994).

In the tourism literature, issues relevant to moral geographies have been framed explicitly as discourses related to "responsible" (Caruana et al., 2014; Goodman & Francis, 2003; Timothy, 2012), "ethical" (Fennell, 2017), and "moral" tourism (Butcher, 2000; Mostafanezhad & Hannam, 2016). Morality in tourism explicitly, or implicitly, is present in issues of social and environmental justice (Jamal, 2019), advocacy for ethical treatment, inclusionary development practices, community empowerment, equity, and rights-based approaches (see examples in Nepal & Saarinen, 2016). Not explicitly defined as moral geographies of space for tourism, but implicit in their interpretations, include literature relevant to volunteer tourism (Mostafanezhad, 2016), pro-poor tourism (Harrison, 2008; Scheyvens, 2011), community-based tourism (Salazar, 2012; Vajirakachorn & Nepal, 2014), inclusive tourism (Scheyvens & Biddulph, 2018), and ecotourism. Ecotourism is closely associated with ethical tourism linked to environmental conservation and community development (Nepal & Saarinen, 2016; Dahal et al., 2014; Lai & Nepal, 2006; Nepal, 2000b; Nyaupane & Thapa, 2004), as well as codes of conduct and guides (e.g., The Good Tourist, The Green Travel Guide, etc.) that promote a moral agenda of advising tourists' personal behaviors (Butcher, 2000; Poudel et al., 2013). In these contexts, moral tourism is characterized by its advocacy of more sensitive behavior toward environments and respect for other ways of life (Butcher, 2000).

Spaces of trekking and mountaineering tourism in the Himalaya underline prevailing values concerning what is "in-place", thus straddling the boundaries of moral geographies of inclusion and exclusion (Cresswell, 1996). Within these boundaries, there is a shared sense of morality among like-minded people. A sense of community and belonging permits and legitimizes certain behaviors that work to regulate human conduct (Rheingold, 1993; Walzer, 1983). The plight of porters in trekking tourism in the Himalayan region illustrates the incompatibility and inconsistency between broader concerns about ethical tourism globally and tensions in host-guest practices on the ground.

## The moral geographies of porters in the Himalaya

Wearing et al. (2007) argue that the mountaineering and trekking tourism industry in the Himalayan region tends to ignore the reality of porter conditions, despite the fact that the nature of the trekking experience theoretically offers tourists the opportunity to better relate to local workers. The issue is much broader and more widely prevalent, as Mowforth and Munt (2015) note—that there is the potential for the fetishistic colonialist mist around the representation of tourism labor to be broken down within the trekking sector. Trekking brings tourists into direct contact with local people, thus forcing the tourism consumer to consider the local relationships behind the product they are consuming (Mowforth & Munt, 2015). The trekking industry in the Himalayan region has attempted to present an idealized representation of porters that is built on colonial images of Himalayan grandeur and the challenges presented to Western tourists in conquering the region's natural environment (Wearing et al., 2007). It is somewhat surprising with a few exceptions that issues of social justice, ethics, labor relations, and local empowerment have not been well documented in the Himalayan tourism literature.

### *The role of porters in trekking tourism*

Locally employed stakeholders in trekking tourism in the Himalaya can be separated into three specific groups: mountaineering operators and trekking lodge owners, trekking guides, and porters. Among these stakeholders, porters occupy the lowest rungs of the socio-economic hierarchy (Barott, 2018). However, even within the porter category, there are three class distinctions, which suggest a further hierarchical stratification of porters based on skills, economic incentives, social prestige, and ethnicity. At the upper echelons of this hierarchy are high-altitude porters who are employed in mountaineering and climbing expeditions. These porters are engaged in high-risk activities due to the nature of their job, which involves hauling climbing equipment and gear at higher altitudes for mountaineering expeditions. They are well paid, compared to other porters, and are admired by others for the social prestige that comes with the job and the social circle to which they belong. Trekking porters are in the middle category. These workers carry heavy equipment, trekking provisions, and tourists' baggage, and accompany trek leaders, guides, and tourists during the trekking journey. The third, and lowest-ranked category, are the so-called merchandise or commercial porters (Nepal et al., 2002), who transport goods for mountain lodges, shops, and other businesses located along the trekking routes. Among the three categories of porters, commercial porters are the most vulnerable due to their inability to set the price of their labor, and other terms of portering (e.g., load weights and occupational safety), and unnecessary competition from porters who are willing to carry for less compensation. They are frequently paid according to weight transported, resulting in heavy loads being carried each trip to maximize their earnings.

Further distinctions can be made of the porters' ethnicities and the evolving nature of labor composition in trekking and mountaineering tourism. For example, during the pioneering years of the 1950s and through the 1970s, most of the porters and mountain guides in the Everest region were represented by the Sherpa community, especially from the Upper Khumbu villages such as Namche Bazaar, Khumjung, and Khunde (Ortner, 1999; von Fürer-Haimendorf, 1964). Post-1990s, portering has been dominated by other ethnic groups such as Rai and Tamang in the Everest region and somewhat mixed ethnic groups in the Annapurna region (Nepal et al., 2002; Ortner, 1999). Elsewhere, these are primarily Drok-pa and Brok-pa in Ladakh (Rajashekariah & Chandan, 2013), and Shimshali porters in northern Pakistan where portering is not confined to a special ethnic group or class of households (Butz, 2006). Specific to the Everest region, today, members of the Sherpa community have moved up the socio-economic ladder and are primarily lodge owners and expedition guides and outfitters. They now employ people from the Rai and Tamang communities. Still, most of the mountaineering expedition porters are Sherpa from high-altitude villages including Thame, Khumjung, Phortse, and Pangboche. While the data is now a bit obsolete, a survey of porters in the Everest region showed that "within a 12-month period in 1996 and 1997, the gate at Everest National Park registered 13,389 entries by trekking-related porters. In addition, 14,279 commercial porters and 2,791 packstock carrying trekking and merchandise loads entered the park" (Nepal et al., 2002, p. 49). That survey was done when the number of trekkers was almost half of what it is today.

Porters are the backbone of trekking and mountaineering tourism in the Himalaya, particularly in popular destinations such as Nepal's Mt. Everest region where motorized transport does not exist. Even in areas with transportation access, trekking tourism relies primarily on the backs of porters. One of the common characteristics of porters in this region is that they are from poor families, lack education and awareness of the tourism industry, and are not aware of workers' rights, advocacy organizations, and labor practices. An earlier study reported 30% of porters in the Mt. Everest region had no education at all (Panzeri et al., 2013). Trekking porters are usually young (early teens to 35 years of age), and are away from home for extended periods of time to earn the most money during peak tourist seasons before returning home for school, farm work, or to take care of their families. For a trekking porter, a typical itinerary in the Mt. Everest region involves being with the trekking party for about 12–14 days, carrying heavy loads, getting minimal pay, and laboring without adequate food and shelter provided by the trekking guide/company. Expecting a generous tip at the end of the trek, many porters go out of their way to assist and support the trekkers in a variety of ways. These may include carrying extra baggage for tourists who are unable to carry their personal baggage and backpacks at higher altitudes (for example, above 4,000 meters), assisting those who become unwell, or simply rising to the occasion when the job demands, sometimes at their own peril.

### Porters' perspectives of self and others

As part of a field trip conducted in May and June 2018 to the Everest region, this author and three Canadian undergraduate students conducted several interviews with trekking porters during their journey to the Everest Base Camp. This section highlights some perspectives from the interviews. Of the 12 porters interviewed, only one was female, 10 were Rai, and two were Tamang. All cited financial need as the primary motivation for working as a porter and indicated that the lack of other employment opportunities was the main reason why they decided to work as porters. Five of the porters were in their late teens, were high school students, and spoke English well. These porters had personal goals to move up the tourism hierarchy and become something more than "just a porter". The current work was seen as exploring the various ways trekking is conducted, getting to know foreign trekkers and their cultural sensibilities, and finding out what it takes to succeed in the trekking business. Some porters had plans to become entrepreneurs themselves, either as the owner of a restaurant in the region or operating a trekking company. Only one of the porters had the desire to work outside tourism, as he expressed interest in becoming a schoolteacher. Of the 12 porters, six were from the same village and had family ties to each other. Anthropological research conducted among the Khumbu Sherpas has shown that family ties and kinship are important social capital in gaining a foothold in trekking and mountaineering-related employment and entrepreneurship (Fisher, 1990; Ortner, 1999).

When asked how they perceived themselves as part of the trekking industry in the Everest region, some porters stated that they are critical to the success of the foreigners' travel experience but felt they were not treated well by their guides and lodge owners. Surya Rai, a 19-year-old porter, had this to say:

> I don't like how we are treated. The Sirdar [trekking guide] says we have insurance, but I haven't seen the papers. I too dream of traveling outside of my village, but don't know how is that even possible with this job. The Sherpas do not like us, they say we are lazy, but you can see how hard we work . . . to them, as if we don't exist. They like trekkers and will do anything for them, not for us. I think whenever a Sherpa sees a tourist, he thinks of how he can make money out of that tourist. . . . But, we have no value for him. Without us, who is going to carry the tourists' bags?

When asked to elaborate on whether or not the porters had the opportunity to interact with the foreigners they were portering for, some said they had occasional conversations with them. Others stated that it depended on an individual trekking guide, who is usually the intermediary or facilitator of interactions between porters and the foreigners. Porters were aware not just of their social distance, such as language, culture, and value (Nyaupane et al., 2015), from the foreign trekkers, but also of their stark economic differences with trekking guides who earned higher wages and received generous tips and gifts. The porters also made reference to the privileged positions of both guides and tourists, noting that they are unable to travel for fun even if they want to. When asked if they considered trekking with guides and tourists as their own form of travel, they stated they are doing it by necessity, certainly not for fun, but admitted that at times they too enjoy the mountain scenery and stop to take photos of each other to share with their friends and relatives back home. Some trekking guides were perceived as deliberately separating the porters from the trekkers, as porters were often discouraged from taking part in informal social interactions during the trek. Butz (2002) reports a similar situation in his study of porters in Shimsal, Pakistan. This, according to the porters, prevents them from learning and practicing their English with the foreigners.

As for economic incentives, for porters the daily wage in trekking is about US$15 in the Nepalese Himalaya. However, when costs for meals and accommodations are factored in, the daily wage amounts to six or eight US dollars, depending on the location of villages for overnight stops; the higher the villages, the more costly meals and accommodations are. It is a challenge for many porters to save the US$100 needed for the cost of guide training courses to enable them to upgrade their jobs. Hence, while many porters aspire to be guides, very few actually make it to the next hierarchical order. This is compounded by the fact that most trekking guides nowadays are based in the cities where opportunities for schooling and training are better. Regardless, when the skills required to be a porter and the economic returns are considered, many porters mentioned that opportunities in trekking tourism are better and their earnings not bad compared to other jobs outside trekking. Even when considering the seasonality of trekking (4–5 months of the year), portering is especially well suited to young men from remote villages who may not have the skills to find well-paying job in other sectors. Hence, the porters this author interviewed expressed satisfaction with their employment situation despite the many challenges they had experienced. The porters further stated that they are happy to provide their services to enhance a positive image of Nepal among the foreign tourists. When asked what positive image they consider important in impressing foreigners, many porters referred to their "hard working" and "honest" characteristics, and "smiley and friendly" dispositions. Even when opportunities to interact with the trekkers were sporadic, most porters thought of the trekkers as "friendly", "happy", and "kind". Some porters stated they had received occasional gifts and money from trekkers, especially when they offered help to trekkers who were experiencing physical hardships and health problems.

The issues identified earlier are not unique to Khumbu or to the Nepalese Himalaya. In fact, Butz's (2002, p. 60) research in Shimshal, northern Pakistan, identified several categories of porter concerns related to income; distribution of opportunities; danger, injury, hardship; excitement, experience, adventure; agricultural labor obligations; culture/identity; autonomy/control; reciprocal learning; intersubjectivity/reciprocity; and representation. He reports that tourist guides and the agencies they work for actively undermine porters' tactical efforts to establish remunerative and self-affirming interactions with tourists, and that guides are perceived as obstacles to a satisfying work life. Similar concerns have been reported in studies in other Himalayan destinations, such as Ladakh (Rajashekariah & Chandan, 2013) and Kullu-Manali, India (Gardner et al., 2002).

### Spaces of social exclusion

Proctor (1998) has suggested that the metaphor of space provides perhaps the most familiar entry of geographers into substantive questions of ethics. As mentioned earlier, one of the strongest areas of attention among geographers is concerned with spatial dimensions of social justice (Harvey, 1993; Smith, 2004). This section discusses the ways in which porters involved in trekking tourism are subjected to social exclusion, often manifested in spatial/physical segregation from tourists, guides and other porters, and other forms of explicit and implicit discrimination.

It should be noted that some porters, for example the Rai from the neighboring non-tourist villages in the Everest region, have historically experienced social isolation within their own homeland. The southern regions inhabited by the Rai have not seen the economic development that the Sherpas have experienced. Today, many Rai people are employed in the Everest region working as porters and kitchen staff in Sherpa-owned lodges, and in the case of Rai women, as maids in affluent Sherpa households. In interviews, some of the Rai porters felt humiliated by their Sherpa employers as "slaves in their own homeland" confirming what Frydenlund (2017) has reported in labor and race relation studies in Nepal (cited in Barott, 2018). Indeed, it has been a reversal of fortunes for the Rai who were more prosperous than Khumbu Sherpa prior to tourism because the Rai occupied better agricultural land and had better employment prospects outside farming. Tourism reversed their fortunes as Sherpa capitalized on rapidly expanding tourism opportunities in Upper Khumbu (Everest region). Tourism has accentuated social stratification in which the Sherpa have enjoyed considerable economic success, while the Rai have experienced further restrictions on natural resource use and consumption, making them ever more vulnerable to livelihood risks (Nepal et al., 2002).

### Spatial segregation

A typical Himalayan trekking journey entails walking for several days, assisted by guides and porters who carry trekkers' baggage and equipment. A normal trekking day for the tourist begins early in the morning, packing trekking gear (sleeping bag, clothing, and personal items), and eating breakfast, followed by listening to the guide's instructions and daily schedules en route before ascending the trail with the guide in the lead. Once trekkers' backpacks are packed and porters have their morning breakfast, the porters head off before the guides and trekkers. Thus, porters are segregated from the rest of the group on the trail, and they sleep, eat, and walk separately even though they are part of the same party and follow the same routine.

Social exclusion and segregation in trekking tourism may thus arise due to spatial employment relations between porters and guides/trekkers. Gough (2010) argues that the morality of social relations is intrinsically geographical, as workers compete individually or collectively for a geographically structured supply of jobs, including the use of social oppression and territorial chauvinism in such competitions. In his study of tourists and porters in Shimsal, Butz (2002, pp. 61–62) observed something similar:

> The road head is the place where guides hire porters and allocate loads, and where porters and tourists encounter one another for the first time. This is a site of control for the guide. He has everyone together where he can keep an eye on them. Unlike porters, he has already established relations with trekkers, and as there are more applicants than porter loads he has considerable coercive influence over the behaviour of porters. The trek unfolds best for the guide if he and the tourists travel together along the trail, as a tightly knit group, spatially separated from the porters.

Similar observations were made in the Everest region where guides encourage porters to go ahead separately before the rest of the trekking group. One of the porters stated that guides separate the porters from the tourists partly assuming that if porters and trekkers are together it might give the porters the opportunity to complain about their situation and ask for tips. While this may or may not be true, this physical separation limits the time and opportunity for the porters to interact and socialize with the tourists during the journey. Guides are thus able to exert control on social and economic relations between the porters and the trekkers. For example, Kamal Rai, a porter, stated: ". . . I can speak some English [currently in Grade 10] so I try to talk to them, but most of the time we start early and arrive earlier [than the tourists], there is not much opportunity [to make a conversation with the tourist]". When the trekkers arrive at the destination for the day, the entire party splits into two distinct groups. Tourists stay and eat in the lodges (or mess tent if tent camping) with the guides, while the porters eat in local restaurants (or at their own campfires if tent camping) exclusively catering to porters. Even when they stay in the same lodge, porters are often relegated to a detached building away from the main lodge building, or usually in a damp and dark corner inside the lodge not suited for accommodating the trekkers. In the morning, after eating breakfast separately, porters depart early, leaving the trekkers behind to follow at a leisurely pace and repeat the previous day's routine.

The author visited a porter shelter-cum-accommodation near a slightly upscale lodge establishment in Deboche in the Everest region. The lodge where the porters were staying was a 20-minute walk away from where the trekkers had spent the night, as cheaper accommodations exclusively for porters are generally away from areas with tourist lodges. The lodge beds were wooden blocks with no mattresses; a comforter was provided and intended to double as a blanket and mattress. The main heat source in the room was fuel wood, which can have negative health implications if there is inadequate ventilation. Sanitary conditions were unsatisfactory due to the absence of running water. There was no facility to wash hands or shower, aside from bathing in the glacier-fed river, or paying a prohibitively high price for a bucket of hot water. In one of the tourist lodges in Chukhung (at 4,700 meters), the Sherpa owner had made arrangements for porters' accommodation, but had strictly enforced a rule that prohibited porters from using the toilets inside the lodge and asked them to use the toilets outside. When porters were asked if this arrangement was acceptable to them, they told the author that as long as they have access to a proper toilet and running water it did not matter to them if the toilet was inside or outside.

*Social discrimination*

As mentioned earlier, porters are subject to racial and social discrimination by the guides, lodge owners, and sometimes even foreigners who have the proclivity to be very generous to a Sherpa but somewhat indifferent to non-Sherpas. This is the main reason why many young Rai engaged in mountaineering and trekking try to pass as Sherpas and falsify their true ethnic identity to negotiate better employment terms, conditions of well-being, and maintain harmonious social relations locally.

In addition to poor living conditions, many porters experience challenging working conditions, including racism from lodge owners. One porter stated that several lodge owners do not let them enter the teahouses where the tourists are because "[the porters] look and smell dirty, which ruins the lodge's image". This has been confirmed in previous studies too, for example Barott (2018, p. 12) found that porters in the Everest region were excluded from guesthouses "because they ate too much dal bhat [rice and lentils—a Nepalese staple food]", and were scorned as "dirty, wasteful, obnoxious, troublesome drinkers and gamblers".

Many porters also complain of inadequate meal portions and nutrition while on the trek. Given the caloric expenditure exacted in carrying heavy baggage along challenging paths for hours each day, daily food intake is insufficient in quantity and nutrient content. Therefore, when asked what changes they would like to see regarding their well-being, all interviewees requested having cheaper and more accessible food with bigger portions. The situation of porters is made more challenging as they also have to deal with the effects of altitude, cold, and severe headaches, sometimes making them unfit to travel higher up.

There have been numerous stories of ill-treatment of porters by trekking companies and guides. Porters have been dismissed from the trek the day before reaching the destination and prohibited from receiving all their earnings, as guides and trek leaders maximize their own income by keeping the tips owed to the porters. Porter deaths from accidents or exposure to extreme weather events or altitudes almost always go unreported or unnamed (Ortner, 1999).

The situation of commercial porters is the worst of the three types of porters. The author noted that most of these workers did not have proper trekking equipment, wore flip-flops for footwear, and were clothed inadequately. The inability to afford proper clothing and equipment is a consequence of their minimal earnings; they are paid extremely low wages by lodge owners and other service providers in the region. As commercial porters are at the bottom of the trekking hierarchy, they are powerless to change their circumstances, and instead must risk their health to increase their income by carrying heavier loads. In many cases, porters spend their nights in flimsy tents and caves along the trekking route, exposing themselves to freezing temperatures. Commercial porters have no insurance, so if they are injured or become ill, they have no access to medical care unless they are in Namche Bazaar, which has a porter shelter that provides free medical assistance. Several locals told stories about porters having to walk for weeks with a broken leg because they had no insurance or alternate access to healthcare.

*Inadequate porter welfare programs*

In the context of Shimsal, Pakistan, Butz (2002) argues that a major impediment to porters' welfare is the tremendous power wielded by nonlocal guides who are the usual mediators between the foreigners and the porters. As guides are formally responsible for hiring and paying porters, and for the well-being and contentment of tourists, it costs porters considerable effort and some risk to establish independent relations with tourists. Interactions with tourists independent of guides' mediation can be a source of greater economic reward, greater safety, and enjoyable experience during the journey for the porters. However, these interactions are determined by variables such as the route and schedule of the trek, who the guide is, what company organized the trek, the characteristics of the tourists, the ratio of porters to trekkers, the language skills of porters, and other factors (Butz, 2002).

Stories about porters left stranded sick, enduring freezing cold weather without adequate clothing, food, and shelter are not new, and have been reported since the early 1990s, mostly as anecdotal documentation. Systematic research on porters' welfare hardly exists in the Himalaya. Similarly, with very few exceptions, research is greatly lacking on porters' perspectives on tourism's benefits, guides/trekkers' perspectives on porters, documentation of discriminatory tourism practices on the ground, and policies and regulations to safeguard porters' livelihoods while providing them with a secure and dignified work environment.

There are several nongovernment organizations focused on improving porter welfare and working conditions. The International Porter Protection Group (IPPG) was established in 1997 with the mandate of improving safety and health conditions by raising awareness of porter-related issues with related stakeholders such as trekking business owners, guides, and travel companies (IPPG, 2020). In the Everest region, the IPPG, until very recently, operated porter shelters and rescue services in Machermo, Gokyo, and Namche Bazaar. Working in collaboration with the Himalayan Trust, a Nepali NGO established to address social and environmental concerns, the IPPG is a volunteer organization that depends on the generosity of tourists, funds raised outside Nepal, and on commitments from volunteers who have to stay in remote locations for extended periods. Similarly, Community Action Nepal (CAN) is a UK-based charity organization, which runs a trekking cooperative that arranges treks while maintaining fair working conditions, particularly for porters. The Kathmandu Environmental Education Project (KEEP) focuses on environmental and cultural conservation in Nepal by mitigating negative tourism impacts and encouraging positive outcomes. It runs the Porters' Clothing Bank, where donated clothes are given to porters in need of proper gear, including hiking boots. KEEP also promotes education regarding porter treatment and working conditions. Similarly, the Trekking Agents Association of Nepal (TAAN) provides occasional workshops and training for porters.

However, most of these agencies that focus on porters' welfare programs are externally based and supported, and are not regulated and supervised by government agencies. To whom these agencies are accountable and to what extent porters trust them is an open question. Local NGOs are also subjected to the whims of local tourism organizations and politicians who are involved in tourism businesses. For example, the porter shelters that IPPG was maintaining in Gokyo and Machermo were forcibly closed down recently by the local municipality (IPPG, 2020). Given the lack of government support extended to these groups for their advocacy for porters' welfare, livable wage, and local awareness of their precarious working and living conditions, porters have to rely on their own ability to negotiate and strategize with trekking companies and their local representatives, as Butz (2002) has documented in the Pakistani Himalaya. Most porters this author talked to showed interest in training in English language, customer service, and awareness of health issues at higher altitudes. Government authorities in the Himalayan region need to pay attention to this critical support group. If porters are the backbones of mountaineering and trekking tourism, as has often been remarked by tourists and local people alike, then governments in the region should prioritize developing appropriate porter welfare policies, establish support institutions, and address social justice issues.

## Conclusion

This chapter focuses on three main issues, including the role of porters in trekking tourism, porters' perspectives of self and others, and social practices that disenfranchise the porters in their daily lived experiences during trekking activities in the Himalaya. These practices create impediments to personal growth and morale of the porters who aspire to move up the social and economic ladder, and long for stronger interpersonal relationships with mountain guides and tourists. Tourism practices accentuate class and race divisions between stakeholders. Most porters view portering as the least desirable, but critical, entry point into trekking tourism. They are in a social and economic bind that is exploitative and affects their economic well-being and social progress. This study implies that social progress for underprivileged and underrepresented groups needs closer attention as tourism labor policies are formulated at the highest levels of government. At the ground level, tourists need to give voice to local concerns about the unfair treatment of porters, and advocate for ethical and responsible practices in tourism.

Notions of morality and justice, as described in this chapter, assume certain "fundamental human needs which social arrangements should meet" (Gough, 2010, p. 137). In the context of Himalayan trekking porters, human needs imply decent wages, adequate food, safe working conditions, security from natural elements, and harmonious relations with other social actors. Social arrangements imply economic (trekking tourism) and social structures that facilitate and strengthen porters' engagement with the tourists, guides, lodge owners, governments, and nongovernment entities in ways that improve their standing in society. Respect for the porters and concern for their welfare are central to just social relations. Moral geographies advocates the sharing of power, as opposed to power over others, because the latter implies lack of respect for others and harms their well-being (Gough, 2010).

Moral geographies provides a useful lens to scrutinize the dialectics of power exercised by individual and collective agencies in a tourism context. It is particularly suited to examine how negotiations for authority, control, subjugation, regulation, and relaxation among a diverse set of tourism actors play out daily in the context of Himalayan trekking. Physical manifestations of how certain groups are treated individually or collectively during the course of a trekking journey provide visual, aural, and cognitive clues to social inclusion and exclusion present in tourism. Where exclusionary practices are socially and culturally justified and normalized, as is the case in the Himalaya, global tourism presents an opportunity to advocate for the region's disempowered and disenfranchised inhabitants. The porters engaged in trekking tourism are one such group whose concerns for a more equitable distribution of economic benefits from tourism have long been ignored. Nowhere are moral concerns about health, safety, security, social belonging, and economic well-being of an underrepresented group more acute than in the mountaineering and trekking industry of the Himalaya. Additional research in other Himalayan countries is needed to understand the Nepalese situation and how it compares to these same conditions among trekking stakeholders in Pakistan, India, Bhutan, and Tibet.

Given recent political changes and social movements that assert indigenous rights and recognition, many highland ethnic communities have been drawn into discourses concerning ethnic identity, social mobility, employment access, and labor hierarchies. Mountaineering and trekking tourism in the Himalayan region has afforded many ethnic communities (e.g., Sherpa in the Mt. Everest region) an identity that is economically advantageous and superior to other highland ethnicities elsewhere. Communities such as the Sherpa, once a dominant ethnic group that historically provided portering services to the mountaineering and trekking industry in the Everest region, have left that occupation for other financially lucrative opportunities. In the Nepalese Himalaya, the Rai and Tamang have replaced the Sherpa as porters in popular tourist destinations, giving them access to better prospects in the tourism industry. However, many porters face social discrimination, experience hazardous work conditions, and are in an exploitative relationship with their local employers, city-based trekking agents, and their local representatives.

The precarious reality of porters in trekking tourism raises many ethical issues. For some tourists, it is a moral dilemma whether to hire or not to hire a porter. Not hiring may actually be worse; for a porter, it means a lost economic opportunity. From the perspective of sustainable tourism, tourists should rightly exercise their power to determine the nature and outcome of social and economic transactions that occur in foreign settings. However, for this to occur, tourists need to be aware of prevailing social issues at the destination and be self-aware of their own decisions and actions. Tourism can be a force for social change, and tourists can mediate that change. It is critical to sensitize local tourism entrepreneurs and guides of good ethical practices internationally. The role of guides is particularly important, as this implies some compromise on their part, and advocacy on behalf of the poorer, neglected stakeholders in tourism. For the porters, gaining ground in trekking tourism means promoting unity and solidarity, and getting organized locally. Demands for better wages and adequate security can only be met if porters organize themselves and seek broader social support beyond the tourism industry.

There is a glaring need for increased awareness of social justice issues among policy and decision makers. For the government, establishing ethical standards in tourism policies and practices, instituting labor reforms within the industry, and sensitizing key stakeholders on the importance of socially just tourism for international goodwill and reputation should be the foundations of promoting sustainable tourism in the region. Moreover, governments in all countries of the Himalaya should develop and promote rights-based approaches to the provision of porter welfare services. For tourism scholars, there is a critical need for a social agenda with advocacy for marginalized peoples.

From a research perspective, ethical concerns have been at the forefront of tourism scholarship for some time. Issues of ethics, morality, and justice have been framed also as the "critical turn" in tourism studies (Atelijevic et al., 2017). This scholarship continues to move forward with more recent calls for research into slavery tourism (Cheer, 2018). While much has been written on the lived experiences of working poor in the tourism industry (see Cañada, 2018), not much research exists on porters in trekking tourism. This chapter provides a limited but crucial view of the moral geographies of tourism in the Himalayan region. As such, it should be considered a stepping-stone to a more systematic in-depth research agenda on this important topic.

## Acknowledgments

The author thanks University of Waterloo undergraduate students Asra Choudry, Emily Shields, and Laura Rodriguez for help with interviews with porters in the Everest region in May 2018. Many thanks to the 12 porters who kindly agreed to be interviewed for this study; the author has changed the identity of the porters quoted directly in the text.

# References

Atelijevic, I., Morgan, N., & Pritchard, A. (Eds.). (2017). *The Critical Turn in Tourism Studies—Creating an Academy of Hope* (2nd ed.). London: Routledge.

Barott, N. (2018). *Uphill Struggle: Impediments and Facilitators to Porter Health in the Khumbu Region* (p. 2985). Independent Study Project (ISP) Collection. Ithaca, NY: Cornell University.

Butcher, J. (2000). *The Moralization of Tourism: Sun, Sand . . . and Saving the World?* London: Routledge.

Butz, D. (2002). Sustainable tourism and everyday life in Shimsal, Pakistan. *Tourism Recreation Research*, 27(3), 53–65.

Butz, D. (2006). Tourism and portering: Labour relations in Shimsal, Gojal, Hunza. In H. Kreutzmann (Ed.), *Karakoram in Transition—Culture, Development and Ecology in the Hunza Valley* (pp. 394–403). Oxford: Oxford University Press.

Cañada, E. (2018). Too precarious to be inclusive? Hotel maid employment in Spain. *Tourism Geographies*, 20(4), 653–674.

Caruana, R., Glozer, S., Crane, A., & McCabe, S. (2014). Tourists' accounts of responsible tourism. *Annals of Tourism Research*, 46, 115–129.

Cheer, J. (2018). Geographies of marginalization: Encountering modern slavery in tourism. *Tourism Geographies*, 20(4), 728–732.

Cresswell, T. (1996). *In Place/Out Place: Geography, Ideology and Transgression*. Minneapolis: University of Minnesota Press.

Dahal, S., Nepal, S.K., & Schuett, M. (2014). Marginalized communities and local conservation institutions: Indicators of participation in Nepal's Annapurna Conservation Area. *Environmental Management*, 53, 219–230.

Fennell, D. (2017). *Tourism Ethics* (2nd ed.). Bristol: Channel View Publications.

Fisher, J.F. (1990). *Sherpas: Reflections on change in Himalayan Nepal*. Berkeley: University of California Press.

Frydenlund, S. (2017). Labor and race in Nepal's indigenous nationalities discourse: Beyond 'tribal' vs 'peasant' categories. *The Journal of the Association for Nepal and Himalayan Studies*, 37(1), 26–37.

Gardner, J., Sinclair, J., Berkes, F., & Singh, R.B. (2002). Accelerated tourism development and its impacts in Kullu-Manali, H.P., India. *Tourism Recreation Research*, 27(3), 9–20.

Goodman, H., & Francis, J. (2003). Ethical and responsible tourism: Consumer trends in the UK. *Journal of Vacation Marketing*, 9(3), 271–284.

Gough, J. (2010). Workers' strategies to secure jobs, their uses of scale, and competing economic moralities: Rethinking the 'geography of justice'. *Political Geography*, 29, 130–139.

Harrison, D. (2008). Pro-poor tourism: A critique. *Third World Quarterly*, 29(5), 851–868.

Harvey, D. (1993). Class relations, social justice and the politics of difference. In M. Keith & S. Pile (Eds.), *Place and the Politics of Identity* (pp. 41–66). New York: Routledge.

Harvey, D. (1996). *Justice, Nature and the Geography of Difference*. Oxford: Blackwell.

Holloway, S.H. (1998). Local childcare cultures: Moral geographies of mothering and the social organization of pre-school education. *Gender, Place and Culture*, 5, 29–53.

IPPG (2020). Retrieved May 6, 2020, from www.theadventuremedic.com/features/ippg-nepal-rescue-posts-forced-to-close/

Jamal, T. (2019). *Justice and Ethics in Tourism*. London: Routledge.

Lai, P., & Nepal, S.K. (2006). Local perspectives of ecotourism development in Tawushan Nature Reserve, Taiwan. *Tourism Management*, 27, 1117–1129.

Matless, D. (1994). Moral geographies in Broadlands. *Ecumene*, 1, 27–56.

Mostafanezhad, M. (2016). Organic farm volunteer tourism as social movement participation: A Polanyian political economy analysis of world wide opportunities on organic farms (WWOOF) in Hawai'i. *Journal of Sustainable Tourism*, 24(1), 114–131.

Mostafanezhad, M., & Hannam, K. (2016). *Moral Encounters in Tourism*. New York: Routledge.

Mowforth, M., & Munt, I. (2015). *Tourism and Sustainability: Development, Globalisation and New Tourism in the Third World* (2nd ed.). London: Routledge.

Nepal, S.K. (2000a). Tourism in protected areas: The Nepalese Himalaya. *Annals of Tourism Research*, 27(3), 661–681.

Nepal, S.K. (2000b). National parks, conservation areas, tourism, and local communities in the Nepalese Himalaya. In R.W. Butler & S.W. Boyd (Eds.), *Tourism and National Parks: Issues and Implications* (pp. 73–94). London: Wiley.

Nepal, S.K., Kohler, T., & Banzhaf, B.R. (2002). *Great Himalaya—Tourism and the Dynamics of Change in Nepal*. Berne: Swiss Foundation for Alpine Research.

Nepal, S.K., & Saarinen, J. (Eds). (2016). *Political Ecology and Tourism*. London: Routledge.

Nyaupane, G.P., & Thapa, B. (2004). Evaluation of ecotourism: A comparative assessment in the Annapurna conservation area project, Nepal. *Journal of Ecotourism*, 3(1), 20–45.

Nyaupane, G.P., Timothy, D.J., & Poudel, S. (2015). Understanding tourists in religious destinations: A social distance perspective. *Tourism Management*, 48, 343–353.

Ortner, S. (1999). *Life and Death on Mt. Everest—Sherpas and Himalayan Mountaineering*. New Delhi: Oxford University Press.

Panzeri, D., Caroli, P., & Haack, B. (2013). Sagarmatha Park (Mt Everest) porter survey and analysis. *Tourism Management*, 36, 26–34.

Poudel, S., Nyaupane, G.P., & Timothy, D.J. (2013). Assessing visitors preference of various roles of tour guides in the Himalayas. *Tourism Analysis*, 18(1), 45–59.

Proctor, J.D. (1998). Ethics in geography: Giving moral form to the geographical imagination. *Area*, 30(1), 8–18.

Rajashekariah, K., & Chandan, P. (2013). *Value Chain Mapping of Tourism in Ladakh*. New Delhi: WWF-India.

Rheingold, H. (1993). *The Virtual Community: Homesteading on the Electronic Frontier*. London: MIT Press.

Salazar, N. (2012). Community-based cultural tourism: Issues, threats and opportunities. *Journal of Sustainable Tourism*, 20(1), 9–22.

Scheyvens, R. (2011). The challenge of sustainable tourism development in the Maldives: Understanding the social and political dimensions of sustainability. *Asia Pacific Viewpoint*, 52(2), 148–164.

Scheyvens, R., & Biddulph, R. (2018). Inclusive tourism development. *Tourism Geographies*, 20(4), 589–609.

Smith, D.M. (1997). Geography and ethics: A moral turn. *Progress in Human Geography*, 21, 583–590.

Smith, D.M. (2000a). *Moral Geographies: Ethics in a World of Difference*. Edinburgh: Edinburgh University Press.

Smith, D.M. (2000b). Social justice revisited. *Environment and Planning A*, 32, 1149–1162.

Smith, D.M. (2004). Morality, ethics and social justice. In P. Cloke, P. Crang & M. Goodwin (Eds.), *Envisioning Human Geographies* (pp. 195–209). London: Routledge.

Timothy, D.J. (2012). Destination communities and responsible tourism. In D. Leslie (Ed.), *Responsible Tourism: Concepts, Theory and Practice* (pp. 72–81). Wallingford: CABI.

Vajirakachorn, T., & Nepal, S.K. (2014). Local perspectives of community-based tourism: Case study from Thailand's Amphawa floating market. *International Journal of Tourism Anthropology*, 3(4), 342–356.

von Fürer-Haimendorf, C. (1964). *The Sherpas of Nepal—Buddhist Highlanders*. New Delhi: Oxford Book Co.

Walzer, M. (1983). *Spheres of Justice: A Defense of Pluralism and Equality*. New York: Basic Books.

Wearing, S., van der Duim, R., & Schweinsberg, S. (2007). Equitable representation of local porters: Towards a sustainable Nepalese trekking industry. *Matkailututkimus (Finnish Journal of Tourism Research)*, 3(1), 72–93.

# Part 3

# Forms of Tourism in the Himalaya

# 8 Tourism for Development in the Himalaya

## Using Ecotourism to Improve Livelihoods

*Lisa Choegyal*

## Introduction

"Making tourism more sustainable is not just about controlling and managing the negative impacts of the industry. Tourism is in a very special position to benefit local communities, economically and socially, and to raise awareness and support for conservation of the environment" (United Nations Environment Program and World Tourism Organization, 2005, p. 2). The UN International Year of Ecotourism 2002 was organized by the United Nations Environment Programme (UNEP) and the World Tourism Organization (UNWTO) to bring focus to tourism's potential benefits (UNWTO, 2002). A number of global events, deliberations and seminars culminated in the Québec Declaration, the benchmark for ecotourism, which was forged at the World Ecotourism Summit in May 2002 (United Nations, 2003). Many destination countries formulated and published national ecotourism strategies as part of the lead-up to the Québec summit, including Bhutan and Nepal, in 2001. An enormous body of literature, studies and other global initiatives followed (e.g., Buckley, 2003, 2009, 2012; Butcher, 2006; MacLaren, 2002), including the promotion of the UN International Year of Sustainable Tourism for Development in 2017 in support of the industry's role in achieving the Sustainable Development Goals (SDGs) (United Nations, 2020).

Although the term 'ecotourism' was first coined in the Americas late last century (Ceballos-Lascurain, 1996), it has since taken on relevance all over the world. Countries in the Himalayan region have embraced and adapted ecotourism as a tool for development and conservation, using their wealth of biodiversity and cultural assets and spectacular natural scenery to grow tourism in a style and scale that engages rural communities. Correctly applied, ecotourism ensures that visitors contribute to local livelihoods and environmental preservation by providing mechanisms to help finance and justify protected areas and resource management in less developed and poverty-prone destinations.

Against the natural and cultural backdrop of the Himalaya, and as ecotourism has grown throughout the region, the sector has developed certain unique characteristics and patterns that may or may not be manifested in other parts of the globe. This chapter examines the intricacies of ecotourism in the Himalaya, with special reference to the efforts being undertaken in the countries of the region to develop

DOI: 10.4324/9781003030126-11

a more ecologically and socially sensitive form of mountain tourism. The chapter first defines terms and provides some historical perspectives, including some of the earlier criticisms levied at the concept of ecotourism, and certain growth trends in global ecotourism. This is followed by an examination of ecotourism in the Himalaya and how the area's unique biophysical systems and cultures have provided an unmatched setting for ecotourism, the national policies and development projects that have facilitated its acceptance and growth, and many of the challenges it faces in Nepal, Bhutan, India, Pakistan and China.

## Meanings and Values of Ecotourism

The concept of ecotourism, a combination of ecology and tourism, first emerged in the 1980s as a response to the unsustainable characteristics of tourism. By 1990, the International Ecotourism Society (TIES) had been founded in North America by Megan Epler-Wood and Martha Honey. This institution defines ecotourism as "responsible travel to natural areas that conserves the environment, sustains the well-being of the local people, and involves interpretation and education" (TIES, 2020, n.p.). As explored later, other destinations, including those in the greater Himalayan region, have adapted the definition of ecotourism to suit their national needs, but all generally support the tenets established by TIES (Cabral & Dhar, 2020; Chan & Bhatta, 2013; Dam, 2013; Fennell, 2020; Kala & Maikhuri, 2011; K.C., 2017; Maharana et al., 2000; Nyaupane & Thapa, 2004; Spenceley, 2020). Most ecotourism experiences include, but are not limited to, adventure activities (e.g., mountaineering, trekking, hiking, water activities), agritourism, visits to rural areas and villages and appreciating certain ecosystems (e.g., rainforest, arctic tundra, deserts).

According to TIES (2020, n.p.), the principles of ecotourism are "about uniting conservation, communities, and sustainable travel. This means that those who implement, participate in and market ecotourism activities should adopt the following ecotourism principles:

- Minimize physical, social, behavioral and psychological impacts.
- Build environmental and cultural awareness and respect.
- Provide positive experiences for both visitors and hosts.
- Provide direct financial benefits for conservation.
- Generate financial benefits for both local people and private industry.
- Deliver memorable interpretative experiences to visitors that help raise sensitivity to host countries' political, environmental, and social climates.
- Design, construct and operate low-impact facilities.
- Recognize the rights and spiritual beliefs of the indigenous people in [the] community and work in partnership with them to create empowerment."

Although the word 'ecotourism' is now widely used in conservation and tourism sectors, definitions vary greatly. Whereas its original focus was nature and biodiversity (Ceballos-Lascurain, 1996; K.C., 2017), ecotourism has grown to encompass heritage and culture in part at least, as well as other assets utilized by tourism. Weaver (2002) suggests that this is especially true in Asia, where ecotourism is often hybridized, linking other resources beyond nature, and tourisms beyond nature-based tourism. In many parts of the world, ecotourism now resembles a principled approach to various tourisms rather than a type of nature-oriented tourism alone.

Many different terminologies are now used, often interchangeably, to reflect similar principles of sustainability within the context of tourism—to promote ecological awareness and conservation, and to empower local communities economically, socially and politically (Rai & Sundriyal, 1997; Scheyvens, 1999; Timothy, 2007). The use of these varied terms (e.g. green tourism, sustainable tourism, community-based tourism, alternative tourism, pro-poor tourism, ethical tourism, regenerative tourism and responsible tourism) can be confusing, but they all boil down to essentially the same goal of promoting the principles of sustainable development in tourism (Fennell, 2020; Koščak & O'Rourke, 2019; Leslie, 2012; Plateau Perspectives, 2020). For the purposes of examining purpose-driven ecotourism in this chapter, some of the terms noted earlier are used interchangeably.

Despite the lofty ideals of purposeful ecotourism, the vision of ecotourism saw considerable criticism in its early years. It was often misinterpreted and incorrectly labeled and marketed by service providers as any sort of tourism that took place in nature (Self et al., 2010; Timothy & White, 1999). This was particularly problematic in developing countries. Thus, its sustainability elements were emphasized far less than its nature-embedded locations. Simultaneously, ecotourism's emphasis on nature protection and community well-being were frequently treated as a product label and overused as a marketing gimmick by tourism promoters to capitalize on people's desires to 'do something good' during their holidays in an increasingly sustainability-aware marketplace (Fennell, 2020; Wall, 1997). Yet, many so-called ecotour operators saw ecotourism as a means of seeking profit over protection and essentially began selling experiences that were largely antithetical to the principles they ostensibly supported (Powell & Ham, 2008; Wheeler, 1993). This led to concerns that ecotourism would eventually become 'mass ecotourism', penetrating the most sensitive environments and cultures en masse, resulting in the same degrading activities and large-scale development that characterized mass tourism in general (Butcher, 2007). Although abused and misused in some locations and at certain times, ecotourism has not realized many of the fears it initially stoked in certain destinations and among certain observers.

These criticisms notwithstanding, ecotourism is now generally regarded as a sustainable approach to tourism planning, development and management that has received widespread acceptance within the industry and among most scholars. This recognition has been brought about by years of work by multiple organizations to legitimize ecotourism and its support of sustainability principles, green certification requirements for tour operators, strong educational efforts in destinations and among service providers and numerous policy initiatives.

## Ecotourism Trends

Although the global market for ecotourism is hard to quantify accurately and data are sometimes conflicting, trends confirm strong consumer enthusiasm and awareness for sustainable styles of nature-based tourism, prompting national tourism organizations and tourism operators to consider it seriously in terms of promotion and planning. A growing body of research indicates the rapid growth and popularity of ecotourism as a travel product and means of supporting the principles of sustainability (e.g. Chakraborty, 2019; Fennell, 2020; Khanra et al., 2021). While examining the unprecedented economic impacts of COVID-19 in a 2020 statement, the UNWTO (2020) emphasizes the importance of ecotourism and estimates that approximately 7% of world tourism relates to wildlife, a segment growing by 3% annually. Post-COVID-19 visitor patterns are likely to reinforce this upward direction as travelers feel the need to get back to nature and further the cause of sustainability.

According to TIES (2007), since the 1990s, ecotourism has been growing 20–34% per year. In 2004, it grew globally three times faster than the tourism industry as a whole and is continuing to see growth at 10–12% per annum in international markets. In recent years, several countries have seen an extraordinary growth in ecotourism, with Iceland, Kenya, Palau, Ethiopia and Nepal being among the most popular. Nepal realized a 24% growth in 2018 arrivals (totaling 1.173 million) over the previous year, most of whom visit the country for its ecotourism opportunities (Allied Market Research, 2021). Sun-and-sand resort tourism has now reached a high level of maturity as a market, and its growth is projected to flatten, while 'experiential tourism', including ecotourism, nature tourism, adventure tourism and rural tourism, are expected to grow rapidly over the next few decades (TIES, 2007).

Many scholars and organizations have predicted that most of tourism's future growth will occur in and around the world's remaining natural areas (Elmahdy et al., 2017; TIES, 2007; Winter et al., 2020), which recent research appears to confirm (e.g., Mandić, 2019). Researchers also predict that, along with this growth in nature tourism, there will be a growth in eco-resorts and green hotels (Erdem & Tetik, 2013) and a significant growth in demand, bringing the value of the world's ecotourism market to approximately US$333.8 billion by 2027 (Allied Market Research, 2021).

In 2007, the International Ecotourism Society (TIES, 2007) noted that ecotourists spend far more money in the destination than cruise passengers or package tour participants. Whereas most mass tourism expenditures leak out of the local economy through international hotel companies, airlines and imported food, sometimes as much as 95% of ecotourists' expenditures remain in the local economy because of its emphasis on local services and products. The economic value and positive fiscal impacts of ecotourism for host communities is one of ecotourism's hallmark defining characteristics (Meleddu & Pulina, 2016).

While value for money, comfort and convenience tend to be the overriding factors in consumer holiday choices, the following findings from earlier studies evaluated by Honey and Chafe (2005) are revealing. In the early 2000s, more than two-thirds of US and Australian travelers, and 90% of British tourists, considered active protection of the environment and support for local communities to be part of a hotel's responsibility. In Europe, 20–30% of travelers were aware of the needs and values of sustainable tourism. Some 10–20% of travelers sought 'green' options; 5–10% demanded green holidays. In a German survey, 65% (39 million) of travelers expect environmental quality; 42% (25 million) felt that it was particularly important to use environmentally friendly accommodations. Nearly half of survey participants in Britain said they would be more likely to use a tourism provider that has a written code to guarantee good working conditions, protect the environment and support local charities in the destination.

The studies examined by Honey and Chafe (2005) also found that 70% of US, British and Australian travelers would be willing to pay up to US$150 more for a two-week stay in a hotel that has a responsible environmental attitude. In the UK, 87% of travelers said their holiday should not damage the environment, and 39% said they were prepared to pay 5% extra for ethical guarantees.

## Ecotourism in the Himalayan Region

Most of the Himalaya are located in Nepal, the Indian states of Jammu and Kashmir, Himachal Pradesh, Uttarakhand, Sikkim, and Arunachal Pradesh and Bhutan. The geographic area considered in this chapter also includes the Tibetan Plateau provinces of China that straddle north of the highest mountain range in the world, namely Tibet Autonomous Region and Qinghai Province. The Himalayan region is endowed with a wealth of diverse resources and tourism assets, and the traditions of its mountain people are generally well suited to hospitality. It is home to many natural and cultural heritage sites inscribed on the World Heritage List. With an ancient heritage dating back many millennia, including some of the world's major Buddhist and Hindu sites, the region contains the world's highest mountain, a vast range of white peaks, a wealth of rare endangered species of flora and fauna, and internationally recognized biodiversity hotspots protecting snow leopards, red pandas and much else. There are many different ethnic and language groups with distinctive indigenous lifestyles. Likewise, the grand mountain scenery provides a memorable backdrop for a wide range of adventure activities (Choegyal, 2019a).

Capitalizing on these assets, many Himalayan countries have embedded the development aspects of ecotourism into their national tourism policies (Choegyal, 2019a; Kala & Maikhuri, 2011; Luo, 2012; Maharana et al., 2000; Rai & Sundriyal, 1997; Seth, 2019; Singh & Mishra, 2004). In addition, private sector operators and local community initiatives in some destinations have greatly contributed to the ecotourism story. For example, wildlife tourism was pioneered by Tiger Tops in Chitwan from 1964, several years before the National Park was officially gazette in 1973 (and later the first in South Asia to be awarded World Heritage status, one of four sites in Nepal).

In China, the language of ecotourism was being used as early as 1996, when the Chinese Academy of Sciences led an ecotourism master plan study for the

Qomolangma Nature Preserve (QNP), with an international team supported by UNDP and managed by the China International Center for Economic and Technical Exchanges. Following a four-week study mission in Tibet, the 300-page report outlined ecotourism development strategies and presented ten project proposals. A QNP Ecotourism Master Plan summary was published in 1998 for potential donors, with recommendations for handicraft projects, capacity building, community and conservation tourism and cross-border linkages in the Himalaya (QNP Working Commission, 1998).

More recently, the potential for tourism to benefit local and national economies has been well recognized in China with an increase of over 50% in the number of domestic trips between 2008 and 2019 (Zhao & Liu, 2020). 'Spending time in nature' consistently appears as the most important driver for Chinese travelers. With international travel curtailed by COVID-19, an estimated 2.38 billion domestic trips were expected in 2020, with 550 million trips during the Golden Week holiday (BBC News, 2020), many of which were expected to take place in natural environments. Qinghai Province, situated to the north of Tibet Autonomous Region, is among the least visited regions of China but is becoming an increasingly popular ecotourism destination, together with the Tibetan Plateau region, especially over the past decade, with a more than threefold increase in tourist trips and a sevenfold increase in tourism income. This growth has provided a substantial source of revenue for residents, especially in communities within conservation zones (Foggin & Yuan, 2020).

Although until rather recently India has not considered tourism a priority sector, the country's National Tourism Policy states: "Ecotourism should be made a priority tourism product for India with the focal points located in the Himalaya and North East States" (Government of India, 2002, p. 16). India recognizes the necessity of environmental sustainability with the Environmental Information System Centers in Sikkim, and the National and State Ecotourism Directorates under the Ministry of Tourism (Government of India, 2020; Government of Sikkim, 2020). The Ecotourism and Conservation Society of Sikkim is an example of an active partnership with the private sector (Ecotourism and Conservation Society of Sikkim, 2020).

In the Indian Himalaya, ecotourism and related adventure activities were established historically in western mountain states, including Uttarakhand, Himachal Pradesh and Jammu and Kashmir, with houseboats, hiking, horse riding, camping and snow skiing being the primary activities. Trekking through high altitude villages of the stark landscape of Ladakh to enjoy the scenery, stay in adapted local houses or view the elusive snow leopard are products developed for both the international and domestic markets (Vannelli et al., 2019). In the east, India's wide canvas stretches from the rolling hills of Darjeeling and Kalimpong to the high mountains in Arunachal Pradesh and Sikkim, established as a domestic destination for trekking and cultural tourism. The entire region is intersected by swift-flowing rivers, which are ideal for white water rafting, kayaking and fishing, and hydro dams offer other water-based tourism opportunities. Darjeeling and Assam are household names in much of the world, and there are opportunities to further capitalize on this profile with the promotion of sustainable tea tourism as an add-on for general interest markets (Indian Himalayan Center for Adventure and Eco Tourism, 2020; Jolliffe, 2007).

Poverty reduction has always been a stated goal in Nepal's tourism policy (Nyaupane & Poudel, 2011). The current Nepal National Tourism Strategic Plan 2015–2024 states: "Goals for the tourism economy: to contribute to greater GDP growth and employment, reduce poverty and increase sustainable access to foreign exchange for national development". Nepal's natural environment goals are: "To minimize the impact of tourism on the environment through a proactive planning and implementation strategy and promotion of good practice" (Government of Nepal, 2019, n.p.). The plan goes on to state that Nepal's exceptional natural and cultural resources are its major tourism assets. Its 20 protected areas cover 23.23% of the country's territory, and a large part of the Himalaya, the highest mountain range in the world, is in Nepal. Eight of the world's 14 mountain peaks higher than 8,000 meters are located in the country. These mountains are the key sources of Nepal's hard and soft adventure tourism activities, not all of which can be considered ecotourism by definition.

Nepal's strengths include institutionalized partnerships between natural resource management agencies, and tourism decision makers, and a legal structure that ensures that tourism benefits are shared with buffer zone communities and user groups. For example, the Nepal Tourism Board, a national tourism organization established in public-private partnership, includes representatives from the Ministry of Forests and Environment and its Department of National Parks and Wildlife Conservation (DNPWC). Nepal has a long tradition of tourism concessions managed by the DNPWC in partnership with the private sector (Tourism Recreation Conservation, 2014). For several decades, international NGOs such as the Zoological Society of London and WWF Nepal have worked with local communities using ecotourism to strengthen conservation (Choegyal, 2019b).

Nepal and China share the summit of Mount Everest, the highest peak of them all at 8,848 meters (29,028 ft). More than 100 expedition groups attempt to climb the peak every year, mainly from Nepal and some from Tibet. Trekking is the single largest tourist activity in Nepal and attracts over 200,000 foreign trekkers annually (Ministry of Culture, Tourism and Civil Aviation, 2020), and an unknown number of domestic visitors. Since Nepal first opened for tourism and mountaineering in the 1950s, the regulatory framework, private and community initiatives, and numerous integrated conservation and development projects designed to protect biodiversity and reduce poverty have been a role model in the region in ecotourism. Although not all of Nepal's trekking and mountaineering sector can be considered ecotourism, given the environmental and social strains effected by these activities, the number of outfitters that abide by the principles of sustainability is growing.

Bhutan is internationally positioned by the Tourism Council of Bhutan as a sustainable cultural tourism destination, measuring success on its Gross National Happiness index. In 2019, Bhutan won the Green Destinations Gold Award at ITB Berlin in Germany, and has been recognized recently as one of the world's most desirable travel destinations (Tourism Council of Bhutan, 2020). Some 72% of the country's land is under forest cover, with ten protected areas (51.44% of the country's territory) forming part of the Eastern Himalayan biodiversity hotspot. Iconic species include the snow leopard, takin and Himalayan wolf.

Despite this natural wealth, Bhutan's tourism is dominated by its cultural sites (*tshechus*, *dzongs* and monasteries), indicating that its nature-based ecotourism potential has yet to be fully achieved (Gyelshen, 2019; Nyaupane & Timothy, 2010). Bhutan's low population density, intact natural landscape and significant biodiversity could provide tremendous scope for nature-based ecotourism as a long-term solution for sustaining protected areas, wildlife and community liveli-hoods. Being a small and remote kingdom, and having unique geopolitical con-ditions, Bhutan's pre-booked tourism regulations are hard to replicate in other destinations in the Himalaya or elsewhere.

The Bhutan National Ecotourism Strategy, developed with WWF Bhutan and the Nature Conservation Division for Bhutan's Department of Tourism in 2001, uses the definition: "Styles of tourism that positively enhance the conservation of the environment and/or cultural and religious heritage, and respond to the needs of local communities" (Bhutan Department of Tourism, 2002, n.p.). The Eco-tourism Strategy emphasizes the slogan 'high value, low volume', which infor-mally means 'high value, low impact' (Nyaupane & Timothy, 2010), reflecting the language of ecotourism and more accurately describing Bhutan's attitude to the rationale for controlling foreign visitor numbers through mandatory high daily travel costs and other protective policies. Bhutan's most recent ecotourism definition is: "High-value low impact travel that supports the protection of natu-ral and cultural heritage, provides positive and enriching experiences for visitors and hosts, assures tangible benefits to local people, and contributes to the Gross National Happiness" (Gyelshen, 2019, n.p.).

Ecotourism activities in Bhutan include the pilot public-private partnership 2004 in Jigme Singye Wangchuck National Park, extended to Merak-Sakteng in 2011. The same year, the Royal Society for Protection of Nature initiated an ecotourism project in Phobjikha, the country's largest wetland famed for its population of black-necked cranes. WWF Bhutan kick-started homestay pro-grams in remote villages of Wangchuk Centennial Park in 2012 (World Wide Fund for Nature, 2012). More recent initiatives include community-based sustainable tourism in Haa and My Gakidh Village in Punakha, with further potential identified along the country's approximately 23 approved trek routes (Gyelshen, 2019).

However, ecotourism requires tangible economic and livelihood benefits for local people, with involvement in decision-making and management of ecotour-ism ventures. In Bhutan's case, such ecotourism options are difficult for rural communities to invest in due to policies that control tourism through Thimphu operators and which target wealthier travelers and often require the use of pre-arranged itineraries. One study concluded, "prerequisite for a substantial promo-tion of ecotourism would be changes in the Bhutanese tourism policy to encourage the diversification of tourism products" (Gurung & Seeland, 2008, p. 489).

## Regional Ecotourism Challenges and Constraints

Despite an impressive range of attractions and consistent tourism arrivals growth, the countries of the Himalayan region have not realized ecotourism's full potential benefits. Source markets widely perceive the region as a difficult destination. Even before the calamity of COVID-19, social and political disturbances along with a succession of natural catastrophes had been detrimental to tourism (Beirman et al., 2018; Nyaupane & Chhetri, 2009; Thapa, 2004).

Regional tourism issues and constraints include product development weaknesses, relatively short stays, low expenditures, lack of awareness among tourists and service providers, historically poor government-resident relations and concerns with pollution and environmental degradation (Chan & Bhatta, 2013; Dam, 2013; Kala & Maikhuri, 2011). There is generally potential to attract a broader spectrum of markets including domestic visitors, higher-value general interest markets and regional cross-border markets. Examples of these pre-COVID-19 overall sector weaknesses in the Himalaya include Tibet's dependency on domestic visitors from other parts of China, India's lack of federal government priority for tourism, Nepal's dependency on growth from low-spending East Asian markets and Bhutan's tourism equilibrium being disrupted by large influxes of Indian visitors (Druk Journal, 2019).

Importantly, all Himalayan countries have severe infrastructure weaknesses, including roads and air connections, electricity, water supply, sanitation, medical and rescue services, wayside amenities and border facilities. Inadequate standards of tourist services, especially related to immigration procedures, lost synergies with neighbors or outright conflict (e.g. India-China, India-Pakistan) and availability of tourist information are additional concerns. Combined, these factors have created an unfavorable climate for private investment, which is vital for robust tourism development in the region (Asian Development Bank, 2005).

In some Himalayan locations, there have long been conflicts between governments and local residents. Early in the process of creating nature reserves and national parks, many Himalayan rural dwellers were expelled from their villages and traditional lands to make way for protected areas and to facilitate the tourism industry that was predicted to follow. Many people were displaced from their traditional livelihoods, uprooting entire communities, often in the name of tourism and environmental protection (Chan & Bhatta, 2013; Kala & Maikhuri, 2011). Studies reveal how some ethnic minorities in and around nature preserves lack the power to benefit financially from tourism and conservation (Dahal et al., 2014). Fortunately, although many mistakes were made in the 1980s and 1990s in many Himalayan nature tourism destinations, through the process of ecotourism and sustainable development in general, many of these initial conflicts have dissipated and have been replaced by increased collaborative efforts that work to empower communities rather than disempower them through tourism (Chan & Bhatta, 2013; Kala & Maikhuri, 2011; K.C., 2017; Luo, 2012).

Another problem in the region is that some extreme activities that are frequently labeled 'ecotourism' do not follow the required standards of sustainability. This is especially problematic in the areas of trekking and mountaineering. Himalayan trekking has become especially popular in recent decades. This has led to severe overcrowding on and around trekking trails and base camps with increased litter, human waste and environmental pressures (Huddart & Stott, 2020), even on Mount Everest in the last few years. With tourism-induced high demand for water and wood for heating and building, many areas saw considerable deforestation and water shortages early in the tourism growth phases (Nyaupane & Thapa, 2004). Trailside deforestation happened quickly, forcing porters to go further afield to locate fuel for campfires and heating tourists' bathwater (Timothy & Boyd, 2015). Today, overcrowding, litter and waste are still problems in Himalayan trekking, but the environmental behaviors of tourists and service providers have improved considerably over the past two decades. Because of this, a study by Nyaupane and Thapa (2004) illustrates that communities in areas where ecotourism is practiced have more positive perceptions of the social and ecological impacts of tourism than communities in non-ecotourism areas. Although these ecological problems are still pervasive in the trekking sector of the Himalaya (Huddart & Stott, 2020), ecotourism in general in recent years has helped encourage more sustainable tourist behavior in the Himalaya and reduce tourism's impact compared to the early years (Poudel & Nyaupane, 2017).

In addition to environmental concerns, the trekking industry still suffers from the exploitation of Himalayan porters. In the beginning, it was the Sherpa people, but today it is other ethnic highland and lowland minorities who work as porters in trekking tourism. Nepal et al., (2021, see Nepal's chapter, this volume) recounts how many minority porters still today suffer from maltreatment and ethically questionable practices by guides and trekking firms. These include among other issues, social exclusion, spatial segregation, social discrimination and inadequate compensation and health care. Although some trekking outfitters require the implementation of ecotourism principles, many still do not. Unfortunately, even today, many Himalayan treks are labeled 'ecotourism' but remain far from the principles of nature protection and the socio-economic well-being of host communities.

Despite these shortcomings, unique natural endowments and a growing public awareness of the principles of ecotourism have gone some way in positioning the Himalaya as a world-class destination for ecotourism and adventure activities (Huddart & Stott, 2020). Nepal and, to a lesser extent, Bhutan, in particular, are internationally acknowledged as ecotourism pioneers. Outstanding natural and biodiversity assets, ancient temples and monasteries, unique cultural landscapes, sacred trails, hidden valleys and lowland jungles appeal to international visitors seeking adventure and ecotourism opportunities (Choegyal, 2008).

## Ecotourism Development Projects

Long recognized by governments, development agencies, resource managers and the private sector, the Himalayan countries are particularly well suited to tourism, with ecotourism presenting opportunities to deliver development benefits to remote rural communities as well as contributing to GDP and foreign exchange revenue. Of all the countries considered here, Nepal has been particularly active in developing the Himalayan ecotourism narrative, harnessing programs from multilateral and bilateral donor agencies, as well as international NGOs, through the Ministry of Culture, Tourism and Civil Aviation and the Nepal Tourism Board (NTB), which for many years had a dedicated Sustainable Tourism Unit to look after ecotourism activities. In recognition of the country's pioneer status, the NTB has hosted numerous ecotourism study tours, which highlight the lessons learned from Nepal's extensive public and private sector ecotourism experience. Managed in partnership with NGOs and the private sector, these have comprised groups of government officials, NGO and industry participants from Afghanistan (2009 & 2011), Bhutan (2012), Cambodia (2007), Ethiopia (2014), Myanmar (2015), Pakistan (2019), the Philippines (2008 & 2014), Tibet (2007), Qinghai (2012) and Vietnam (2011). Study tour activities have included field visits and structured briefings from relevant government and NGO projects, and private sector and community ecotourism partnerships.

As early as 2000, the Asian Development Bank (ADB) had recognized the benefits of ecotourism as a potential tool for poverty reduction. The research, planning, infrastructure and software activities of the ADB Ecotourism Project 2000–2001 included designing key mountain airport improvements to facilitate trekking and other tourism access. Although implementation was delayed by the security situation of that time, many recommendations were absorbed into subsequent ADB initiatives. A useful output of the project was the National Ecotourism Strategy and Marketing Program of Nepal, published in an official booklet in December 2001 (Nepal Tourism Board, 2001). For the first time, the Great Himalaya Trail (GHT) was described in a Government of Nepal document (see below).

Perhaps the most influential project with the most significant impact on ecotourism throughout the Himalayan region was the Tourism for Rural Poverty Alleviation Program (TRPAP), designed to mainstream the concept of pro-poor tourism, which had been developed by academics elsewhere in the world (e.g., Ashley et al., 2000, 2001; Goodwin, 1998). Applying ecotourism principles to Himalayan conditions, many of today's ecotourism practitioners cut their teeth during the course of this ambitious five-year project.

Based in the Nepal Tourism Board, TRPAP was a partnership funded by the UK Department of International Development (DFID), SNV Netherlands Development Organisation (SNV) and UNDP 2002–2007, which was considered a successful experimental design (Bhattarai et al., 2006). Working closely with villagers and local governments, TRPAP developed tourism in six districts of Taplejung (Kangchenjunga), Solukhumbu (Everest), Rasuwa (Langtang), Chitwan, Rupandehi (Lumbini) and Dolpa. These districts recognized tourism as a core economic sector, and the program focused on raising awareness of tourism issues, facilitating organizational responses through community participation. It successfully formed a network of community and user groups, enabling local people to engage in tourism. For the first time, district and village tourism plans were prepared through grassroots consultation with the community, and were included in District Periodic Plans.

TRPAP sought to contribute to poverty alleviation in Nepal by mainstreaming pro-poor sustainable tourism policies, and developing strategies and innovative models that were pro-women, pro-environment and pro-communities. Social mobilization and strengthened community enterprises included handicrafts and souvenir organizations; tour, trekking and wildlife guide services; monastery, temple and village visits; local transportation; and conservation awareness, much of it focusing on the red panda and Sarus crane. The DNPWC was a key partner in TRPAP, and the tourism component of the Sagarmatha National Park and Buffer Zone Management Plan 2006–2011 was a significant project output, applying international best practice to embrace Everest's local, national and international considerations as a World Heritage Site. Although activities were hampered by civil disturbances and a resultant lack of foreign tourist flows, many of the lessons learned during TRPAP were compiled into a book and have been absorbed into Nepal's tourism development practices and adopted elsewhere (Dhakal et al., 2007).

The GHT was conceived as an international multi-country trek route in the Himalaya, the bulk of its extent lying within the borders of Nepal (Upadhayaya, 2018). The GHT attracts much interest from development agencies as a vehicle to raise awareness of climate change and deliver benefits deep into the countryside. Proponents were led by DFID UKAID, but included Himalayan Climate Initiative, the International Centre for Integrated Mountain Development, SNV, The Mountain Institute, WWF Nepal and other organizations. The Nepal Tourism Board established a GHT unit; trek companies promoted the route (e.g., World Expeditions, 2020), and a number of websites were developed in several languages.

DFID funded the Great Himalaya Trail Development Program (GHTDP) Phase 1: 2010–2012 to harness market-led tourism to improve livelihoods, create employment and bring sustainable development opportunities to remote communities through the creation of an iconic and globally significant new tourism product for Nepal throughout the country. It provided a significant boost to Nepal's tourism industry and associated support sectors, channeling tourists and pro-poor tourism investment to underdeveloped districts, and stimulating a range of private sector business, employment and production opportunities for impoverished mountain communities.

GHTDP activities focused on settlements along trekking routes in five districts: Humla and Dolpa, Gorkha (including Manaslu), lower Solukhumbu (south of Lukla) and Taplejung (including Kanchenjunga) (Choegyal, 2011; Upadhayaya, 2018). The sites were selected for their appeal to markets and suitability to best achieve the project's purpose of spreading tourism benefits to new trek areas, particularly to women and poor people from excluded groups. Historically, 95% of all trekkers to Nepal are concentrated in Annapurna, Langtang and Sagarmatha, with only a few thousand tourists venturing further afield. Managed by SNV Nepal and UNDP, DFID financed three interrelated programs: marketing and promotion; assisting new and enhanced micro and small to medium enterprises along the trail, with market linkages to strengthen Nepal tourism businesses; and building institutional capacity at central and district levels (Choegyal, 2011).

The GHT was the focus of activities in the DFID-funded Samarth-Nepal Market Development Program (NMDP) 2013–2017, which followed from the GHTDP, including the publication of the Great Himalaya Trails Standards Handbook (Saintz & Hugo, 2017) and developing a website (see www.greathimalayatrails.com). Taking a facilitative role, the NMDP aimed to reduce poverty in Nepal by increasing the incomes of farmers and small-scale entrepreneurs by taking a Making Markets Work for the Poor (M4P) approach. Tourism activities included sector planning, trail improvements, small-scale infrastructure, capacity building, promotion and marketing. During Samarth-NMDP, and in partnership with the International Finance Corporation of the World Bank Group and Investment Board of Nepal, the Ministry of Culture, Tourism and Civil Aviation commissioned and adopted Integrated Tourism Destination Development Plans for Pokhara, Eastern Nepal and Western Nepal (Ministry of Culture, Tourism and Civil Aviation, 2018), all of which were ecotourism relevant.

Of particular interest to ecotourism were the results of Samarth-NMDP's Review and Assessment of the Tourism Permitting Systems and Revenue Streams, completed in 2016 with recommendations for the Ministry of Finance (Choegyal et al., 2016). The study examined Nepal's well-established system of permits, royalties, fees and licenses for tourism-related activities, particularly those associated with mountain and wildlife tourism. Concerns had been raised that the existing permit system did not provide tourists with sufficient safety and support, was overly complex and intimidating for visitors, and that local tourism destinations and communities may not have received sufficient assistance to improve livelihoods or maintain the local environment for tourism. The study found scope for improvement overall, and the goal was set to initiate a smart permit system that increases government revenue, enhances the tourism experience and benefits the local people and resources in the destination.

The study by Choegyal and her colleagues (2016) projected that 27% of all revenue collected from tourism permits could be redirected to local development and destination management in rural areas, an increase over the previous ten-year average of 20% per year, if the study recommendations were applied. These included switching protected area fees to a more equitable merit-based user-pay system calculated on a per-person, per-day basis to improve visitors' experiences and satisfaction in Nepal's national parks and conservation areas. Increased revenue generated would boost the national treasury and assist protected areas, thereby also benefiting buffer zones, adjacent committees and local destination communities with their share. As a mechanism for the future, incentives, seasonal packages and a progressive fee system could help stimulate multi-park visitation and manage visitors, either to encourage more visitation or control numbers and peak-period crowding as required.

In addition to the national projects mentioned earlier, ecotourism has been identified as a common potential strength of the entire Himalayan region, agreed by participating South Asian Subregional Economic Cooperation (SASEC) Working Group countries Bangladesh, Bhutan, India, Nepal and Sri Lanka, and facilitated by the ADB from 2004 to 2010. With an overall objective to spread the benefits of pro-poor tourism, particularly to remote and impoverished areas and building on the region's comparative advantage of natural assets, several ecotourism programs and projects were developed under the following sub-themes: Trekking in the Himalaya, Ecotourism in the Ganga-Brahmaputra and Adventure Tourism in South Asia. Coordinated marketing material reflected these themes, under Ecotourism Pioneers and Nature Culture Adventure branding, and ecotourism training was delivered to stakeholders in participating countries under a human resource development component. In addition to ecotourism's strengths in the region, religious tourism, including Hindu pilgrimage and Buddhist circuits were deemed relevant and complementary (Asian Development Bank, 2005).

Many of Nepal's recent ecotourism efforts include improving facilities and services designed to lengthen visitor stays, increase yield and engage local communities in tourism's benefits. The Nepal Ecotourism Road Circuit packaged nature, culture and community attractions between Kathmandu, Pokhara, Tansen Palpa, Lumbini and Chitwan into one loop, upgrading the Tansen Palpa historic hill town with water supply systems, heritage restoration and training in homestays and handicrafts (Asian Development Bank, 2008).

The SASEC tourism studies were evolved by the ADB and the national governments into the South Asia Tourism Infrastructure Development Program (SATIDP), which from 2010 to 2015 implemented projects in Bangladesh, India and Nepal. These included a visitor center and connectivity enhancement with roads and Bhairahawa airport upgrading to regional international status serving Lumbini, the birthplace of Lord Buddha, and also destination improvements to visitor experiences at Lumbini to safeguard the outstanding universal values of the site, and fostering community participation and engagement in conservation and tourism. Under SATIDP, Nepal's Ministry of Culture, Tourism and Civil Aviation produced the comprehensive Greater Lumbini Tourism Promotion Plan 2015–2024. This is an action plan of tourism activities for the six Terai and middle hill districts of Palpa, Gulmi, Arghakhanchi, Rupandehi, Kapilvastu and Nawalparasi, aiming to improve local impact and community participation and conservation and environmental safeguards, among other objectives (Ministry of Culture, Tourism and Civil Aviation, 2015)

## A Note on Post-pandemic Ecotourism

Recently, the UNWTO (2020) recognized the impacts of the 2020–2021 pandemic on global tourism, and particularly its effects on nature-based ecotourism. The UNWTO lamented that the sudden collapse of tourism has severely reduced funding for biodiversity conservation, which is alarming given that a large percentage of global tourism relates to wildlife and other natural realms. Likewise, the pandemic has put many jobs at risk, raised incidents of poaching, looting and the consumption of bushmeat owing to the absence of tourists and protected area staff. The impact on biodiversity and ecosystems is particularly critical in less developed countries, where tourism revenues enable marine and terrestrial conservation efforts. Many examples of community involvement in nature tourism show how communities, including indigenous peoples, have been able to protect their cultural and natural heritage while creating wealth and improving their well-being. The impact of COVID-19 on tourism places further pressure on heritage conservation and on the cultural and social fabric of communities, particularly among indigenous people and ethnic minorities. Many intangible heritage practices, such as festivals and traditional gatherings, have been halted or postponed, and with the closure of handicraft markets, indigenous women's revenues have been severely curtailed. Approximately 90% of countries have closed their World Heritage Sites, resulting in deep socio-economic consequences for communities that rely on tourism.

There is no doubt that the COVID-19 outbreak has affected Himalayan tourism in equally drastic ways, with millions of jobs lost and few government safety nets available. The wide-reaching impacts in the Himalaya, including zero international tourism arrivals in many areas during most of 2020, have reinforced the vital importance that ecotourism plays in supporting livelihoods and biodiversity protection, even in the most remote communities.

## Conclusion

The Himalaya Mountains have become one of the world's leading venues for, and proponents of, ecotourism with a strong record of project experience, development studies, planning and policy progress, although this varies from country to country. The trend toward more sustainable forms of tourism during the past two decades has resulted in a well-trained cadre of ecotourism professionals and leaders that help guide tourism's continued growth in the region. Despite ecotourism's rocky start in this part of Asia and some enduring environmental and social concerns, increased numbers of communities have benefited socioeconomically from ecotourism. Likewise, the overall valuation of the montane ecosystem has increased considerably so that now protective legislation is in place, enforcement is more reliable and more service providers, community members, policy makers and other stakeholders support the principles of sustainable tourism.

As a natural and cultural realm, the Himalaya have certain physical and human characteristics that make ecotourism there unique. These include a greater emphasis on cultural tourism in the mix, issues associated with lowland ethnic minorities working in the highlands, delicate montane ecosystems and certain socio-political histories that have created image problems in the past but which are currently being overcome. The Himalaya are one of the great destination regions of the world. As Bhutan, China/Tibet, India, Nepal and Pakistan continue on their upward trajectory in ensuring that the principles of sustainable development are supported by the growth of tourism, they will continue to carve out their unique position in the global ecotourism marketplace.

## References

Allied Market Research (2021). Global ecotourism market: Opportunities and forecast, 2021–2027. Retrieved February 15, 2021, from www.alliedmarketresearch. com/eco-tourism-market-A06364#:~:text=Request%20Now%20!,dominance%20 throughout%20the%20forecast%20period.

Ashley, C., Boyd, C., & Goodwin, H. (2000). Pro-poor tourism: Putting poverty at the heart of the tourism agenda. *ODI: Natural Resource Perspectives*, 51, 1–6.

Ashley, C., Roe, D., & Goodwin, H. (2001). *Pro-Poor Tourism Strategies: Making Tourism Work for the Poor—A Review of Experience*. London: Overseas Development Institute.

Asian Development Bank (2005). *South Asia Subregional Economic Cooperation, Tourism Development Plan*. Manila: Asian Development Bank and South Asia Subregional Economic Cooperation Tourism Working Group.

Asian Development Bank (2008). *South Asia Subregional Economic Cooperation (SASEC) Tourism Development Project: Final Report, Volume 1, Main Report*. Manila: Asian Development Bank and METCON Nepal.

BBC News (2020). China holiday: Millions on the move for golden week. *BBC News*, October 1. Retrieved February 19, 2021, from www.bbc.com/news/world-asia-china-54377722.

Bhattarai, S., Adhikari, S.R., & Bamford, D. (2006). *Tourism for Rural Poverty Alleviation Programme (TRPAP), Final Evaluation Report*. Kathmandu: United Nations Development Programme.

Bhutan Department of Tourism (2002). *Bhutan National Ecotourism Strategy*. Thimphu: Ministry of Trade & Industry and World Wildlife Fund.

Beirman, D., Upadhayaya, P.K., Pradhananga, P., & Darcy, S. (2018). Nepal tourism in the aftermath of the April/May 2015 earthquake and aftershocks: Repercussions, recovery and the rise of new tourism sectors. *Tourism Recreation Research*, 43(4), 544–554.

Buckley, R.C. (2003). *Case Studies in Ecotourism*. Wallingford: CABI.

Buckley, R.C. (2009). *Ecotourism: Principles and Practices*. Wallingford: CABI.

Buckley, R.C. (2012). Ecotourism. In S. Geall, J. Liu, & S. Pellissery (Eds.), *Berkshire Encyclopedia of Sustainability. China, India and East and Southeast Asia: Assessing Sustainability* (pp. 92–96). Great Barrington, MA: Berkshire Publishing.

Butcher, J. (2006). The United Nations international year of ecotourism: A critical analysis of development implications. *Progress in Development Studies*, 6(2), 146–156.

Butcher, J. (2007). *Ecotourism, NGOs and Development: A Critical Analysis*. London: Routledge.

Cabral, C., & Dhar, R.L. (2020). Ecotourism research in India: From an integrative literature review to a future research framework. *Journal of Ecotourism*, 19(1), 23–49.

Ceballos-Lascurain, H. (1996). *Tourism, Ecotourism, and Protected Areas: The State of Nature-Based Tourism Around the World and Guidelines for Its Development*. Gland, Switzerland: IUCN.

Chakraborty, A. (2019). Does nature matter? Arguing for a biophysical turn in the ecotourism narrative. *Journal of Ecotourism*, 18(3), 243–260.

Chan, R., & Bhatta, K. (2013). Ecotourism planning and sustainable community development: Theoretical perspectives for Nepal. *South Asian Journal of Tourism and Heritage*, 6(1), 69–96.

Choegyal, L. (2008). *Tourism, Eastern Himalayas Ecoregion Tourism Scoping Study Report*. Kathmandu: WWF International Living Himalayas Network Initiative.

Choegyal, L. (2011). The great Himalaya trail: A new Nepal tourism product with both trek marketing and development rationale. *Nepal Tourism & Development Review*, 1(1), 71–76.

Choegyal, L. (2019a). Tourism in the Himalayas. *Druk Journal*, 5(2), n.p. (online journal)

Choegyal, L. (2019b). *WWF Nepal Ecotourism Strategy, Final Report*. Kathmandu: WWF Nepal.

Choegyal, L., Marshall, T., Pandey, R.J., & Thakali, S. (2016). *Review and Assessment of the Tourism Permitting System and Revenue Streams in Nepal*. Kathmandu: Samarth.

Dahal, S., Nepal, S.K., & Schuett, M. (2014). Marginalized communities and local conservation institutions: Indicators of participation in Nepal's Annapurna conservation area. *Environmental Management*, 53, 219–230.

Dam, S. (2013). Issues of sustainable ecotourism development in Sikkim: An analysis. *South Asian Journal of Tourism and Heritage*, 6(2), 32–48.

Dhakal, D.P., Khadka, M., Sharma, S., & Choegyal, L. (2007). *Lessons Learned: Nepal's Experience Implementing Sustainable Rural Tourism Development Model*. Kathmandu: UNDP.

Druk Journal (2019). Editorial: A deep dive into tourism in Bhutan. *Druk Journal*, 5(2), n.p. (online journal)

Ecotourism and Conservation Society of Sikkim (2020). About us. Retrieved December 1, 2020, from www.indiamart.com/ecotourism-conservation-societygangtok/aboutus.html

Elmahdy, Y.M., Haukeland, J.V., & Fredman, P. (2017). *Tourism Megatrends: A Literature Review Focused on Nature-Based Tourism*. Ås: Norwegian University of Life Sciences.

Erdem, B., & Tetik, N. (2013). An environmentally-sensitive approach in the hotel industry: Ecolodges. *International Journal for Responsible Tourism*, 2(2), 22–40.

Fennell, D.A. (2020). *Ecotourism* (5th ed.). London: Routledge.

Foggin, J.M., & Yuan, C. (2020). *Promoting Conservation & Community Development through Ecotourism: Experiences from Valued Conservation Landscapes on the Tibetan Plateau (Qinghai Province, China)*. Bishkek, Kyrgyzstan: Plateau Perspectives.

Goodwin, H. (1998). Sustainable tourism and poverty elimination. Discussion paper prepared for the DFID/DETR Workshop on Sustainable Tourism and Poverty, October 13.

Government of India (2002). *National Tourism Policy*. New Delhi: Department of Tourism, Ministry of Tourism and Culture.

Government of India (2020). Environmental information system. Retrieved December 5, 2020, from www.envis.nic.in/

Government of Nepal (2019). *Nepal National Tourism Strategic Plan 2015–2024*. Kathmandu: Ministry of Culture, Tourism and Civil Aviation.

Government of Sikkim (2020). About ENVIS. Retrieved December 5, 2020, from www.sikkimforest.gov.in/About_ENVIS.htm

Gurung, D.B., & Seeland, K. (2008). Ecotourism in Bhutan: Extending its benefits to rural communities. *Annals of Tourism Research*, 35(2), 489–508.

Gyelshen, N. (2019). Concept of ecotourism. *Druk Journal*, 5(2), n.p. (online journal)

Honey, M., & Chafe, Z. (2005). *Consumer Demand and Operator Support for Socially and Environmentally Responsible Tourism*. Washington, DC: Center on Ecotourism and Sustainable Development and the International Ecotourism Society.

Huddart, D., & Stott, T. (2020). *Adventure Tourism: Environmental Impacts and Management*. Cham, Switzerland: Springer.

Indian Himalayan Center for Adventure and Eco Tourism (2020). Eco-tourism in Sikkim. Retrieved December 5, 2020, from www.ihcaesikkim.org/eco-tourism

Jolliffe, L. (Ed.). (2007). *Tea and Tourism: Tourists, Traditions and Transformations*. Clevedon: Channel View Publications.

Kala, C.P., & Maikhuri, R.K. (2011). Mitigating people-park conflicts on resource use through ecotourism: A case of the Nanda Devi biosphere reserve, Indian Himalaya. *Journal of Mountain Science*, 8(1), 87–95.

K.C., A. (2017). Ecotourism in Nepal. *The Gaze: Journal of Tourism and Hospitality*, 8, 1–19.

Khanra, S., Dhir, A., Kaur, P., & Mäntymäki, M. (2021). Bibliometric analysis and literature review of ecotourism: Toward sustainable development. *Tourism Management Perspectives*, 37, 100777.

Koščak, M., & O'Rourke, T. (Eds.). (2019). *Ethical and Responsible Tourism: Managing Sustainability in Local Tourism Destinations*. London: Earthscan.

Leslie, D. (Ed.). (2012). *Responsible Tourism: Concepts, Theory and Practice*. Wallingford: CABI.

Luo, H. (2012). On the development model and strategy of Tibet ecotourism. *Journal of Tibet University*, 27(4), 33–38.

MacLaren, F.T. (2002). The international year of ecotourism in review. *Journal of Sustainable Tourism*, 10(5), 443–448.

Maharana, I., Rai, S.C., & Sharma, E. (2000). Valuing ecotourism in a sacred lake of the Sikkim Himalaya, India. *Environmental Conservation*, 27(3), 269–277.

Mandić, A. (2019). Nature-based solutions for sustainable tourism development in protected natural areas. *Environment Systems and Decisions*, 39, 249–268.

Meleddu, M., & Pulina, M. (2016). Evaluation of individuals' intention to pay a premium price for ecotourism: An exploratory study. *Journal of Behavioral and Experimental Economics*, 65, 67–78.

Ministry of Culture, Tourism and Civil Aviation (2015). *Greater Lumbini Tourism Promotion Plan: Towards Making the Lord Buddha's Birthplace and Associated Sites a Regional Tourism Hub (2015–2024 AD)*. Kathmandu: Ministry of Culture, Tourism and Civil Aviation.

Ministry of Culture, Tourism and Civil Aviation (2018). Reports. Retrieved December 1, 2020, from www.tourism.gov.np/publications/6

Ministry of Culture, Tourism and Civil Aviation (2020). *Nepal Tourism Statistics 2019*. Kathmandu: Ministry of Culture, Tourism and Civil Aviation.

Nepal, S.K., Kohler, T., & Banzhaf, B.R. (2002). *Great Himalaya - Tourism and the Dynamics of Change in Nepal*. Berne: Swiss Foundation for Alpine Research.

Nepal Tourism Board (2001). *National Ecotourism Strategy and Marketing Program of Nepal*. Kathmandu: Nepal Tourism Board.

Nyaupane, G.P., & Chhetri, N. (2009). Vulnerability to climate change of nature-based tourism in the Nepalese Himalayas. *Tourism Geographies*, 11(1), 95–119.

Nyaupane, G.P., & Poudel, S. (2011). Linkages among biodiversity, livelihood, and tourism. *Annals of Tourism Research*, 38(4), 1344–1366.

Nyaupane, G.P., & Thapa, B. (2004). Evaluation of ecotourism: A comparative assessment in the Annapurna conservation area project, Nepal. *Journal of Ecotourism*, 3(1), 20–45.

Nyaupane, G.P., & Timothy, D.J. (2010). Power, regionalism and tourism policy in Bhutan. *Annals of Tourism Research*, 37(4), 969–988.

Plateau Perspectives (2020). Choosing ecotourism in Kyrgyzstan. Retrieved November 30, 2020, from https://stories.plateauperspectives.org/choose-ecotourism

Poudel, S., & Nyaupane, G.P. (2017). Understanding environmentally responsible behaviour of ecotourists: The Reasoned Action Approach. *Tourism Planning & Development*, 14(3), 337–352.

Powell, R.B., & Ham, S.H. (2008). Can ecotourism interpretation really lead to pro-conservation knowledge, attitudes and behavior? Evidence from the Galapagos Islands. *Journal of Sustainable Tourism*, 16(4), 467–489.

QNP Working Commission (1998). *Qomolangma Nature Preserve: Ecotourism Masterplan Summary*. Beijing: QNP Working Commission, CICETE and UNDP.

Rai, S.C., & Sundriyal, R.C. (1997). Tourism and biodiversity conservation: The Sikkim Himalaya. *Ambio*, 26(4), 235–242.

Saintz, G., & Hugo, L. (2017). *Great Himalaya Trails: Trail Standards Handbook*. Kathmandu: Samarth.

Scheyvens, R. (1999). Ecotourism and the empowerment of local communities. *Tourism Management*, 20(2), 245–249.

Self, R.M., Self, D.R., & Bell-Haynes, J. (2010). Marketing tourism in the Galapagos Islands: Ecotourism or greenwashing? *International Business & Economics Research Journal*, 9(6), 111–125.

Seth, C.M. (2019). Developing eco-tourism in the Himalayan state of Jammu and Kashmir, India. In U. Stankov, S.N. Boemi, S. Attia, S. Kostopoulou, & N. Mohareb (Eds.), *Cultural Sustainable Tourism: A Selection of Research Papers from IEREK Conference on Cultural Sustainable Tourism* (pp. 193–201). Cham, Switzerland: Springer.

Singh, R.B., & Mishra, D.K. (2004). Green tourism in mountain regions-reducing vulnerability and promoting people and place centric development in the Himalayas. *Journal of Mountain Science*, 1(1), 57–64.

Spenceley, A. (2020). *Tools and Resources for Nature-Based Tourism*. Washington, DC: World Bank Group.

Thapa, B. (2004). Tourism in Nepal: Shangri-La's troubled times. *Journal of Travel & Tourism Marketing*, 15(2–3), 117–138.

TIES (2007). *TIES Global Ecotourism Fact Sheet*. Washington, DC: The International Ecotourism Society.

TIES (2020). What is ecotourism? Retrieved December 5, 2020, from https://ecotourism.org

Timothy, D.J. (2007). Empowerment and stakeholder participation in tourism destination communities. In A. Church & T. Coles (Eds.), *Tourism, Power and Space* (pp. 199–216). London: Routledge.

Timothy, D.J., & Boyd, S.W. (2015). *Tourism and Trails: Cultural, Ecological and Management Issues*. Bristol: Channel View Publications.

Timothy, D.J., & White, K. (1999). Community-based ecotourism development on the periphery of Belize. *Current Issues in Tourism*, 2(2/3), 226–242.

Tourism Council of Bhutan (2020). *Bhutan Tourism Monitor, 2019*. Thimphu: Tourism Council of Bhutan. Retrieved December 1, 2020, from www.tourism.gov.bt/uploads/attachment_files/tcb_K11a_BTM%202019.pdf

Tourism Recreation Conservation (2014). *Nepal National Park Concessioning Framework: Scoping Report and Project Plan*. Kathmandu: International Finance Corporation Nepal, World Bank Group.

United Nations (2003). Québec declaration on ecotourism defines basis for its international development. Retrieved December 1, 2020, from www.un.org/press/en/2002/unep113.doc.htm

United Nations (2020). Sustainable development goals: Sustainable tourism. Retrieved December 1, 2020, from https://sustainabledevelopment.un.org/topics/sustainabletourism

United Nations Environment Program and World Tourism Organization (2005). *Making Tourism More Sustainable: A Guide for Policy Makers*. Paris and Madrid: UNEP and UNWTO.

UNWTO (2002). International year of ecotourism, 2002. Retrieved December 1, 2020, from www.unwto.org/international-year-ecotourism-2002

UNWTO (2020). Secretary general's policy brief on tourism and COVID-19. Retrieved December 2, 2020, from www.unwto.org/tourism-and-covid-19-unprecedented-economic-impacts

Upadhayaya, P.K. (2018). Sustainable management of trekking trails for the adventure tourism in mountains: A study of Nepal's great Himalaya Trails. *Journal of Tourism and Adventure*, 1(1), 1–31.

Vannelli, K., Hampton, M.P., Namgail, T., & Black, S.A. (2019). Community participation in ecotourism and its effect on local perceptions of snow leopard (*Panthera uncia*) conservation. *Human Dimensions of Wildlife*, 24(2), 180–193.

Wall, G. (1997). Sustainable tourism-unsustainable development. In S. Wahab & J.J. Pigram (Eds.), *Tourism Development and Growth: The Challenge of Sustainability* (pp. 33–49). London: Routledge.

Weaver, D. (2002). Asian ecotourism: Patterns and themes. *Tourism Geographies*, 4(2), 153–172.

Wheeller, B. (1993). Sustaining the ego. *Journal of Sustainable Tourism*, 1(2), 129–139.

Winter, P.L., Selin, S., Cerveny, L., & Bricker, K. (2020). Outdoor recreation, nature-based tourism and sustainability. *Sustainability*, 12(1), 81.

World Expeditions (2020). The great Himalaya trail. Retrieved December 1, 2020, from http://trekthegreathimalayatrail.com/

World Wide Fund for Nature (2012). Ecotourism in Bhutan. Retrieved December 3, 2020, from https://wwf.panda.org/wwf_news/?203428/Ecotourism-in-Bhutan

Zhao, Y., & Liu, B. (2020). The evolution and new trends of China's tourism industry. *National Accounting Review*, 2(4), 337–353.

# 9 Tourism and Himalayan Indigenous Communities

*Anup K C and Mariela Fernandez*

## Introduction

Tourism can contribute to the sustainable development of Indigenous communities (Coria & Calfucura, 2012; Su et al., 2016). When Indigenous people participate in and/or control the tourism development processes in their communities, they stand a better chance of working in higher-level management positions, being entrepreneurs, and experiencing a reduction in poverty. Indigenous communities may also experience an increase in infrastructure development and provision of public services (Anand et al., 2012; Bhalla et al., 2016; Bhatt, 2012; Long et al., 2018; Manyara & Jones, 2007; Sharma & Sarmah, 2019; Sood et al., 2017; Walter et al., 2018). Additionally, Indigenous communities can utilize tourism as a mechanism to preserve their cultures.

Although tourism can contribute to the sustainable development of Indigenous communities, it may not always be supported by these communities for various reasons, including safety and security issues (Hounnaklang, 2016), strong and opposing religious beliefs, lack of awareness (Sood et al., 2017), and disproportionate benefit-sharing (Yang et al., 2013). Accordingly, the impacts of tourism may need to be examined to ensure that tourism benefits Indigenous communities. To explore this issue further, this chapter focuses on the Pan-Himalayan region, which includes the Himalaya and its adjacent regions covering the northeastern part of Afghanistan, northern Pakistan, northern India, Nepal, Bhutan, and southwest China (Editors, 2011). This region has many Indigenous groups, and tourism is an important socioeconomic force among them (Zhang et al., 2017). For instance, the number of ethnic groups in China is 56 (Zhang et al., 2017), with Tibetans being the main Indigenous group of the Himalayan region of China. In Nepal, there are 59 Indigenous societies (Bhusal, 2007; KC, 2016), 705 in India (MoTA, 2014), and many other groups in Bhutan, Pakistan, and Afghanistan. There are no integrated reviews on the involvement of Indigenous communities in tourism activities, its impacts, and issues in the Himalaya. Thus, this chapter examines three major questions. First, what are the different forms of tourism adopted by Pan-Himalayan Indigenous communities? Second, how does tourism affect these communities? Third, what tourism issues face Himalayan Indigenous communities?

DOI: 10.4324/9781003030126-12

## Indigenous People and Their Involvement in Tourism

Indigenous people, also known as First Peoples, Aboriginal Peoples, or Native Peoples, are pre-colonial communities with historical continuity of pre-invasion that are different from those of the dominant communities (United Nations, 2019). In the context of the Pan-Himalaya, different names are given to the Indigenous people, including tribal peoples, hill tribes, scheduled tribes, *Adivasis*, or *Janajatis*. Indigenous people in the Himalaya were colonized anciently by various groups including Dravidian, Indo-Iranian, Tibeto-Burman, and Austro-Asiatic communities. However, colonization mostly refers to European colonization from the sixteenth to twentieth centuries. During colonialization, Indigenous people faced discrimination by foreign elites and experienced a myriad of challenges (Kumar, 2018).

Settler colonialism led to the large-scale displacement of Indigenous people from their traditional lands and settlements so that the colonizers could reap the rewards of exploiting natural resources for economic reasons (Colchester, 2004; Kumar, 2018). The process led to colonizers' political and cultural dominance, during which many ethnic groups found themselves excluded from using natural and cultural resources and participating in the political arena, as most leadership positions were filled by the non-natives, with only the lower positions available to Indigenous people. Given their marginalization in many spheres, many native peoples experienced discrimination by the more dominant ethnic groups in society and colonialists, and they were relegated to lives of poverty and struggles for basic human rights. This pattern is still apparent today in many regions (Kumar, 2018).

Many international organizations have worked with Indigenous people to ensure basic human rights. Notable organizations include the International Labor Organization (ILO), which organized a convention for the rights of Indigenous people in 1989, known as ILO-Convention 169, and the United Nations Working Group (established in 1982) responsible for the Declaration on the Rights of Indigenous People (UNDRIP) in 2007 (Kumar, 2018). The constitutions of Nepal, India, and Pakistan support equal rights for all citizens, including Indigenous communities (Kumar, 2018), but the implementation of these legal rights is questionable as people continue to be affected by acts of discrimination. To ensure equal rights for its Indigenous populations, Nepal ratified both the ILO declaration and the UNDRIP declaration (Dunford, 2019). Afghanistan, Pakistan, China, and India have not ratified the latest ILO 169 but have adopted the UNDRIP's action. Bhutan has not ratified either the ILO or the UNDRIP declarations (Kumar, 2018). Indigenous people have an equal right to maintain, protect, and develop their cultures and traditions. In some cases, tourism has helped in the realm of social and economic enhancement, and cultural protection (Carr et al., 2016).

## Indigenous People and Tourism in the Pan-Himalaya

The Himalayan Indigenous communities have long been involved in different forms of tourism, including nature-based, religious, and adventure tourism

(K C, 2017b). More recently, community-based homestay tourism, which integrates nature and culture and is operated by Indigenous communities, is growing in popularity (Acharya & Halpenny, 2013; Bhalla et al., 2016). These forms of tourism are described below as pertaining to the Indigenous groups in the region.

## Nature-Based Tourism

Nature-based tourism is extremely popular in the Himalaya because of the unique physiography, rich biodiversity, and scenic beauty. The majority of countries in the region have become popular nature destinations and have been well researched in this regard, including Bhutan (Brunet et al., 2001; Gurung & Seeland, 2008; Nyaupane & Timothy, 2010), China (Catibog-Sinha & Wen, 2008; Genzong et al., 2007; Nianyong & Zhuge, 2001; Shi et al., 2015; Zhao & Jia, 2008), India (Anand et al., 2012; Bhalla et al., 2016; Serenari et al., 2012; Sood et al., 2017), Nepal (Baral et al., 2012; K C et al., 2015; Nyaupane & Thapa, 2004; Neupane et al., 2021), and Pakistan (Adnan Hye & Ali Khan, 2013; Baloch, 2007), and, to a much lesser degree, Afghanistan (Hall & Page, 2017; Simms et al., 2011). Afghanistan suffers from the severe effects of war and political instability, which prevents widespread tourism in that country. The high diversity of flora and fauna (Adnan Hye & Ali Khan, 2013; Brunet et al., 2001), the designation of valued protected areas (Neupane et al., 2021; Nianyong & Zhuge, 2001), the presence of freshwater bodies (Debarbieux et al., 2014), and the scenic montane landscapes (Adhikari & Fischer, 2011; Walter et al., 2018) serve as the main nature-based tourist assets in the region. Nature photography, visiting parks and protected areas, and wildlife viewing are among the most popular nature-based activities in the region (Acharya & Halpenny, 2013; Lipton & Bhattarai, 2014; Panta & Thapa, 2018).

Protected areas are the major draw of nature-based tourism (Dangi & Gribb, 2018; Nyaupane & Poudel, 2011). Indigenous communities have historically used protected areas for their livelihoods. When protected areas were established, communities lost access to their traditional lands. During the 1990s, however, several governments shifted from exclusionary policies to inclusionary policies regarding native communities and natural resources, thereby encouraging their participation and co-management of natural resources. This allowed communities to participate in tourism activities and benefit from conservation (K C et al., 2021). Local people have participated in the conservation and management of natural resources (participatory conservation) because the increase in wildlife numbers had resulted in more human-wildlife conflicts. To reduce these clashes and maximize the benefits of nature-based tourism in protected areas, certain international nongovernmental organizations (NGOs) have helped create buffer zones and buffer zone management regulations through consultations with Indigenous communities as one of the main stakeholders of conservation and development (Bhattarai et al., 2017). Protected area authorities and NGOs have worked together to involve Indigenous communities in conservation and tourism development (K C et al., 2021).

### Religious Tourism

The Himalaya range is replete with holy shrines, temples, and sacred mountains. Mount Kailash in Tibet is a popular destination for pilgrims of many faiths (Debarbieux et al., 2014), and there are many more sacred pilgrimage sites in that part of China (Zhang et al., 2017). Nepal is home to many sacred sites that are visited by millions of pilgrims and cultural tourists each year (Bhusal, 2007; K C, 2016), including Muktinath, Pashupati Nath, Gosaikunda, and Lumbini— a very important pilgrimage destination for Buddhists, who believe this to be the site of Buddha's birth (Nyaupane et al., 2015). Many of these sites and the pilgrimage trails are located in areas populated by Indigenous people, a large portion of whom are involved in satisfying the needs of pilgrims and other religious tourists. Uttarakhand, India, is a popular pilgrimage and religious tourism destination (Debarbieux et al., 2014). The northern part of Pakistan has strong potential for religious tourism, mainly related to Sikhism, Islam, and Buddhism (Baloch, 2007).

There are many religious sites in Bhutan that attract domestic and international tourists (Brunet et al., 2001). Some sacred sites are located in the lowlands or foothills with easy transport access, while others are in the mountains where trekking is necessary. The development of road transportation and cable car or gondala has increased the number of pilgrims and other tourists in the region (Beazley & Lassoie, 2017; Bleie, 2003). Although these accessibility developments have brought in some unintended and unwelcomed consequences (discussed below), these innovations have increased the economic benefits of tourism to the Indigenous communities. In Lumbini, Nepal, one of the world's great Buddhist pilgrimage destinations, the local population is comprised of Muslims and Hindus. Non-Buddhist inhabitants have influenced the conservation and management of the Lumbini site. Many local people have different motives and do not necessarily support heritage preservation activities that limit their economic participation. Also, the government and NGOs influence area management traditionally without consulting the local indigenous people (Nyaupane, 2009).

### Adventure Tourism

Adventure tourism in the mountains, which includes trekking and mountaineering, is popular in the Himalaya (Baloch, 2007). Eight of the 14 highest mountains in the world are in Nepal, including Mt. Everest. Mt. Everest is globally known for mountaineering where Sherpa communities have long played a significant role in the adventure experience (Musa et al., 2004; Nyaupane, 2015). Today, the Sherpa have moved up the employment ladder and tend to be lead porters, guides, and managers, while other Indigenous people from lower elevations have stepped in to work in the porter positions that were filled mostly by Sherpas in the earlier years of trekking tourism. The Sherpa have lived in the High Himalaya for generations and are therefore better acclimated to high-altitude living, while the newer Indigenous laborers in recent years are not as used to the extreme elevations and sometimes suffer ill health (see altitude sickness chapter by Pun et al. in this volume). Everest area natives operate family-owned trekking companies and hotels along the trekking routes in Nepal (Nyaupane, 2015). Other popular trekking destinations in Nepal include the Annapurna region (Baral et al., 2012), the Manaslu region (K C & Thapa Parajuli, 2014), the Langtang region, the Khaptad region, and the Kanchanjunga region, where Indigenous communities provide food and lodging services. The northern area of Pakistan is popular among trekkers and mountaineers (Adnan Hye & Ali Khan, 2013). Gilgit-Baltistan is a premier mountain destination owing to the presence of the world's second highest peak, K2 (Imran et al., 2014). Tourists also engage in trekking in the religious and pilgrimage areas in Uttarakhand, India, and Mount Kailash in Tibet (Debarbieux et al., 2014). Trekking tourism is also growing in Bhutan, but owing to that kingdom's strict minimum daily tourism tariff, it is a more expensive option than the other Himalayan states (Gurung & Seeland, 2008).

### Community-Based Homestay Tourism

Among the different forms of tourism in Himalayan Indigenous communities, community-based homestay tourism is one of the most important because it directly benefits the native people. Research has shown that locally controlled homestay programs benefit Indigenous communities through economic, social, and political empowerment (Ismail et al., 2016; Timothy & White, 1999). In community-based homestay tourism, guests are hosted by locals in their homes (Bhatt, 2012; Jamaludin et al., 2012; Shukor et al., 2014). The homes are not hotels, motels, or commercialized residences (Ismail et al., 2016; Sood et al., 2017), and they are mostly located away from urban areas with limited accommodation facilities (Shukor et al., 2014). As a family business, family members are engaged in cooking food, managing their operations, and hosting guests (Jiang et al., 2018; Tavakoli et al., 2017). Most of them do not employ additional staff, other than family members (McIntosh et al., 2011). In some cases, people living in the neighborhood help provide entertainment (Tavakoli et al., 2017).

These homestays were initiated for a different purpose, but their concept developed as numbers of domestic and international researchers and visitors grew. Initially, the first homestays were used to host students and volunteers from outside the region. Encouraged by the earnings from hosting students, the Indigenous communities improved their houses and facilities to attract more guests. Later, local governments and international organizations saw homestays as a means of economically empowering Indigenous communities and provided technical and financial support to develop and sustain these homestays (Ali et al., 2014; Hounnaklang, 2016). Homestays became popular lodging options in the Himalaya, with the exception of Bhutan, where budget independent travel is not permitted, although even that country has seen some success with its limited homestay program along the Nabji-Korphu trail, which opened in 2006 (Gurung & Seeland, 2008). Their guests tend to be better educated, having a keen interest in local cultures and lifestyles. The guesthouses attract budget-conscious tourists and international backpackers who like to stay with the families and become immersed in Indigenous cultures and ways of life (Degang & Xiaoting, 2006; Ismail et al., 2016; Shi et al., 2015; Sood et al., 2017), including participating in cultural festivals, wearing traditional attire, eating and cooking traditional foods, observing and participating in farming methods, and even attempting to learn some of the Indigenous languages (Chang, 2006; Jiang et al., 2018). There are several successful examples of Indigenous homestay programs in the Himalaya, including many in Jammu and Kashmir, Himachal Pradesh, Uttarakhand, and Sikkim, India (Bhalla et al., 2016).

In the Kullu district of the Indian Himalaya, community homestay-based tourism is encouraged by the government through tax exemptions on income for five years and providing subsidies for electricity and water (Sood et al., 2017). In Nepal, homestay tourism aims to educate tourists about local cultures (Biswakarma, 2015) and is popular in several parts of the country (Acharya & Halpenny, 2013; Dahal et al., 2020; Walter et al., 2018; Neupane et al., 2021). In the mountainous area of China, this type of tourism has attracted significant tourist attention and benefited many Indigenous communities. It has enhanced the tourism management capacity of local farmers and put to good use fallow or vacant rural spaces, local farm products, and surplus labor (Long et al., 2018; Su et al., 2016; Wang et al., 2018).

In all of the types of tourism described earlier, nature and culture work together to attract tourists in the Himalaya (Adnan Hye & Ali Khan, 2013; Baloch, 2007; Baral et al., 2012; Brunet et al., 2001; Gurung & Seeland, 2008; Maikhuri et al., 2000; Nianyong & Zhuge, 2001). Understanding tourism preferences in this region can help Indigenous providers enhance the satisfaction of their guests, which in turn contributes to the development, management, and sustainability of Indigenous tourism endeavours (Pabel et al., 2017).

## Impacts of Tourism on Indigenous Communities

Indigenous tourism can empower communities, protect biodiversity, and improve people's livelihoods (Coria & Calfucura, 2012). It can provide many economic, social, cultural, and environmental benefits to native communities (Ruhanen & Whitford, 2019; Sharma & Sarmah, 2019; Shukor et al., 2014). Small tourism enterprises run by Indigenous communities are key in achieving the goals of sustainable development of these communities and are most impactful when they receive initial external support from local governments or NGOs (Zhang & Zhang, 2018). To protect the limited resources from rapid population growth and unrestrained tourism development, sustainability principles have been widely employed (Samsudin & Maliki, 2015; see also Choegyal's chapter in this volume). In some cases, tourism has been unable to provide all the expected benefits for Indigenous communities owing to a lack of local human, financial, and social capital, unbalanced distribution of economic benefits, and land insecurity (Coria & Calfucura, 2012). These issues in the Himalaya are explained in detail in the following sections.

In the Kumaon Division of Uttarakhand, India, homestays have promoted the protection of local cultural activities, heritage, natural capital, and equitable access for households across economic classes. By no means is it perfect, but it has also supported the conservation of forests, natural resources, and the local environment (Bhalla et al., 2016). Tourism in Ladakh, India, has addressed local livelihood needs through biodiversity conservation by bringing environmental, economic, and livelihood benefits to the people. It has helped instill pro-environment behavior, sustainable mountain tourism enterprises, and more equitable development in the region (Anand et al., 2012). Despite all of its good intentions, Indigenous tourism has caused both positive and negative impacts on the environment, society, and economy in many parts of the region (Acharya & Halpenny, 2013; K C et al., 2015; Walter et al., 2018).

### Environmental Impacts

Tourism operated by Indigenous communities has encouraged conservation and sustainable tourism development (Shukor et al., 2014). Because of the widely accepted concept of Indigenous knowledge as a key tool for ecological management since native peoples have been the sustainable stewards of the land for centuries (Johnston, 2003; Ruhanen & Whitford, 2019), native control helps protect undisturbed and disturbed sites and ecologically fragile landscapes (Hounnaklang, 2016). The 'Plant a Tree' campaign is one of the programs that encouraged tourists to appreciate the environment and help locals appreciate environmental ethics and awareness, and helps protect the forests (Kunjuraman & Hussin, 2017). It encourages people to implement good practices for maintaining better environmental quality (Samsudin & Maliki, 2015) and enhancing village image and scenery (Shukor et al., 2014). In some areas, solid waste problems have improved where communities collect a small waste management fee (Anand et al., 2012; Dahal et al., 2020; Sharma & Sarmah, 2019; Tran & Walter, 2014). There are many different examples of environmental impacts faced by the Indigenous communities of the Himalaya.

In the Mawlynnong village, India, a Khasi tribal women's group is involved in tourism activities; one of their main goals is to prevent waste and dust accumulation. Organic waste is collected and converted into compost. Local people maintain the roads by picking up waste, sweeping, and weeding (Sharma & Sarmah, 2019). In the Kumaon division of Uttarakhand, homestays are eco-friendly, and local people engage in wildlife census, ecotourism impact monitoring, and training and workshops (Bhalla et al., 2016). In Ladakh, the Indigenous Changpas use green, efficient, and renewable energy sources in their tourism development efforts. Solid waste and water sources are managed efficiently. The Changpas protect the fragile ecosystem and biodiversity by not allowing visitors to camp in restricted areas (key wildlife feeding and breeding areas). Likewise, different areas are designated for pack animal grazing to reduce the adverse environmental impacts (Anand et al., 2012). In various localities of Nepal, Indigenous communities are making efforts to maintain clean and green environments for tourism (Acharya & Halpenny, 2013; K C et al., 2015; Walter et al., 2018).

Despite these examples of positive environmental outcomes, tourism consumes scarce resources, thereby competing with the needs of native peoples, produces waste by-products, and causes different types of pollution (K C, 2016). Conservation activities have recently increased human-wildlife conflicts in protected areas and the surrounding communities where Indigenous people live (Nyaupane & Poudel, 2011; Stone, 2015). Unmanaged tourism causes adverse impacts on the natural environment and cultural landscape (Hounnaklang, 2016; Samsudin & Maliki, 2015). The movement of vehicles and fuel burning cause air pollution (Hounnaklang, 2016; Shukor et al., 2014). Other problems include an increase in water pollution and noise pollution from boat engines and cars in addition to a decrease in fish, shrimp, and firefly populations (Hounnaklang, 2016). Increasing numbers of tourist vehicles cause traffic congestion and parking problems (Dangi & Gribb, 2018; Shukor et al., 2014). Nyaupane and his colleagues (2014) reported that competitiveness in the trekking sector is lowering the standards of trekking services, which in turn increases environmental problems in the Khumbu region of Nepal.

### Social Impacts

Tourism has benefitted Indigenous societies in a number of ways, including enhancing communication and collaboration. It can increase grassroots participation in development and potentially reduce hostility between park management and the local communities, as the communities themselves become positioned to make decisions and determine much of their own future (Stone, 2015). Tourism enhances the exchange of knowledge and cross-cultural interaction, and provides diverse experiences for the community (Jamaludin et al., 2012). Through a participatory approach, it has the ability to empower Indigenous communities socially, economically, politically, and psychologically and to preserve their valuable cultural heritage (Masud et al., 2017; Sita & Nor, 2015; Su et al., 2016; Timothy, 2007). Participatory tourism development also encourages youths to become involved in the industry as a means of valuing their heritage as they work to protect their inherited cultural landscapes for future generations (Samsudin & Maliki, 2015). Although many scholars have noted the potential detrimental effects of tourism on cultural change and commodification if it is not managed and monitored correctly (Lundup, 2018; Mu et al., 2019), there is a general consensus that tourism can be a mechanism for preserving Indigenous cultures, especially among younger generations (Grimwood et al., 2019; Ruhanen & Whitford, 2019). Besides working in small enterprises, youth commonly perform traditional music, dances, customs, and other cultural activities for tourists' enjoyment (Sita & Nor, 2015).

In most Indigenous Himalayan societies, there is a work division among males and females in which males welcome the guests and provide basic information about the village, while females cook and manage other household activities (Mura, 2015). But, unlike many other economic activities, sustainable forms of tourism help provide women equal access to material resources, such as food, income, shelter, and medical care. It has the potential to empower women and provide new sources of income, such as through hosting tourists, guiding tours, cooking, and performing music and dance (Nyaupane & Poudel, 2011). It reduces the physical burdens on women and increases their socioeconomic productivity, compared to many traditional roles in wet rice cultivation and shellfish harvesting. Through tourism, women often acquire better roles and opportunities in education and knowledge enterprises, especially as they are invited to attend training seminars and workshops on planning, communication, leadership, accounting, managing small businesses, hosting, and tour guiding. Females in tourism areas are also more involved in decision-making, policy formulation, planning, and administering tourism projects (Acharya & Halpenny, 2013; McMillan et al., 2011; Scheyvens, 2000).

Even though destination inhabitants may derive economic benefits from their involvement in tourism, many of them, and those not involved, frequently have to bear the social burdens of tourism. One issue noted in native communities is the noise made by visitors and their activities during the night, as well as a general lack of privacy. Tourists often peer voyeuristically into Indigenous homes, temples, gardens, and other private spaces in an attempt to catch a glimpse of 'authentic' native life (Gu & Wong, 2006; Hounnaklang, 2016).

Many local entrepreneurs feel a sense of inequality in the distribution of guests and tourism benefit-sharing (Hounnaklang, 2016). Unfortunately, tourism sometimes exacerbates the existing power relations and widens the disparity between the rich and the poor, emphasizing inequities in power-sharing and decision-making among Indigenous community members where the elite families and communities primarily benefit from tourism (Coria & Calfucura, 2012; Stone, 2015). The people who receive the most benefits from tourism are able to build better homes and better satisfy other needs for transportation, communication, health, sanitation, and electricity. Because of their capital accumulation, wealthier families are able to renovate their houses and improve tourism facilities, while other families must use their limited incomes on basic needs such as food and clothes (Tran & Walter, 2014). Many tourism projects in the Global South have also benefitted outsiders more than Indigenous populations (Manyara & Jones, 2007). These negative and positive social impacts are also apparent in the Himalaya (Apollo, 2016).

In Ladakh, tourism has been shown to enhance the entrepreneurial capacity of Indigenous women, helping to lessen their financial dependence on their husbands (Bhatt, 2012). There is also an increase in the participation and broader representation of the community. Tourism has begun to empower women and increased their incomes (Anand et al., 2012; Caparrós, 2018). In the Kumaon division of Uttarakhand, India, Indigenous people are now collaborating, engaging in partnerships, and becoming involved in hosting cultural events (Bhalla et al., 2016). The people of Mawlynnong village, India, have gained business experience from tourism and have become more efficient in their development efforts (Sharma & Sarmah, 2019). In China, the mountain people are interacting more with tourists, and the participation of Indigenous communities has increased substantially.

Tibetans and others are now promoting their local products as souvenirs and introducing their cultures, foods, architecture, artifacts, and music to tourists (Su et al., 2016). Despite these positive aspects, tourism has brought about significant changes in Indigenous religions and ways of living, and alcoholism and drug abuse (Poudel, 2014). On the Annapurna Sanctuary Trail in Nepal, incidents of increased crime and adverse impacts on culture and social life have been documented (Nyaupane & Thapa, 2004). In other regions, prostitution has grown as an outcome of tourism (Nyaupane et al., 2006). These occurrences have been difficult for many Indigenous groups to handle as they see their cultures diminish and many of their traditions disappear, despite the economic benefits that come with tourism (Sood et al., 2017).

### Economic Impacts

Tourism is a significant source of income for most countries (Ogucha et al., 2015) and the main revenue source in many regions (Gu & Wong, 2006). Tourism is often adopted by Indigenous people for its positive effects on the livelihoods and employment in economically marginal or impoverished regions (K C & Thapa Parajuli, 2014; Samsudin & Maliki, 2015). Tourism frequently promotes regional development, creates new enterprises and stimulates entrepreneurial thinking, and enlarges the market for locally-produced goods and services, including agricultural products (Ogucha et al., 2015). Many Indigenous people are involved in making and selling representative handicrafts (Gioli et al., 2019; Singh & Mishra, 2004). Tourism growth raises revenue for governments, corporations, and private businesses (Yang et al., 2013). From tourism, local institutions and NGOs acquire funds to operate their development efforts (Shukor et al., 2014; Stone, 2015). In many ways, tourism has been successful in providing income for impoverished Indigenous communities throughout the world (Hounnaklang, 2016; Shukor et al., 2014). Local earnings and savings can remain local and utilized for investments in other development projects (Stone, 2015). Such efforts can help provide business opportunities for local inhabitants and reduce some of the inequalities between the rich and the poor (Jamal et al., 2011) and between female and male entrepreneurs (Gu & Wong, 2006).

In Ladakh, tourism as an alternative to a limited resource situation, has created many new jobs, improved livelihoods, and helped reduce poverty in Indigenous communities. Youths in Ladakh are involved as porters, cooks, and trekking guides. Women are economically more empowered as they are earning incomes from successful homestays (Anand et al., 2012; Bhatt, 2012). In India's Kullu district, homestays have provided new economic opportunities for rural Indian communities who are more fully employed and involved in making and selling traditional Kullu clothing and handicrafts (Sood et al., 2017). In Mawlynnong, India, tourism provides new sources of income and contributes to the well-being of the Indigenous community. In the Binsar Wildlife Sanctuary in the Indian Himalaya, tourism has provided income-generating opportunities. There, the Indigenous people sell locally produced honey and juice from the rhododendron flower

(Bhalla et al., 2016). There are many examples of tourism providing employment and entrepreneurial opportunities in all countries of the region.

Despite these positive perspectives, Indigenous communities often face a lack of capital, inconsistent or seasonal incomes, and increased debt (Hounnaklang, 2016). In many cases, tourism income is only supplementary and does not provide for all of the economic needs of Indigenous families (Tran & Walter, 2014). As previously noted, sometimes tourism favors outside investors rather than native communities (Coria & Calfucura, 2012), ensuring that elites receive the greatest share of benefits compared to destination residents, who are the ones who suffer the costs of tourism, often increasing conflict between the haves and the have-nots (Coria & Calfucura, 2012; Yang et al., 2013). With growing tourist arrivals, prices usually rise, increasing the cost of living, sometimes prohibitively. So far, in the Himalaya, the economic benefits are seen more or less as balanced against the costs.

In the Chinese Himalaya, like other parts of the country, tourism is often controlled by outside interests in the form of national tourism development corporations and non-native migrants who frequently come in to buy local homes and businesses to convert them into tourism services (Nyaupane et al., 2006; Qu et al., 2019). In Nepal's Khumbu region, there is evidence of increasing competition, commercialization, income inequality, and tourism-induced increased cost of living (Nyaupane et al., 2014). Unfair competition in Indigenous communities can lead to adverse economic conditions. For example, unregistered homestays raise challenges for registered homestays as the unofficial operations do not follow local homestay guidelines. When visitors stay in the unofficial homestays and have negative experiences, such as environmentally unfriendly services or actions, it taints the image of the community and the registered inns that are adhering to the rules.

## Tourism Issues in Pan-Himalayan Indigenous Communities

### Risk of Natural Disasters

Inhabitants of the Himalaya face a constant threat of landslides, floods, glacial lake overflows, earthquakes, and dry spells. As most Indigenous communities live in remote areas, they are occasionally affected by natural disasters. In Nepal, avalanches and icefall took the lives of 16 Sherpas in the Mt. Everest region in 2014 (Beaumont, 2015). In 2015, a devastating earthquake and its aftershocks destroyed the tourism infrastructures of Indigenous communities in Barpak; lodges run by Indigenous communities on many trekking routes were destroyed (K C, 2017a). Snow avalanches in the Langtang region after the 2015 earthquake buried 116 houses and killed 308 people (176 residents, 80 foreigners, and 10 army personnel) (Callaghan & Thapa, 2015). Post-earthquake avalanches in the Mt. Everest region claimed the lives of 17 people and injured 61 others (Beaumont, 2015). Severe floods in Uttarakhand, India, killed more than 6,000 pilgrims in 2013, and there is an ever-present threat of flash flooding from glacial lakes in the region (Debarbieux et al., 2014). In 2005, an earthquake in Pakistan had damaged the tourism infrastructure in the most popular tourist locations in the country's northern mountain areas (Adnan Hye & Ali Khan, 2013). All of these disasters disproportionately affect Indigenous communities and their tourism livelihoods, primarily because they live in vulnerable areas, and are less equipped to deal with such crises in terms of inaccessibility, insufficient medical care and emergency services, and evacuation readiness.

### A Changing Climate

Climate change has had a notable impact on tourism in the Himalaya (K C, 2017a). The increasing scarcity of water is affecting tourism, agriculture, and health security. Indigenous lodging establishments have been adversely affected by water shortages as they are unable to provide for the water needs of their guests (Lama, 2010). Increasing number of extreme weather events are also an indication of climate change (Scott et al., 2019). In 2014, more than 32 people were killed by a sudden snowstorm in Nepal's Annapurna region (Burke & Walker, 2014). In 2010, flights were canceled due to bad weather for several days at the Lukla airport, Nepal, which affected more than 1,000 tourists (Sangraula, 2010). Sudden changes in the weather regularly affect mountain flights and cause uncertainty for all flights in the higher altitudes. This sometimes traps tourists for several days. In the short term, rising temperatures may increase access for trekking tourism, but in the long run, it will have adverse effects on tourism (K C & Thapa Parajuli, 2015). Ice melt in the Himalaya reduces the aesthetic appeal of the mountains (Debarbieux et al., 2014). Unusual weather events and various other crises (e.g., warming) are expected to increase with changes in the Himalayan climate, directly and indirectly affecting Indigenous people and their involvement in tourism (Nyaupane & Chhetri, 2009).

Given the high elevations of the Himalaya, tourism activities operated by Indigenous communities are mostly seasonal. There is a pervading fear that climate change will affect seasonality in the highlands, growing the off-season and shortening the high season. The highly seasonal nature of Himalayan tourism means that many Indigenous entrepreneurs get further into debt as they are unable to pay their bank loans during certain times of the year, and it is difficult to recuperate the costs of their investments in their businesses (Su et al., 2016). If tourist numbers decline, native people will need to migrate from their villages, at least seasonally if not permanently, to find work in agriculture, trade, or day labor

### Political Instability

Political instability in Nepal, Pakistan, Afghanistan and parts of India have affected Indigenous tourism in the past. Before 2006, Nepal's Maoist insurgency affected tourism in some remote areas of the country (Regmi & Walter, 2017). Unstable governments in the past were unable to form or enforce long-term tourism policies. Recent government-formed tourism policies are not very supportive of the private sector (Bhandari, 2019). The threat of terrorism in Pakistan and Afghanistan has kept tourists away for many years, despite the fact that those countries are home to remarkable natural landscapes and dynamic Indigenous cultural heritage (Adnan Hye & Ali Khan, 2013; Raza & Jawaid, 2013). The ongoing armed conflicts between India and Pakistan in Kashmir, as well as that between India and China, have adversely affected tourism in an area with otherwise significant tourism potential (Islam, 2014; Nyaupane & Budruk, 2009; Timothy, 2019).

### *Cultural Issues*

Indigenous communities have an uneasy time reconciling tourism and culture (Hounnaklang, 2016). Many fear the entry of tourists and entrepreneurs from outside the community could introduce culture shock and cultural change to these marginalized communities (Yang et al., 2013) and result in disrespect for their traditions, such as beliefs in the traditional system of goddesses and gods, caste-based hierarchies, the taboo of hiding women during their monthly periods, and the gendered division of labor between males and females. These cultural concerns keep some Indigenous people from becoming involved in tourism (Sood et al., 2017).

### *Infrastructure Issues*

To run tourism enterprises in remote areas and Indigenous communities is challenging due to poor infrastructure, poor promotional efforts, a lack of trained human resources, inadequate finances, safety and security concerns, poor local leadership, and a lack of tourism management experiences (Coria & Calfucura, 2012; Kunjuraman & Hussin, 2017). In some areas, haphazard road construction in the area of religious sites has increased visitation and overcommercialization, as well as destroyed many historic pilgrimage trails and trade routes (Beazley & Lassoie, 2017). While trekking in Nepal's protected areas (e.g., Annapurna Conservation Area, Manaslu Conservation Area, Gaurishankar Conservation Area, and Panchase Protected Forest), the authors observed that many trekking routes have been disturbed and even destroyed by road construction. Relatedly, increased dust from vehicular movements has raised visibility and health concerns among trekkers. The scenic beauty of the hills is also affected by road construction, which also affects the tourism activities operated by Indigenous communities. Increased road access to formerly remote areas has transformed traditional tourism activities (e.g., trekking and pilgrimage) into modern adventure sport and leisure in the Annapurna Conservation Area, Nepal (Lama & Job, 2014). It has also enabled much higher numbers of domestic tourists throughout the entire Himalayan region.

## The Need to Improve Himalayan Indigenous Tourism

If Indigenous tourism becomes economically unsustainable due to low tourist numbers during crises, support from the government and external parties (financial or technical) will be important. If the flow of tourists is high, the number of visitors should be limited at a steady pace through prior authorization or quotas (Coria & Calfucura, 2012). Thus, careful tourism planning and management is critical in Indigenous contexts because of the fragile cultures they value and the volatile geographies they inhabit (Bhalla et al., 2016; Coria & Calfucura, 2012). Indigenous communities should always be empowered in a position to negotiate their own interactions with tourists and how they wish to supply goods and services in tourism (Zhang et al., 2017). To sustain tourism in Indigenous communities, governments should implement policies to inform people of ethnic minorities about the importance of their culture, not just for tourism but also for the sake of protecting their heritage. Quick profiteering is dangerous, and native peoples should be cautioned to focus on conserving culture and nature, and interacting with tourists in socially and ecologically sustainable ways (Zhang et al., 2017). Promotional activities that highlight native people's authenticity and uniqueness, such as special cultural festivals, should be a high priority. Domestic tourists are often more sensitive to cultural differences and cultural frailties. Likewise, they are often big spenders who desire to be immersed in traditional communities. To attract domestic visitors, advertisements in local medias is helpful.

Additional research is needed to identify appropriate cultural resources for Indigenous tourism and to understand what elements of native cultures might be inappropriate to share with outsiders (Chang, 2006). This should come from the communities themselves within a participatory development framework. The effectiveness of the Indigenous homestay programs in India, Nepal, China, and Bhutan needs additional research attention to understand its value and impacts on the host communities in accordance with the principles of sustainable tourism development (Sharma & Sarmah, 2019). Also, research on the quality of service provided by these communities and the satisfaction rate of tourists should be conducted to help the communities better understand what creates satisfied customers (Jamal et al., 2011; Su et al., 2016).

## Conclusion

Like other regions of the world, Indigenous tourism is gaining popularity in the Himalaya due to the unique cultural backgrounds and lifestyles of the Indigenous Himalayan communities. The isolation created by the mountains has preserved native cultures and lifestyles for thousands of years, so many of them remain fairly authentic and in rather pristine environments. People visit Indigenous communities to enjoy authentic culture, local food, customary clothing, and traditional lifestyles. These, together with the dramatic mountain scenery backdrop, create an appeal unrivaled anywhere else on the earth. Not all Indigenous communities find it easy to receive tourists and develop tourism owing to some of their cultural beliefs, lack of awareness and confidence, and low economic background.

There are many manifestations of Indigenous tourism in the Himalaya. Nature-based tourism, religious tourism, adventure tourism, and community-based homestay tourism are some of the most important ones. The isolation and remoteness of many Indigenous communities have made homestay tourism quite popular and successful in recent years.

Tourism has both positive and negative ecological, social, and economic impacts on every destination, but these are especially acute in the remote native communities of the Himalaya. The region is also especially prone to the risks of natural disasters, climate change, the seasonal nature of visitation, and political instability. Sustainably managed destinations tend to have the most positive environmental outcomes, while unfettered growth in mismanaged destinations experience the most adverse impacts. So far, tourism has increased many native people's awareness of the value of the lands and cultures they inherited, and sustainable tourism has empowered Himalayan Indigenous communities to begin controlling tourism development, participating in its benefits, and collaborating with stakeholders to fund and manage the burgeoning industry in the region. One of the most glaring realities of Himalayan tourism is the inequities between the impoverished and the elites, many of whom have strong connections with external stakeholders. Growing incomes, increased employment, and better public services are some positive impacts, unbalanced economic opportunities and inflated prices are the most notable negative economic outcomes. Efforts should be made to continue planning and managing Indigenous tourism in ways that will benefit them most, reduce their suffering, and make them more resilient against tourism-induced change.

## References

Acharya, B.P., & Halpenny, E.A. (2013). Homestays as an alternative tourism product for sustainable community development: A case study of women managed tourism product in rural Nepal. *Tourism Planning & Development*, 10(4), 367–387.

Adhikari, Y.P., & Fischer, A. (2011). Tourism: Boon for forest conservation, livelihood, and community development in Ghandruk VDC, Western Nepal. *The Initiation*, 4, 35–45.

Adnan Hye, Q.M., & Ali Khan, R.E. (2013). Tourism-led growth hypothesis: A case study of Pakistan. *Asia Pacific Journal of Tourism Research*, 18(4), 303–313.

Ali, A., Anuar, M.M., & Ahmad, N. (2014). Continuance intention in using homestay terminology in the names of private lodging houses. *International Journal of Business and Society*, 15(3), 379–394.

Anand, A., Chandan, P., & Singh, R.B. (2012). Homestays at Korzok: Supplementing rural livelihoods and supporting green tourism in the Indian Himalayas. *Mountain Research and Development*, 32(2), 126–136.

Apollo, M. (2016). The clash—social, environmental and economical changes in tourism destination areas caused by tourism the case of Himalayan villages (India and Nepal). *Current Issues of Tourism Research*, 5(1), 6–19.

Baloch, Q.B. (2007). Managing tourism in Pakistan (a case study of Chitral valley). *Journal of Managerial Sciences*, 2(2), 169–190.

Baral, N., Stern, M.J., & Hammett, A.L. (2012). Developing a scale for evaluating ecotourism by visitors: A study in the Annapurna conservation area, Nepal. *Journal of Sustainable Tourism*, 20(7), 975–989.

Beaumont, P. (2015). Deadly Everest avalanche triggered by Nepal earthquake. *The Guardian*. Retrieved from www.theguardian.com/world/2015/apr/25/deadly-everest-avalanche-triggered-by-nepal-earthquake

Beazley, R.E., & Lassoie, J.P. (2017). *Himalayan Mobilities: An Exploration of the Impact of Expanding Rural Road Networks on Social and Ecological Systems in the Nepalese Himalaya*. Cham, Switzerland: Springer.

Bhalla, P., Coghlan, A., & Bhattacharya, P. (2016). Homestays' contribution to community-based ecotourism in the Himalayan Region of India. *Tourism Recreation Research*, 41(2), 213–228.

Bhandari, K. (2019). International development ideology and two tourism policies of Nepal. *Environment and Planning C: Politics and Space*, 37(3), 558–576.

Bhatt, S. (2012). Community based homestays: Innovation in tourism. *Context*, 9(2), 77–83.

Bhattarai, B.R., Wright, W., & Poudel, B.S. (2017). Shifting paradigms for Nepal's protected areas: History, challenges and relationships. *Journal of Mountain Science*, 14(5), 964–979.

Bhusal, N.P. (2007). Chitwan national park: A prime destination of eco-tourism in Central Tarai region, Nepal. *The Third Pole*, 5(7), 70–75.

Biswakarma, G. (2015). On the dimensionality of measuring tourist satisfaction towards homestay. *International Journal of Hospitality and Tourism Systems*, 8(2), 51–63.

Bleie, T. (2003). Pilgrim tourism in the Central Himalayas. *Mountain Research and Development*, 23(2), 177–184.

Brunet, S., Bauer, J., De Lacy, T., & Tshering, K. (2001). Tourism development in Bhutan: Tensions between tradition and modernity. *Journal of Sustainable Tourism*, 9(3), 243–263.

Burke, J., & Walker, P. (2014). Nepal blames cheap tourists for falling victim to snowstorm in Himalayas. *The Guardian*. Retrieved from www.theguardian.com/world/2014/oct/17/nepal-cheap-tourists-snow-disaster

Callaghan, A., & Thapa, R. (2015). An oral history of Langtang, the valley destroyed by Nepal earthquake *Outside*. Retrieved from www.outsideonline.com/2016856/oral-history-langtang-valley-destroyed-nepal-earthquake.

Caparrós, B.M. (2018). Trekking to women's empowerment: A case study of a female-operated travel company in Ladakh. In S. Cole (Ed.), *Gender Equality and Tourism: Beyond Empowerment* (pp. 57–66). Wallingford: CABI.

Carr, A., Ruhanen, L., & Whitford, M. (2016). Indigenous peoples and tourism: The challenges and opportunities for sustainable tourism. *Journal of Sustainable Tourism*, 24(8–9), 1067–1079.

Catibog-Sinha, C., & Wen, J. (2008). Sustainable tourism planning and management model for protected natural areas: Xishuangbanna Biosphere Reserve, South China. *Asia Pacific Journal of Tourism Research*, 13(2), 145–162.

Chang, J. (2006). Segmenting tourists to aboriginal cultural festivals: An example in the Rukai Tribal area, Taiwan. *Tourism Management*, 27(6), 1224–1234.

Colchester, M. (2004). Conservation policy and indigenous peoples. *Environmental Science and Policy*, 7(3), 145–153.

Coria, J., & Calfucura, E. (2012). Ecotourism and the development of Indigenous communities: The good, the bad, and the ugly. *Ecological Economics*, 73, 47–55.

Dahal, B., K C, A., & Sapkota, R.P. (2020). Environmental impacts of community-based home stay ecotourism in Nepal. *The Gaze: Journal of Tourism and Hospitality*, 11(1), 60–80.

Dangi, T.B., & Gribb, W.J. (2018). Sustainable ecotourism management and visitor experiences: Managing conflicting perspectives in Rocky Mountain national park, USA. *Journal of Ecotourism*, 17(3), 338–358.

Debarbieux, B., Oiry Varacca, M., Rudaz, G., Maselli, D., Kohler, T., & Jurek, M. (Eds.). (2014). *Tourism in Mountain Regions: Hopes, Fears and Realities.* Geneva: University of Geneva, Department of Geography and Environment.

Degang, W., & Xiaoting, H. (2006). Coincidence and upgrade: A typical case study of rural ecotourism development. *Chinese Journal of Population Resources and Environment*, 4(1), 45–53.

Dunford, M.R. (2019). Indigeneity, ethnopolitics, and taingyinthar: Myanmar and the global Indigenous peoples' movement. *Journal of Southeast Asian Studies*, 50(1), 51–67.

Editors, F. (2011). Flora of the pan-Himalayas: General guidelines. *Journal of Systematics and Evolution*, 49(6), 617–624.

Genzong, X., Penghua, Q., & Shaoxia, T. (2007). Nature reserve and ecotourism development in China's Wuzhishan mountain region. *Chinese Journal of Population Resources and Environment*, 5(1), 74–83.

Gioli, G., Thapa, G., Khan, F., Dasgupta, P., Nathan, D., Chhetri, N., Adhikari, L., Mohanty, S.K., Aurino, E., & Scott, L.M. (2019). Understanding and tackling poverty and vulnerability in mountain livelihoods in the Hindu Kush Himalaya. In P. Wester, A. Mishra, A. Mukherji, & A.B. Shrestha (Eds.), *The Hindu Kush Himalaya Assessment: Mountains, Climate Change, Sustainability and People* (pp. 421–456). Cham, Switzerland: Springer.

Grimwood, B.S., Muldoon, M.L., & Stevens, Z.M. (2019). Settler colonialism, Indigenous cultures, and the promotional landscape of tourism in Ontario, Canada's 'near North'. *Journal of Heritage Tourism*, 14(3), 233–248.

Gu, M., & Wong, P.P. (2006). Residents' perception of tourism impacts: A case study of homestay operators in Dachangshan Dao, north-east China. *Tourism Geographies*, 8(3), 253–273.

Gurung, D.B., & Seeland, K. (2008). Ecotourism in Bhutan. *Annals of Tourism Research*, 35(2), 489–508.

Hall, C.M., & Page, S.J. (2017). Developing tourism in South and Central Asia: Introduction. In C.M. Hall & S.J. Page (Eds.), *The Routledge Handbook of Tourism in Asia* (pp. 223–240). London: Routledge.

Hounnaklang, S. (2016). Concepts, issues and the effectiveness of alternative tourism management in Thailand: A case study of Plai Pong Pang homestay, Amphoe Ampawa, Samut Songkram Province. *International Journal of Arts & Sciences*, 9(3), 337–348.

Imran, S., Alam, K., & Beaumont, N. (2014). Environmental orientations and environmental behaviour: Perceptions of protected area tourism stakeholders. *Tourism Management*, 40, 290–299.

Islam, A. (2014). Impact of armed conflict on economy and tourism: A study of state of Jammu and Kashmir. *IOSR Journal of Economics and Finance*, 4(6), 55–60.

Ismail, M.N.I., Hanafiah, M.H., Aminuddin, N., & Mustafa, N. (2016). Community-based homestay service quality, visitor satisfaction, and behavioral intention. *Procedia—Social and Behavioral Sciences*, 222, 398–405.

Jamal, S.A., Othman, N., & Muhammad, N.M.N. (2011). Tourist perceived value in a community-based homestay visit: An investigation into the functional and experiential aspect of value. *Journal of Vacation Marketing*, 17(1), 5–15.

Jamaludin, M., Othman, N., & Awang, A.R. (2012). Community based homestay programme: A personal experience. *Procedia—Social and Behavioral Sciences*, 42, 451–459.

Jiang, S., Wu, C., & Tseng, K. (2018). Utilization of cloud computing app for homestay operation—design and analysis. *International Journal of Organizational Innovation*, 11(2), 31–44.

Johnston, A.M. (2003). Self-determination: Exercising indigenous rights in tourism. In S. Singh, D.J. Timothy & R.K. Dowling (Eds.), *Tourism in Destination Communities* (pp. 115–133). Wallingford: CABI.

K C, A. (2016). Ecotourism and its role in sustainable development of Nepal. In L. Butowski (Ed.), *Tourism: From Empirical Research Towards Practical Application* (pp. 31–59). Rijeka, Croatia: InTech.

K C, A. (2017a). Climate change and its impact on tourism in Nepal. *Journal of Tourism and Hospitality Education*, 7, 25–43.

K C, A. (2017b). Ecotourism in Nepal. *The Gaze: Journal of Tourism and Hospitality*, 8, 1–19.

K C, A., Ghimire, S., & Dhakal, A. (2021). Ecotourism and its impact on Indigenous people and their local environment: Case of Ghalegaun and Golaghat of Nepal. *GeoJournal*, 86, 2747–2765. doi:10.1007/s10708-020-10222-3

K C, A., Rijal, K., & Sapkota, R.P. (2015). Role of ecotourism in environmental conservation and socioeconomic development in Annapurna conservation area, Nepal. *International Journal of Sustainable Development & World Ecology*, 22(3), 251–258.

K C, A., & Thapa Parajuli, R.B. (2014). Tourism and its impact on livelihood in Manaslu conservation area. *Environment Development and Sustainability*, 16, 1053–1063.

K C, A., & Thapa Parajuli, R.B. (2015). Climate change and its impact on tourism in the Manaslu conservation area, Nepal. *Tourism Planning & Development*, 12(2), 225–237.

Kumar, S. (2018). Indigenous people in South Asia and international law. In J. Kaul, & A. Jha (Eds.), *Shifting Horizons of Public International Law: A South Asian Perspective* (pp. 61–83). New Delhi: Springer.

Kunjuraman, V., & Hussin, R. (2017). Challenges of community-based homestay programme in Sabah, Malaysia: Hopeful or hopeless? *Tourism Management Perspectives*, 21, 1–9.

Lama, A.K. (2010). *Vulnerability of Nature-Based Tourism to Climate Change: Stakeholders' Perceptions of and Response to Climate Change in the Lower Mustang Region of the Annapurna Conservation Area.* Unpublished master's thesis, Lincoln University, New Zealand.

Lama, A.K., & Job, H. (2014). Protected areas and road development: Sustainable development discourses in the Annapurna conservation area, Nepal. *Erdkunde*, 68(4), 229–250.

Lipton, J.K., & Bhattarai, U. (2014). Park establishment, tourism, and livelihood changes: A case study of the establishment of Chitwan national park and the Tharu people of Nepal. *American International Journal of Social Science*, 3(1), 12–24.

Long, F., Liu, J., Zhang, S., Yu, H., & Jiang, H. (2018). Development characteristics and evolution mechanism of homestay agglomeration in Mogan Mountain, China. *Sustainability*, 10(9), 2964.

Lundup, T. (2018). Tourism and Ladakhi culture in transition. *Indian Anthropologist*, 48(1), 47–59.

Maikhuri, R.K., Rana, U., Rao, K.S., Nautiyal, S., & Saxena, K.G. (2000). Promoting ecotourism in the buffer zone areas of Nanda Devi biosphere reserve: An option to resolve people—policy conflict. *International Journal of Sustainable Development & World Ecology*, 7(4), 333–342.

Manyara, G., & Jones, E. (2007). Community-based tourism enterprises development in Kenya: An exploration of their potential as avenues of poverty reduction. *Journal of Sustainable Tourism*, 15(6), 628–644.

Masud, M.M., Aldakhil, A.M., Nassani, A.A., & Azam, M.N. (2017). Community-based ecotourism management for sustainable development of marine protected areas in Malaysia. *Ocean and Coastal Management*, 136, 104–112.

McIntosh, A.J., Lynch, P., & Sweeney, M. (2011). "My home is my castle": Defiance of the commercial homestay host in tourism. *Journal of Travel Research*, 50(5), 509–519.

McMillan, C.L., O'Gorman, K.D., & MacLaren, A.C. (2011). Commercial hospitality: A vehicle for the sustainable empowerment of Nepali women. *International Journal of Contemporary Hospitality Management*, 23(2), 189–208.

MoTA (2014). *Report of the High Level Committee on Socioeconomic, Health and Educational Status of Tribal Communities of India*. New Delhi: Ministry of Tribal Affairs, Government of India.

Mu, Y., Nepal, S.K., & Lai, P.H. (2019). Tourism and sacred landscape in Sagarmatha (Mt. Everest) National Park, Nepal. *Tourism Geographies*, 21(3), 442–459.

Mura, P. (2015). Perceptions of authenticity in a Malaysian homestay. *Tourism Management*, 51, 225–233.

Musa, G., Hall, C.M., & Higham, J.E.S. (2004). Tourism sustainability and health impacts in high altitude adventure, cultural and ecotourism destinations: A case study of Nepal's Sagarmatha national park. *Journal of Sustainable Tourism*, 12(4), 306–331.

Neupane, R., K C, A., Aryal, M., & Rijal, K. (2021). Status of ecotourism in Nepal: A case of Bhadaure-Tamagi village of Panchase area. *Environment, Development and Sustainability*, 23, 15897–15920.

Nianyong, H., & Zhuge, R. (2001). Ecotourism in China's nature reserves: Opportunities and challenges. *Journal of Sustainable Tourism*, 9(3), 228–242.

Nyaupane, G.P. (2009). Heritage complexity and tourism: The case of Lumbini, Nepal. *Journal of Heritage Tourism*, 4(2), 157–172.

Nyaupane, G.P. (2015). Mountaineering on Mt Everest: Evolution, economy, ecology, and ethics. In G. Musa, J. Higham, & A. Thompson-Carr (Eds.), *Mountaineering Tourism* (pp. 265–271). London: Routledge.

Nyaupane, G.P., & Budruk, M. (2009). South Asian heritage tourism conflict, colonialism, and cooperation. In D.J. Timothy & G.P. Nyaupane (Eds.), *Cultural Heritage and Tourism in the Developing World a Regional Perspective* (pp. 127–145). London: Routledge.

Nyaupane, G.P., & Chhetri, N. (2009). Vulnerability to climate change of nature-based tourism in the Nepalese Himalayas. *Tourism Geographies*, 11(1), 95–119.

Nyaupane, G.P., Lew, A.A., & Tatsugawa, K. (2014). Perceptions of trekking tourism and social and environmental change in Nepal's Himalayas. *Tourism Geographies*, 16(3), 415–437.

Nyaupane, G.P., Morais, D.B., & Dowler, L. (2006). The role of community involvement and number/type of visitors on tourism impacts: A controlled comparison of Annapurna, Nepal and Northwest Yunnan, China. *Tourism Management*, 27(6), 1373–1385.

Nyaupane, G.P., & Poudel, S. (2011). Linkages among biodiversity, livelihood, and tourism. *Annals of Tourism Research*, 38(4), 1344–1366.

Nyaupane, G.P., & Thapa, B. (2004). Evaluation of ecotourism: A comparative assessment in the Annapurna conservation area project, Nepal. *Journal of Ecotourism*, 3(1), 20–45.

Nyaupane, G.P., & Timothy, D.J. (2010). Power, regionalism, and tourism policy in Bhutan. *Annals of Tourism Research*, 37(4), 969–988.

Nyaupane, G.P., Timothy, D.J., & Poudel, S. (2015). Understanding tourists in religious destinations: A social distance perspective. *Tourism Management*, 48, 343–353.

Ogucha, E.B., Riungu, G.K., Kiama, F.K., & Mukolwe, E. (2015). The influence of homestay facilities on tourist satisfaction in the Lake Victoria Kenya tourism circuit. *Journal of Ecotourism*, 14(2–3), 278–287.

Pabel, A., Prideaux, B., & Thompson, M. (2017). Tourists' preferences with Indigenous tourism experiences in the wet tropics of Queensland, Australia. *Journal of Hospitality and Tourism Management*, 31, 142–151.

Panta, S.K., & Thapa, B. (2018). Entrepreneurship and women's empowerment in gateway communities of Bardia National Park, Nepal. *Journal of Ecotourism*, 17(1), 20–42.

Poudel, J. (2014). Socio-cultural impact in tourism: A case study of Sauraha, Nepal. *Journal of Advanced Academic Research*, 1(2), 47–55.

Qu, C., Timothy, D.J., & Zhang, C. (2019). Does tourism erode or prosper culture? Evidence from the Tibetan ethnic area of Sichuan Province, China. *Journal of Tourism and Cultural Change*, 17(4), 526–543.

Raza, S.A., & Jawaid, S.T. (2013). Terrorism and tourism: A conjunction and ramification in Pakistan. *Economic Modelling*, 33, 65–70.

Regmi, K.D., & Walter, P. (2017). Modernisation theory, ecotourism policy, and sustainable development for poor countries of the global south: Perspectives from Nepal. *International Journal of Sustainable Development & World Ecology*, 24(1), 1–14.

Ruhanen, L., & Whitford, M. (2019). Cultural heritage and Indigenous tourism. *Journal of Heritage Tourism*, 14(3), 179–191.

Samsudin, P.Y., & Maliki, N.Z. (2015). Preserving cultural landscape in homestay programme towards sustainable tourism: Brief critical review concept. *Procedia—Social and Behavioral Sciences*, 170, 433–441.

Sangraula, B. (2010). 1,000 tourists stranded in Lukla due to week-long flight shutdown. My Republica, Kathmandu, Nepal.

Scheyvens, R. (2000). Promoting women's empowerment through involvement in ecotourism: Experiences from the Third World. *Journal of sustainable tourism*, 8(3), 232–249.

Scott, D., Hall, C.M., & Gössling, S. (2019). Global tourism vulnerability to climate change. *Annals of Tourism Research*, 77, 49–61.

Serenari, C., Leung, Y., Attarian, A., & Franck, C. (2012). Understanding environmentally significant behavior among whitewater rafting and trekking guides in the Garhwal Himalaya, India. *Journal of Sustainable Tourism*, 20(5), 757–772.

Sharma, N., & Sarmah, B. (2019). Consumer engagement in village eco-tourism: A case of the cleanest village in Asia—Mawlynnong. *Journal of Global Scholars of Marketing Science*, 29(2), 248–265.

Shi, L., Zhao, H., Li, Y., Ma, H., Yang, S., & Wang, H. (2015). Evaluation of Shangri-la County's tourism resources and ecotourism carrying capacity. *International Journal of Sustainable Development & World Ecology*, 22(2), 103–109.

Shukor, M.S., Mohd Salleh, N.H., Othman, R., & Mohd Idris, S.H. (2014). Perception of homestay operators towards homestay development in Malaysia. *Jurnal Pengurusan*, 42, 3–17.

Simms, A., Moheb, Z., Salahudin, A.H., Ali, I., & Wood, T. (2011). Saving threatened species in Afghanistan: Snow leopards in the Wakhan Corridor. *International Journal of Environmental Studies*, 68(3), 299–312.

Singh, R.B., & Mishra, D.K. (2004). Green tourism in mountain regions-reducing vulnerability and promoting people and place centric development in the Himalayas. *Journal of Mountain Science*, 1(1), 57–64.

Sita, S.E.D., & Nor, N.A.M. (2015). Degree of contact and local perceptions of tourism impacts: A case study of homestay programme in Sarawak. *Procedia—Social and Behavioral Sciences*, 211, 903–910.

Sood, J., Lynch, P., & Anastasiadou, C. (2017). Community non-participation in homestays in Kullu, Himachal Pradesh, India. *Tourism Management*, 60, 332–347.

Stone, M.T. (2015). Community-based ecotourism: A collaborative partnerships perspective. *Journal of Ecotourism*, 14(2–3), 166–184.

Su, M.M., Long, Y., Wall, G., & Jin, M. (2016). Tourist-community interactions in ethnic tourism: Tuva villages, Kanas scenic area, China. *Journal of Tourism and Cultural Change*, 14(1), 1–26.

Tavakoli, R., Mura, P., & Rajaratnam, S.D. (2017). Social capital in Malaysian homestays: Exploring hosts' social relations. *Current Issues in Tourism*, 20(10), 1028–1043.

Timothy, D.J. (2007). Empowerment and stakeholder participation in tourism destination communities. In A. Church & T. Coles (Eds.), *Tourism, Power and Space* (pp. 199–216). London: Routledge.

Timothy, D.J. (2019). Tourism, border disputes and claims to territorial sovereignty. In R.K. Isaac, E. Çakmak, & R. Butler (Eds.), *Tourism and Hospitality in Conflict-Ridden Destinations* (pp. 25–38). London: Routledge.

Timothy, D.J., & White, K. (1999). Community-based ecotourism development on the periphery of Belize. *Current Issues in Tourism*, 2(2/3), 226–242.

Tran, L., & Walter, P. (2014). Ecotourism, gender and development in northern Vietnam. *Annals of Tourism Research*, 44, 116–130.

United Nations (2019). Indigenous peoples at the UN. Retrieved from www.un.org/development/desa/Indigenouspeoples/about-us.html

Walter, P., Regmi, K.D., & Khanal, P.R. (2018). Host learning in community-based ecotourism in Nepal: The case of Sirubari and Ghalegaun homestays. *Tourism Management Perspectives*, 26, 49–58.

Wang, W., Liu, J., Kozak, R., Jin, M., & Innes, J. (2018). How do conservation and the tourism industry affect local livelihoods? A comparative study of two nature reserves in China. *Sustainability*, 10(6), 1–16.

Yang, J., Ryan, C., & Zhang, L. (2013). Social conflict in communities impacted by tourism. *Tourism Management*, 35, 82–93.

Zhang, J., Xu, H., Xu, H., & Xing, W. (2017). The host-guest interactions in ethnic tourism, Lijiang, China. *Current Issues in Tourism*, 20(7), 724–739.

Zhang, L., & Zhang, J. (2018). Perception of small tourism enterprises in Lao PDR regarding social sustainability under the influence of social network. *Tourism Management*, 69, 109–120.

Zhao, J., & Jia, H. (2008). Strategies for the sustainable development of Lugu Lake region. *International Journal of Sustainable Development & World Ecology*, 15(1), 71–79.

# 10 Volunteer Tourism, Poverty, and People on the Margins

## Altruism, Ego-enhancement, and Voyeurism in the Himalaya

*Dallen J. Timothy*

## Introduction

For the first time in history, in 2012 over one billion international trips were taken worldwide. Since that year, until the 2020–2022 COVID-19 pandemic, the world has seen remarkable growth, with 1.5 billion international trips taken in 2019, a growth of more than 50% since 2010. One of the most remarkable aspects of this growth is the multitude of motivations that drive people to travel (Kozak, 2002; Lin & Nawijn, 2020). It can safely be assumed that there are as many motives or nuanced combinations of motives for taking a trip as there are trips being taken. Whereas many motivations are relatively common, others are very personal and unique to the individual. Some of the most typical motivations are rest and relaxation, experiencing new or different cultures, learning, pursuing a hobby, visiting friends and relatives, spiritual communion with deity or nature, and multitudes of others. Most tourist journeys are motivated by self-serving or egoistic purposes that aim to achieve one's personal satisfaction or pleasure.

One particular set of motivations that has existed for many decades, even centuries, but has only recently been recognized as rousing an identifiable and marketable type of tourism, is altruism and selfless service. A desire to travel in an effort to give of oneself for the greater global good is known as volunteer tourism (Smith & Holmes, 2009; Wearing, 2001; Wearing & McGehee, 2013a). This form of tourism is characterized by individuals traveling to engage in service to other people, the planet, or to institutions in need of help. Examples include helping to construct a house or dig a well in a poor village, rebuild a school or church following a natural disaster, plant trees in an area of heavy deforestation, plant shrubs and grass in areas of heavy erosion, repair trails and fences in a national park, provide free dental work or healthcare in a remote community, help serve meals at a food bank, volunteer at an archaeological dig, or assist a rescue institution in its efforts to care for destitute populations (Fredheim, 2018; McLennan, 2014; Mostafanezhad, 2016; Smithson et al., 2018; Stoddart & Rogerson, 2004; Timothy, 2020, 2021).

Volunteering in archaeological sites, other heritage places, or natural environments takes place in both developed and developing countries, but most volunteer tourism opportunities in other areas take place in countries of the Global

DOI: 10.4324/9781003030126-13

South and are staffed by tourists from the Global North. The Himalayan region is an increasingly popular destination for volunteer vacationers, whose activities adhere to a wide range of service opportunities. This chapter examines the role of volunteer tourism in the Himalaya, highlighting its most common manifestations, such as orphanage work and disaster relief, and the benefits of these activities in the region. The chapter then describes some of the criticisms and pitfalls of volunteer tourism, including the exploitation of children, the voyeuristic nature of philanthropic travel, and the complications associated with transitory medical clinics.

## Volunteer Tourism

Many studies have examined individual motivations for undertaking volunteer experiences abroad, some of which have looked at different kinds of experiences and activities, and what motivates people to give their time and means to do something for the greater good of society or the planet. On the surface, what makes volunteer tourism unique is travelers' altruistic motives to 'give something back', as well as their willingness to pay for their own service experiences. Although it is still a relatively small segment of global tourism, philanthropic travel has grown a lot in recent years, especially as people realize they can combine their humanitarian activities with leisure pursuits. While altruism and the desire to 'make a difference' originally lay at the root of volunteering abroad, the growth of the phenomenon has brought about evolutionary changes as it has become more 'touristified' (Brown, 2005; Coghlan & Fennell, 2009; Mustonen, 2007).

Mustonen (2007) notes that there are many stimuli for volunteer travel, from pure altruism on one end of the spectrum to pure ego-enhancement on the other end. The growth of volunteer tourism has blurred the lines between leisure holidays and altruistic endeavors, creating many combinations of travel experiences that overlap (Mustonen, 2006). Many people see volunteering as an opportunity to become immersed in a foreign culture, to network socially, to learn from the experience of being away from home, and to undergo some degree of independence and hardship (Brown, 2005). People often participate as a means of strengthening their resumes for future educational or employment purposes (Hindman, 2014). Some university programs require students to participate in public service, which volunteering abroad may satisfy. According to one study (Coatsworth et al., 2017), Australian volunteer student nurses working in Thailand saw their service as a means of expanding their career options but also as a way of effecting change in the communities they serve. Their clinical placement overseas helped them develop better nursing skills, enabled them to work in cross-cultural situations, and expanded their ability to empathize with their patients and promote compassionate care. Toward the egoistic end of the spectrum, other participants see these experiences as a rite of passage or an opportunity to strengthen their bragging rights (Calkin, 2014). There has long been a close relationship between youth travel, such as gap year journeys and backpacking trips, and volunteering abroad

(Ooi & Laing, 2010). Volunteer tourism is frequently treated as a mechanism for youth to access less-developed regions and is often an appendage to backpacker travel (Timothy & Zhu, 2022).

Although much of their lodging and food provision is rather spartan, volunteer tourists usually remain in the destination longer and undertake other activities beyond their philanthropic endeavors, which makes them a commercially desirable market segment for many developing countries (Wearing et al., 2020). There is a growing body of research on the outcomes of volunteer experiences for the tourists. For many participants, it is life-changing. It can transform the travelers' values and behaviors, particularly with regard to their attitudes toward material possessions and the value of human life. It also influences their lifestyle choices and future travel decisions (Zahra, 2011; Zahra & McIntosh, 2007).

Many development specialists see volunteer tourism as a legitimate form of pro-poor tourism and an instrument of sustainable development, because it is believed to benefit communities and people in the destination, as well as the volunteers themselves (Mody et al., 2014; Scheyvens, 2007; Singh, 2006; Upadhayaya, 2011; Wearing et al., 2020). Singh (2006) provides an example from the Indian Himalaya where volunteers from several developed countries helped build 40 latrines and waste processing stations, funded by UNICEF, the Indian government, and the volunteers themselves. Such activities have been shown to benefit marginal regions and promote a sense of solidarity between the hosts and the guests (Singh & Singh, 2004). Scholars have also argued that volunteer tourism may be one of the most suitable sorts of tourism to build greater understanding between cultures (Brown, 2005).

Academic research and industry promotional efforts reveal a significant geographical concentration of volunteer opportunities. Humanitarian and pro-environmental activities promoted by volunteer NGOs are geographically concentrated in a relatively small handful of places (Keese, 2011; Tomazos & Butler, 2009). In the Himalayan region, Nepal and India are among the most voluntoured countries in the world. The following sections examine patterns and issues associated with this increasingly popular form of tourism in the Himalaya.

## Volunteer Tourism in the Himalaya

Volunteering has a long history in the Himalaya and is well documented in countries such as Bhutan and Nepal (Choden, 2003; Hacker et al., 2017; Hayward & Colman, 2012). It takes the form of 'self-service' and communal cooperation, working together to solve issues that affect an entire community. It is therefore natural that Himalayan communities would welcome the volunteer efforts of foreign or domestic helpers. A unique perspective on humanitarian tourism in Nepal is the popularity of volunteering in the country among the Nepali diaspora. Many Nepali emigrants with a professional background regularly use part of their homeland visits to engage in social justice, conservation, infrastructure development, educational, healthcare, and elder care efforts, as well as contribute to community development projects (Khanal, 2013).

The majority of marketed volunteer opportunities for Westerners in the Himalaya are in India and Nepal, which are among the most popular destinations in the world for volunteer tourism (Grout, 2009; Wickens, 2011). It is far less common to find such opportunities in Afghanistan, Pakistan, China (Tibet), and Bhutan in the international marketplace, although they do exist in limited numbers (Callanan & Thomas, 2005; Grout, 2009; Tomazos & Butler, 2009). Table 10.1 illustrates a handful of current volunteer tourism opportunities in the Himalaya. There are two overarching temporal perspectives on volunteering in the Himalaya: ongoing long-term projects and those that occur as needed, including in emergency or one-off situations.

*Table 10.1* Examples of commercially available volunteer opportunities in the Himalaya, 2019–2021

| Countries | Location | Organization | Volunteer Opportunities |
|---|---|---|---|
| India | Himachal Pradesh | RCDP International Volunteer | Working with at-risk children |
| India | Himachal Pradesh | RCDP International Volunteer | Teaching English |
| India | Himachal Pradesh | Volunteering Solutions | Renovating children's daycare centers |
| India | Uttarakhand | Worldpackers | Teaching English to children |
| India | Uttarakhand | Worldpackers | Teaching sports |
| India | Uttarakhand | Worldpackers | Child care |
| Nepal | various locations | Volunteering Journeys | Women's entrepreneurship |
| Nepal | various locations | Volunteering Journeys | Health and sanitation |
| Nepal | various locations | Volunteering Journeys | Environmental conservation |
| India | Garhwal | Volunteerforever | Teaching English |
| India | Uttarakhand | GivingWay | Assist teachers at local schools and teach English |
| India | Ladakh, Sikkim, Uttarakhand | Help—Himalayan Education Lifeline Programme | Teaching English |
| Nepal | Kathmandu Valley, western Nepal | Help—Himalayan Education Lifeline Programme | Teaching English |
| Nepal | western Nepal | Help—Himalayan Education Lifeline Programme | Need physicians and dentists to volunteer at clinics |
| India | Himachal Pradesh | New Hope Volunteers | Teaching English to children and working with schools |
| India | Himachal Pradesh | New Hope Volunteers | Street Children Project—cooking, child care, teaching about hygiene, reading, sport, computers, art, etc. |

Source: Compiled by the author from multiple sources

## Long-Term Volunteer Opportunities

The most common ongoing volunteer opportunities can be seen in Table 10.1, including working with children and youth by teaching and working with teachers in the classroom, helping at-risk and street children to establish a better life and a good educational foundation, and child care services (Wickens, 2011). Other common opportunities involve working with adults, such as assisting women in their entrepreneurial pursuits. Several companies consistently seek dentists, physicians, and nurses who can devote their time and skills to help care for the needy. Environmental conservation opportunities are also available in Nepal and India (Mody et al., 2014; Singh, 2006; Singh & Singh, 2004).

Besides teaching, the two main traditional volunteer activities in the region are working with children in orphanages and medical services and other healthcare in 'health camps'. Orphanage volunteers regularly care for children, babysit, prepare meals, read stories, teach, participate in structural repairs, and play games. Fewer skills are needed for these orphanage activities than in medical camps, which makes orphanages an extremely popular volunteer option, especially in Nepal. Orphanage work is often marketed and sold as 'working with at-risk children'.

Health camps are short-term, usually one day to a week, medical setups that are established regularly in rural areas. Some camps focus on general healthcare, or they might emphasize certain specialties that are urgently needed, such as dental, eye care, family planning, mental health, or minor surgery clinics in underprivileged areas (Citrin, 2010; Thara et al., 2008). They are usually sponsored by NGOs, but are also known to be set up by national governments and private hospitals, and they normally rely on volunteer physicians, nurses, and dentists who donate their time and means to participate in assisting the community selected to host the clinic. These camps also frequently rely on non-skilled, untrained volunteers to assist the doctors and nurses in their efforts to treat the impoverished populations (Mohammed, 2021). These health camps have become especially routine in Nepal and India, but are found also in other Himalayan states, although these have not received much research attention. Many rural parts of the Himalaya have come to rely on these medical camps for their periodic healthcare needs (Citrin, 2010; Thara et al., 2008).

## As-Needed Volunteering: Emergencies and Disaster Recovery Efforts

A form of volunteer tourism, which is not unique to the Himalaya, but which is fairly common is disaster recovery (Lin et al., 2018; Wearing et al., 2020). The Himalaya are prone to a variety of natural disasters, with the most common being landslides, floods, avalanches, and periodic earthquakes (see Nyaupane's chapter on vulnerability in this volume), although earthquakes tend to receive the most global media coverage given their size, impact, and irregularity. There is a long history of people traveling to assist immediately following a natural disaster (Michel, 2007; Welty & Bolton, 2017), which has happened in the Himalaya, especially after major earthquakes, as was evident following the Uttarkashi earthquake (1991), Kashmir earthquake (2005), Sikkim earthquake (2011), and the Nepal earthquake (2015) (Kandangwa & Gurung, 2016; Wearing et al., 2020).

In addition to the efforts of local volunteers and rescue teams, the work of international volunteers has been hailed as an important part of post-disaster recovery from two different perspectives. First, oftentimes thousands of volunteers appear from across the globe within a few days of a disaster with the immediate aim of helping clean up, care for the injured or ill, or serve food and water for those in need. In the case of the 2015 Nepal earthquake, volunteers from across the globe arrived to help the Nepali people within days and weeks in the aftermath of the earthquake. Later, follow-up volunteers come to help rebuild disaster-resistant infrastructure and homes, and contribute to entrepreneurial development (Wearing et al., 2020). The second way in which volunteer tourism helps is when it is not necessarily associated directly with the disaster but is considered a relatively inelastic form of tourism that helps restart or stimulate the tourism economy following a crisis. Thus, other types of volunteer efforts (e.g., orphanage work or environmental cleanup) continue when other types of tourism (e.g., trekking and mountaineering) decline because of the disaster. Thus, volunteer tourism is commonly seen as a post-recovery stimulant for economic development (Wearing et al., 2020).

Volunteer tourism in Nepal in 2015 and 2016 after the devastating earthquake of April 25, 2015, and its aftershocks was regarded as a major apparatus for post-disaster tourism recovery, and nearly one-third of all tours booked immediately after May 2015 consisted of volunteer and philanthropic opportunities provided by NGOs or tour operators, or people came individually (Beirman et al., 2018). Some of these volunteer arrivals were spurred in part due to media and television images of Britain's Prince Harry visiting Nepal in 2016 to help clean up devastated villages in the region. Most of the volunteer opportunities focused on reconstructing homes and schools, conducting safety audits of trails, helping to restore damaged tourist attractions, assisting in health camps, tending children, teaching, and working with alternative energy sources—all of which enhanced the quality of life for people and helped the country recover from the devastating event (Beirman et al., 2018). In the beginning, most volunteers came to help in the immediate aftermath and were skilled in areas such as engineering, construction, medicine, and emergency management, while those who came later were less skilled but anxious to help however they could (Beirman et al., 2018).

Wearing et al. (2020) provide considerable insight into the foreign volunteering that took place after the 2015 Nepal earthquake. They identified three key types of volunteer tourism in the aftermath of the disaster. The first was comprised of volunteer programs initiated and managed by national and international organizations, government agencies, NGOs, and other nonprofit associations that focused on post-disaster recovery and development. Second, many individuals arrived in the country independently and approached community leaders or were sought by local communities to help. The third type of volunteer tourist experience was comprised of programs that were embedded within extant trekking packages. Many trekking outfitters included volunteer opportunities in their itineraries. In

2015–2016, rescue teams from all three of these sources were made up of people from more than 40 countries (Wearing et al., 2020).

## Controversies and Concerns

All tourism forms have both positive and negative outcomes, which are usually classified as sociocultural, economic, and environmental. Volunteer tourism is no exception, and it has received considerable criticism in the past three decades commensurate with its growth. Much of this growing criticism argues that volunteer tourism often does more social harm than good (Durham, 2017; Henry & Mostafanezhad, 2019; Hindman, 2014; Mustonen, 2006, 2007; Punaks & Feit, 2014; Schuurman, 2018; Telfer & Sharpley, 2015; Wymer et al., 2010). Nonetheless, most researchers concede that the results of volunteer tourism have had mixed successes and failures (Telfer & Sharpley, 2015). Some of the successes were outlined earlier.

Most volunteer tourism takes place in the Global South, with the majority of volunteers coming from the Global North in an ostensibly altruistic effort to do something good. "It is now fashionable to participate in volunteer tourism, as its altruistic veneer seems to suggest that participants are selfless servants" (Timothy, 2020, p. 89), but critics argue that, like many other so-called 'alternative tourisms', volunteering is simply becoming another form of mass tourism that is packaged in bulk and consumed by mass markets (Guttentag, 2015; Wearing et al., 2016). Volunteer tourism, therefore, resembles other tourisms with their negative social, ecological, and economic impacts, and their neocolonialist undertones in which altruistic endeavors benefit the destination a little bit but are equally harmful as voluntourists "flaunt paternalistic and inequitable associations that sometimes disempower local communities and keep them in a constant state of dependency" (Timothy, 2020, p. 89; see also Henry & Mostafanezhad, 2019 and Wearing & McGehee, 2013b). Although their intentions appear to be good on their own part, the foreign volunteers may see themselves, or behave with racial superiority undertones, as foreign 'white saviors' who have come to rescue the 'brown locals', who simply cannot survive on their own until the next cohort of saviors arrives (Bandyopadhyay, 2019; Bandyopadhyay & Patil, 2017; Easton & Wise, 2015). This has been shown to deepen the divide between the haves from the Global North and the have-nots from the Global South and cause the locals to outsmart the volunteers by taking advantage of their ostensible goodwill (Calkin, 2014; Easton & Wise, 2015; Timothy, 2020).

Traditionally in the Himalaya, communities and individuals could rely on one another in difficult times to donate money or food to those in need, provide emotional and spiritual support, and volunteer time and physical labor on behalf of others and the broader community, including during the harvest season (Choden, 2003; Khanal, 2013). Some evidence suggests that foreign voluntourism appears to have changed the indigenous nature of community volunteerism somewhat by creating an overarching atmosphere of dependency and indifference. Hacker et al. (2017) argue that traditional self-service and community approaches in many parts of the Himalaya have been replaced by a dependence on foreign volunteers and the NGOs that encourage volunteer tourism. Likewise, partially due to tourism, many local volunteers now expect to be paid for their humanitarian endeavors (Hacker et al., 2017)—a foreign concept and an abhorrent practice unknown in the region until recently.

In the Himalaya, many volunteer programs have been overcommercialized to the point where the actual philanthropic element is far secondary to the holiday experience. According to one online promotional narrative, the experience of volunteering in the Himalaya leans acutely toward personal gain. Porter (2019) describes the reasons for volunteering in the Himalaya as the region's breathtaking beauty, fascinating culture that transcends borders, local hospitality that wows, the unforgettable experience, and rewarding work. All of these hint strongly at ego-enhancement more than altruism. According to Calkin (2014), volunteer opportunities in India entice participants with dreams of adventure and relaxation, images of beautiful mountain scenery, and, somewhat secondarily, an opportunity to help vulnerable children.

One of the main criticisms of volunteer tourism in the Himalaya has been a lack of ethics, such as the prioritization of travelers' and tour operators' needs over the needs of the communities they are purporting to help (Hindman, 2014). There have been reported instances of some NGOs selling tours and trekking expeditions, which goes far beyond their humanitarian mandates, and some trekking companies have charged their clients additional fees to participate in volunteer activities along trekking routes. The volunteer sector is not well regulated or enforced in countries such as Nepal, India, and Pakistan (Beirman et al., 2018), which has led to the exploitation of both the tourists and the destination residents.

Though one of the most pervasive humanitarian activities in the region, orphanage volunteer tourism is one of the most derogatory, especially when the rights of children and their parents are infringed upon, which is not infrequent (Durham, 2017). Punaks and Feit (2014) describe how orphanage volunteer tourism encourages the trafficking of children, unbeknownst to most volunteers. The orphanage volunteering industry is so pervasive and powerful in countries such as Nepal and India that many children, who do have living parents, are institutionalized for commercial purposes. Orphanage staff have been known to go from village to village recruiting children to populate the institutions. Sometimes, parents are duped into believing that their children are going away to gain an education (Durham, 2017). Often, parents are deceived by unscrupulous, fake orphanages "that mine rural villages for children and pay parents to whisk their sons and daughters away to what's promised as a better life" (Blumenfeld, 2019, p. 119). In some cases, fake orphanages are set up complete with resident non-orphaned children, as a means of swindling the do-gooders and frequently the companies that sell them the experience (Easton & Wise, 2015). Some children's homes are populated by trafficked false orphans, and the children who live in them are coerced into lying about their personal stories of parental loss and destitution (Blumenfeld, 2019; Durham, 2017).

Durham (2017) maintains that volunteer tourism results in many children being unnecessarily institutionalized, including falsified paperwork that 'proves' they are orphans. In particular, Durham (2017) outlines three main types of harm that accrue to children through philanthropic orphanage tourism. The first is direct impacts, especially sexual and physical abuse. Foreigners are rarely screened or vetted to participate in orphanage volunteering, and this opportunity to be around children has found favor among many pedophiles, even convicted sex offenders, from the West. Many volunteer activities provide opportunities for participants to spend time alone with children. The second type of harm is marked by "well-meaning foreigners . . . putting children at great harm through complicit actions in supporting a system that is known to be harmful" (Durham, 2017, p. 58). This includes volunteer aid, donor sponsorship, and funding new initiatives, among others. Finally, the cumulative impacts of volunteer activities in orphanage settings results in increased trafficking, migration (sometimes forced), homelessness, disrupted families, a sustained cycle of poverty, and the expansion of child sexual exploitation.

Orphanage conditions and institutional living perpetuated in part by foreign volunteerism have also been shown to hinder children's physical, emotional, and mental development. Children in orphanages, whether truly orphaned or not, may form emotional bonds with their foreign visitors, "only to suffer abandonment issues when volunteers move in and out on conveyor belts, departing with empty promises to someday return" (Blumenfeld, 2019, p. 119). For many children, this effects an inability to form emotional attachments, and slows brain development, language learning, and coping skills (Durham, 2017).

Some countries in the region lack child protection laws and policies against maltreatment, child trafficking, and sexual offences. Thus, many predators are empowered by volunteer opportunities, and many humanitarian aid programs enroll people who have no training or background checks. Nepal is one of Asia's poorest countries, and multitudes of NGOs operate with limited government oversight. This is likely because of the aid that comes with NGOs, and for the most part, national institutions and local communities trust foreigners and their institutions. Nonetheless, "the absence of strict regulations means aid groups can be used as cover for human traffickers and predatory behavior by humanitarian workers" (Blumenfeld, 2019, p. 119).

Citrin (2010) provides a blatant critique of the Himalayan health camps that are so popular among healthcare volunteers in Nepal and India. He argues that these camps often give a false reliance on pills and surgeries without addressing the underlying problems that cause the need for medications and medical procedures, such as lack of education and understanding about nutrition, inadequate food hygiene, and unclean water. Likewise, the fleeting clinics do not address the issue of native healthcare traditions, as foreign practitioners nearly always reject traditional medicine outright, which alienates many Indigenous people who are reluctant to take advantage of the opportunity to participate. These camps tend to be transitory and occur in "brief but intense periods of intentional interaction" (Citrin, 2010, p. 41) with little follow-up care for the patients. Moreover, many of them rely on volunteers with little or no medical training, even to administer vaccinations or divvy out medications. Medicines, according to Citrin (2010), are often touted as a cure-all in the health camps and pills are distributed generously, often leading the poor to disregard the doctors' instructions, instead choosing to sell the medications to other people or local pharmacies for cash. Also, like other forms of volunteering, medical humanitarians frequently use these opportunities to pad their resumes with "drive-by medicine" in a convenient and scenic vacation destination (Citrin, 2010, p. 40).

Finally, volunteer opportunities abroad often don the appearance of cultural voyeurism. Medical volunteers often use the developing world as a laboratory for experimenting on illnesses that have been eradicated in their home countries, or they use critical care as an opportunity to get the all-important selfie with their inflicted patients to bolster their social media roles as "aid cowboys" (Citrin, 2010). Many philanthropic tourists treat their experience voyeuristically as an opportunity to see and photograph the aftereffects of a natural disaster or to document their saintly intrusions into impoverished villages with only a "tokenistic few days of volunteer work" (Wearing et al., 2020, p. 7) as part of a larger holiday experience. This phenomenon is boldly expressed in the words of Blumenfeld (2019, p. 115): "Today we live at a time when, for some, the volunteer activity is all about producing a suitable selfie to share on social media". Likewise,

> the gaps in culture, background, and privilege are apparent on social media, where some participants post preening 'selfies' with indigenous children and use hashtags like #InstagrammingAfrica to share a filtered version of their glamorous lives. It all seems to devalue the original purpose of volunteering abroad, and makes one wonder if these participants' motives were charitable at all.
>
> (Wesby, 2015, quoted in Blumenfeld, 2019, p. 115)

## Conclusion

Volunteer tourism has been an important activity in the Global South for many decades, although its recognition as a marketable and manageable form of tourism has only come to the fore in the past quarter century. Traveling for charitable purposes is becoming increasingly popular among youth and the wealthy, often as a means of demonstrating care for the earth or underprivileged societies. Regardless of its origins of selfless service, humanitarian tourism has become overcommercialized and an increasingly prodigious manifestation of mass tourism with all its negative and positive consequences.

In the Himalayan region, volunteer tourism is most prominent in Nepal and India, although it can be found less prominently in all countries of the region. The most common humanitarian activities include medical care, orphanage work, teaching, and disaster relief. While all of these have salient positive outcomes for the communities that host volunteers, they also all contribute to the image problems associated with volunteer tourism. Most research in the region indicates that volunteer tourism generally benefits the visitors more than it benefits the residents. In the Himalaya, guest volunteering tends to have an exploitative character and promotes unbalanced neocolonial relationships between the haves (tourists and tour companies) and the have-nots (communities). It has also changed the nature of traditional mutual aid and volunteerism in Himalayan communities, accelerated a sense of dependency between the helped and the helpers, provided avenues for increased child abuse and trafficking, and delivered fleeting and cursory medical assistance that lacks follow-up care, often by untrained caregivers, and which typically disregards local traditions. These volunteer opportunities are frequently manipulated as pleasure vacations with a little bit of humanitarian work attached and are the settings for voyeuristic actions that embellish CVs, provide photo opportunities to garner 'likes' and followers on social media, and stroke the egos of the volunteers.

Regardless of these pitfalls, some of which are not unique to the region, volunteer tourism has the potential to be a benefit to communities in India, Nepal, Pakistan, Afghanistan, China, and Bhutan. Many humanitarian guests are motivated by proper altruism, and their efforts should not be ignored by critics. Successful and effective volunteer tourism requires evidence-based policies and the political and social will to enforce them. This form of tourism requires special management and sensitive marketing considerations. Careful planning is needed to develop philanthropic activities that are mutually beneficial to the tourists, the host communities, and the tour operators and organizations that facilitate their travel (Wearing et al., 2020).

## References

Bandyopadhyay, R. (2019). Volunteer tourism and "The white man's burden": Globalization of suffering, white savior complex, religion and modernity. *Journal of Sustainable Tourism*, 27(3), 327–343.

Bandyopadhyay, R., & Patil, V. (2017). 'The white woman's burden': The racialized, gendered politics of volunteer tourism. *Tourism Geographies*, 19(4), 644–657.

Beirman, D., Upadhayaya, P.K., Pradhananga, P., & Darcy, S. (2018). Nepal tourism in the aftermath of the April/May 2015 earthquake and aftershocks: Repercussions, recovery and the rise of new tourism sectors. *Tourism Recreation Research*, 43(4), 544–554.

Blumenfeld, J. (2019). *Travel With Purpose: A Field Guide to Voluntourism*. Lanham, MD: Rowman & Littlefield.

Brown, S. (2005). Travelling with a purpose: Understanding the motives and benefits of volunteer vacationers. *Current Issues in Tourism*, 8(6), 479–496.

Calkin, S. (2014). Mind the 'gap year': A critical discourse analysis of volunteer tourism promotional material. *Global Discourse*, 4(1), 30–43.

Callanan, M., & Thomas, S. (2005). Volunteer tourism: Deconstructing volunteer activities within a dynamic environment. In M. Novelli (Ed.), *Niche Tourism: Contemporary Issues, Trends and Cases* (pp. 183–200). Amsterdam: Elsevier.

Choden, T. (2003). *Traditional Forms of Volunteerism in Bhutan*. Thimphu: The Centre for Bhutan Studies.

Citrin, D. (2010). The anatomy of ephemeral healthcare: "Health Camps" and short-term medical voluntourism in remote Nepal. *Studies in Nepali History and Society*, 15(1), 27–72.

Coatsworth, K., Hurley, J., & Miller-Rosser, K. (2017). A phenomenological study of student nurses volunteering in Nepal: Have their experiences altered their understanding of nursing? *Collegian*, 24(4), 339–344.

Coghlan, A., & Fennell, D. (2009). Myth or substance: An examination of altruism as the basis of volunteer tourism. *Annals of Leisure Research*, 12(3–4), 377–402.

Durham, J. (2017). *Protecting the Voluntoured: An Exploratory Human Rights Impact Assessment for Ethical Voluntourism in Nepal*. Unpublished doctoral dissertation. Global Campus.

Easton, S., & Wise, N. (2015). Online portrayals of volunteer tourism in Nepal: Exploring the communicated disparities between promotional and user-generated content. *Worldwide Hospitality and Tourism Themes*, 7(2), 141–158.

Fredheim, L.H. (2018). Endangerment-driven heritage volunteering: Democratisation or 'changeless change'. *International Journal of Heritage Studies*, 24(6), 619–633.

Grout, P. (2009). *The 100 Best Volunteer Vacations to Enrich Your Life*. Washington, DC: National Geographic.

Guttentag, D. (2015). Volunteer tourism: Insights from the past, concerns about the present and questions for the future. In T.V. Singh (Ed.), *Challenges in Tourism Research* (pp. 112–118). Bristol: Channel View Publications.

Hacker, E., Picken, A., & Lewis, S. (2017). Perceptions of volunteering and their effect on sustainable development and poverty alleviation in Mozambique, Nepal and Kenya. In J. Butcher & C.J. Einolf (Eds.), *Perspectives on Volunteering: Voices from the South* (pp. 53–74). Cham, Switzerland: Springer.

Hayward, K., & Colman, R. (2012). *The Economic Value of Voluntary Work in Bhutan*. Thimpu: National Statistics Bureau, Government of Bhutan.

Henry, J., & Mostafanezhad, M. (2019). The geopolitics of volunteer tourism. In D.J. Timothy (Ed.), *Handbook of Globalisation and Tourism* (pp. 295–304). London: Edward Elgar.

Hindman, H. (2014). The re-enchantment of development: Creating value for volunteers in Nepal. In M. Mostafanezhad & K. Hannam (Eds.), *Moral Encounters in Tourism* (pp. 47–58). London: Routledge.

Kandangwa, S., & Gurung, S. (2016). *Preparing to Volunteer in Disaster Situation: A Study of Volunteers' Experiences in the April 2015 Nepal Earthquake*. Unpublished bachelor's thesis. Diaconia University of Applied Sciences, Finland.

Keese, J.R. (2011). The geography of volunteer tourism: Place matters. *Tourism Geographies*, 13(2), 257–279.

Khanal, P. (2013). Diaspora volunteering and development in Nepal. In T.T. Yong & M. Mizanur (Eds.), *Diaspora Engagement and Development in South Asia* (pp. 162–175). London: Palgrave Macmillan.

Kozak, M. (2002). Comparative analysis of tourist motivations by nationality and destinations. *Tourism Management*, 23(3), 221–232.

Lin, Y., Kelemen, M., & Tresidder, R. (2018). Post-disaster tourism: Building resilience through community-led approaches in the aftermath of the 2011 disasters in Japan. *Journal of Sustainable Tourism*, 26(10), 1766–1783.

Lin, Y., & Nawijn, J. (2020). The impact of travel motivation on emotions: A longitudinal study. *Journal of Destination Marketing & Management*, 16, 100363.

McLennan, S. (2014). Medical voluntourism in Honduras: 'Helping' the poor? *Progress in Development Studies*, 14(2), 163–179.

Michel, L.M. (2007). Personal responsibility and volunteering after a natural disaster: The case of Hurricane Katrina. *Sociological Spectrum*, 27(6), 633–652.

Mody, M., Day, J., Sydnor, S., Jaffe, W., & Lehto, X. (2014). The different shades of responsibility: Examining domestic and international travelers' motivations for responsible tourism in India. *Tourism Management Perspectives*, 12, 113–124.

Mohammed, C.A. (2021). Interprofessional health camps to improve health outcomes in rural areas of Himachal Pradesh: A novel intervention. *Health Sciences*, 10(3), 1–7.

Mostafanezhad, M. (2016). *Volunteer Tourism: Popular Humanitarianism in Neoliberal Times*. London: Routledge.

Mustonen, P. (2006). Volunteer tourism: Postmodern pilgrimage? *Journal of Tourism and Cultural Change*, 3(3), 160–177.

Mustonen, P. (2007). Volunteer tourism—Altruism or mere tourism? *Anatolia*, 18(1), 97–115.

Ooi, N., & Laing, J.H. (2010). Backpacker tourism: Sustainable and purposeful? Investigating the overlap between backpacker tourism and volunteer tourism motivations. *Journal of Sustainable Tourism*, 18(2), 191–206.

Porter, B. (2019, April 18). 5 mountain-sized reasons to volunteer in the Himalayas. *Go Abroad*. Retrieved from www.goabroad.com/articles/volunteer-abroad/volunteer-in-himalayas-reasons-why

Punaks, M., & Feit, K. (2014). Orphanage voluntourism in Nepal and its links to the displacement and unnecessary institutionalisation of children. *Institutionalised Children Explorations and Beyond*, 1(2), 179–192.

Scheyvens, R. (2007). Exploring the tourism-poverty nexus. *Current Issues in Tourism*, 10(2–3), 231–254.

Schuurman, J.C. (2018). *Voluntourism in Kathmandu, Nepal*. Unpublished master's thesis. Utrecht University, the Netherlands.

Singh, S. (2006). Tourism in the sacred Indian Himalayas: An incipient theology of tourism? *Asia Pacific Journal of Tourism Research*, 11(4), 375–389.

Singh, S., & Singh, T.V. (2004). Volunteer tourism: New pilgrimages to the Himalayas. In T.V. Singh (Ed.), *New Horizons in Tourism: Strange Experiences and Stranger Practices* (pp. 181–194). Wallingford: CABI.

Smith, K., & Holmes, K. (2009). Researching volunteers in tourism: Going beyond. *Annals of Leisure Research*, 12(3–4), 403–420.

Smithson, C., Rowley, J., & Fullwood, R. (2018). Promoting volunteer engagement in the heritage sector. *Journal of Cultural Heritage Management and Sustainable Development*, 8(3), 362–371.

Stoddart, H., & Rogerson, C.M. (2004). Volunteer tourism: The case of Habitat for Humanity South Africa. *GeoJournal*, 60(3), 311–318.

Telfer, D.J., & Sharpley, R. (2015). *Tourism and Development in the Developing World* (2nd ed.). London: Routledge.

Thara, R., Padmavati, R., Aynkran, J.R., & John, S. (2008). Community mental health in India: A rethink. *International Journal of Mental Health Systems*, 2(1), 1–7.

Timothy, D.J. (2020). Archaeological heritage and volunteer tourism. In D.J. Timothy & L.G. Tahan (Eds.), *Archaeology and Tourism: Touring the Past* (pp. 87–105). Bristol: Channel View Publications.

Timothy, D.J. (2021). *Cultural Heritage and Tourism: An Introduction* (2nd ed.). Bristol: Channel View Publications.

Timothy, D.J., & Zhu, X. (2022). Backpacker tourist experiences: Temporal, spatial and cultural perspectives. In R.J. Sharpley (Ed.), *Routledge Handbook of the Tourist Experience* (pp. 249–261). London: Routledge.

Tomazos, K., & Butler, R. (2009). Volunteer tourism: The new ecotourism? *Anatolia*, 20(1), 196–211.

Upadhayaya, P. (2011). Peace through tourism: A critical look at Nepalese tourism. *Nepal Tourism and Development Review*, 1(1), 15–40.

Wearing, S.L. (2001). *Volunteer Tourism: Experiences That Make a Difference*. Wallingford: CABI.

Wearing, S.L., Beirman, D., & Grabowski, S. (2020). Engaging volunteer tourism in post-disaster recovery in Nepal. *Annals of Tourism Research*, 80, 102802.

Wearing, S.L., Benson, A.M., & McGehee, N. (2016). Volunteer tourism and travel volunteers. In D.H. Smith, R.A. Stebbins, & J. Grotz (Eds.), *The Palgrave Handbook of Volunteering, Civic Participation, and Non-Profit Associations* (pp. 275–289). Basingstoke: Palgrave Macmillan.

Wearing, S.L., & McGehee, N.G. (2013a). *International Volunteer Tourism: Integrating Travellers and Communities*. Wallingford: CABI.

Wearing, S.L., & McGehee, N.G. (2013b). Volunteer tourism: A review. *Tourism Management*, 38, 120–130.

Welty, E., & Bolton, M. (2017). The role of short term volunteers in responding to humanitarian crises: Lessons from the 2010 Haiti earthquake. In M. Holenweger, M.K. Jager, & F. Kernic (Eds.), *Leadership in Extreme Situations* (pp. 115–130). Cham: Springer.

Wesby, M. (2015, August 18). The exploitative selfishness of volunteering abroad. *Newsweek*. Retrieved from www.newsweek.com/exploitative-selfishness-volunteering-abroad-363768

Wickens, E. (2011). Journeys of the self: Volunteer tourists in Nepal. In A.M. Benson (Ed.), *Volunteer Tourism: Theory Framework to Practical Applications* (pp. 66–76). London: Routledge.

Wymer Jr, W.W., Self, D.R., & Findley, C.S. (2010). Sensation seekers as a target market for volunteer tourism. *Services Marketing Quarterly*, 31(3), 348–362.

Zahra, A. (2011). Volunteer tourism as a life-changing experience. In A.M. Benson (Ed.), *Volunteer Tourism: Theory Framework to Practical Applications* (pp. 114–125). London: Routledge.

Zahra, A., & McIntosh, A.J. (2007). Volunteer tourism: Evidence of cathartic tourist experiences. *Tourism Recreation Research*, 32(1), 115–119.

# 11 Pilgrimage Tourism and Sacred Places in the Himalaya

*Agnieszka Gawlik, Michal Apollo, Viacheslav Andreychouk and Yana Wengel*

## Introduction

According to the World Tourism Organization a decade ago, 300–600 million tourists visit the world's religious sites each year (UNWTO, 2011). Religious travel is not a new phenomenon; pilgrimage is one of the oldest forms of tourism and has been performed since ancient times (Barber, 1993; Jackowski & Smith, 1992; Raj & Griffin, 2015; Timothy & Olsen, 2006). Religious tourism is one of the fastest-growing tourism segments in the travel industry today (Dowson et al., 2019; Olsen & Timothy, 2021). Non-pilgrim tourists typically visit religious sites out of curiosity, to learn something new, or to appreciate a region's cultural heritage (Shackley, 2001). Pilgrimages and other forms of religious tourism are common in all of the world's major faiths.

For centuries, the Himalaya have been a venue for pilgrimage and continue to be a popular religious destination today (Singh et al., 2022). The Himalayan pilgrimages are somewhat unique in that typically more than one religion (e.g. Hinduism and Buddhism) venerate the same sacred spaces, they often take place at very high elevations, are particularly arduous because of the physical geography of the region, and they exact considerable impacts on the delicate montane ecosystems of the region. This chapter focuses on Himalayan pilgrimages, which are an ancient and continuing religious tradition among many Himalayan cultures (Eck, 2012) and have grown to incorporate pilgrims and non-pilgrim tourists from around the world. Pilgrimage tourism, which is a source of benefit to local economies, is simultaneously a burden and a blessing. After describing pilgrimage tourism and its impacts and resources generally, this chapter examines the unique characteristics of the Himalaya as a pilgrimage tourism destination, namely issues related to ecological and cultural fragility, accessibility, safety and security, impact management, and cultural solidarity.

## Pilgrimage Tourism and Its Impacts

The term 'pilgrimage' is often defined as a journey to a holy place undertaken for religious or spiritual reasons, or a journey to a place with special significance (Barber, 1993; Olsen & Timothy, 2021). Singh (2013) suggests that pilgrimages

DOI: 10.4324/9781003030126-14

evoke deep feelings, faith, beliefs, respect for the divine, and above all, humility within the devotees. Many sacrosanct places have become world-renowned destinations for religious pilgrims and spirit-seekers, as well as ordinary tourists, and are almost exclusively dependent upon pilgrims for their tourism-based economies.

Like other forms of tourism, pilgrimage has both positive and negative impacts on the society and the natural environment (Apollo, 2015, 2017a; Gawlik, 2022; Łoś, 2012; Sati, 2013; Shinde, 2007; Rybina & Lee, 2021; Shinde & Olsen, 2020). Pilgrimage tourism is known to influence destination cultures (Joseph & Kavoori, 2001; Rybina & Lee, 2021). When pilgrims visit holy shrines, they arrive with their own cultures and traditions and influence the local people, which may lead to cultural erosion due to changes in traditional ideas and values, norms and identities, as well as food habits, lifestyles, family relations, values systems, and individual and community behaviours (Apollo, 2014; Rybina & Lee, 2021). In the case of sacred places far from civilization, modern lifestyles brought by pilgrims often clash with the archaic and culturally pristine values of hosting communities, sometimes accelerating sociocultural changes and deepening social distances between the hosts and the guests.

From an ecological perspective, overcrowding, overdevelopment, pollution, and the destruction of flora and fauna are common effects of religious tourism (Alley, 1998; Apollo, 2014, 2017a, 2017b; Sati, 2014, 2015; Timothy & Nyaupane, 2009). Pilgrims contribute a great deal to environmental degradation by dumping waste and litter in open spaces and water bodies, and increased transportation causes air and water pollution. Likewise, building infrastructure and pilgrim services, such as rest houses and restaurants, often has a negative impact, especially in fragile mountain landscapes (Sati, 2015), as does clearing land for temporary accommodations. Construction for pilgrimage in highland areas leads to large-scale deforestation, and soil compaction near pilgrimage routes increases land degradation and erosion. The pressure at shrines and temples is heaviest, although it is evident in every phase of the experience. The work of Apollo and his colleagues (Apollo, 2016; Apollo et al., 2020a) reports that most Hindu sacred places in the Himalaya are highly polluted, unsafe, unhygienic, and environmentally unsound.

A combination of social and environmental issues is the spread of infectious diseases. Mass religious gatherings have long been blamed for transmitting communicable diseases among attendees, which affects people onsite and often spreads globally as the illnesses accompany the pilgrims' home afterwards (Apollo et al., 2020a; Chatterjee, 2007; Gautret & Steffen, 2016; Sridhar et al., 2015). Major environmental concerns include increased pollution and lack of hygiene, which can increase the risk of diseases.

The positive effects of pilgrimage tourism focus mainly on economic development in the consecrated locale, along pilgrimage routes, and in the broader destination region (Dasgupta et al., 2006; Olsen & Trono, 2018; Sati & Kumar, 2004). Over time, many pilgrimage sites have also become general tourist destinations, largely through their faith affiliations and also because of their natural scenery, the cultural landscapes they create, and the heritage they represent. In some cases, community solidarity is enhanced, resulting in increased cooperation between residents and visitors for environmental and cultural protection (Verma & Sarangi, 2019). Many pilgrimage sites have been a source of pride and solidarity at the local, regional, and national levels (Nyaupane & Budruk, 2009).

## Religion as a Journey: Motivations and the 'Expectation of Experience'

Pilgrimage is one of the clearest outward manifestations of faith and a cultural tourism phenomenon. Once defined exclusively as a journey prompted by religious motivations, pilgrimage today is defined differently as being a traditional religious quest or a modern secular journey (Bixby, 2006; Collins-Kreiner, 2016; Timothy & Olsen, 2006). Contemporary religious tourists may be a mix of traditional faith-seekers and 'pilgrims of modernity', as they choose to visit holy sites for non-religious reasons. Thus, travelling to a pilgrimage centre itself is not enough to characterize a journey as a pilgrimage. Likewise, most religious pilgrimages today are motivated by a mix of faith-seeking and hedonistic leisure pursuits (Lochrie et al., 2019; Nyaupane et al., 2015), something which Gladstone (2005) noted is becoming increasingly popular in Hindu pilgrimages. Similarly, many vacationers are now choosing to visit sacred sites while on leisure holidays (Shinde, 2003; Singh, 1997).

Many contemporary 'pilgrims' are driven by non-religious motives (Cohen, 2003; Shuo et al., 2009; Nyaupane et al., 2015) and are known in the literature as 'secular pilgrims'. Sacred site visitors who identify as pilgrims have greater religious motives, while those who identify as tourists typically exhibit recreational, cultural, or adventure motivations (Nyaupane et al., 2015). Secular pilgrims seek enlightening experiences, encounters with otherness, or special occurrences that are deeply meaningful, but they do not necessarily see these as spiritual or religious. Secular pilgrims sometimes define their journeys as somehow spiritual in nature, such as when in the presence of unique and pristine natural features or while visiting the gravesite of a famous athlete they admire (Digance, 2006; Olsen, 2021). For the most part, however, most secular pilgrims easily fit within various classifications such as cultural tourists and adventure tourists.

More general 'spiritual' motivations and gaining new experiences, rather than religion, have driven secular pilgrimage for many years in Western countries (Abbate & Di Nuovo, 2013; Amaro et al., 2018). This trend appears to be manifesting with increasing frequency in the Global South (Singh, 2004), including in the Himalaya. For this reason, pilgrimage can no longer be described simply as religiously-motivated travel to sacred places (Wright, 2008). For example, along with a substantial increase in the volume of visitors to sacred Hindu sites, qualitative changes have become visible in the very essence of pilgrimage (Shinde, 2007). The modern pilgrimage is more hedonic in character and includes changing patterns of visits, less engagement with rituals, the commercialization of pilgrimage into packaged tours, increased marketing, and an expanding consumerist behaviour among all types of visitors (Gladstone, 2005; Guha & Gandhi, 1995; Qurashi, 2017; Singh, 2002). Several authors recognize that what was once traditional pilgrimage has transformed into a global consumer phenomenon (Nyaupane & Timothy, 2010; Qurashi, 2017; Triantafillidou et al., 2010).

## Pilgrimage Tourism in the Himalaya

### *Population, Culture, and Religion of the Himalaya*

High mountain tourism has grown rapidly in recent decades (Apollo, 2017b). The Himalaya are home to nearly 53 million people and received 46.8 million visits a decade ago in 2011 (Apollo, 2016, 2017b). A considerable part of that number were domestic visitors (45.3 million), with a large share being pilgrims visiting sacred temples and shrines in the High Himalaya, including Amarnath (3,888 metres), Manimahesh Lake (4,080 metres), and the complex of four temples at Chota Char Dham (3,048–3,553 metres).

The Himalayan population is divided into four distinct major ethnic groups with many different languages and cultures: the Indic peoples, the Tibetan peoples, the Afghan-Iranian peoples, and the Burmese/Southeast Asian peoples. However, in the High Himalaya three major civilizations converge. Tibetan culture based on monastic Lamaism and Buddhism is dominant in the high-elevation regions and the northern tier of the range. Indic culture, which largely practices Hinduism, is distributed northwards from India and dominates the southern tier of the range, the lower elevations, and the foothills. Islam is found in the westernmost part of the Himalaya in Pakistan, Afghanistan, and parts of India (Apollo, 2017c; Karan, 1987; Zurick & Pacheco, 2006). The Himalayan region's cultural diversity is augmented by numerous tribal traditions and animistic beliefs (Apollo, 2015).

Mountains have a special place in the religious doctrines and practices of many faiths (Jimura, 2016; Labor, 2018; Timothy, 2013, 2021; Wang et al., 2016) and are among the oldest and least physically accessible sacred venues for pilgrimage. Himalayan pilgrimages have been taking place for many centuries. Historically, the Dravidians and the Aryans proclaimed the Himalaya to be the abode of Shiva. Before the sixth century BC, holy sanctuaries were identified primarily on the basis of their association with unique geophysical features, such as rivers, hot springs, exceptional topographic forms, or snowy peaks. Initially, these places were natural, devoid of idols and temples (R.P.B. Singh, 2006). Today, Hindus believe that the Himalaya are the holy abode of deity and the centre of spirit-body unity. Buddhists worship the mountains as the centre of the spiritual universe.

Many venerated sites are located high in the Himalaya and require prolonged and arduous treks through deep valleys and gorges and up steep slopes. Many pilgrims suffer from acute altitude sickness, with a large number having died during their travels, but this is often seen as part of the necessary sacrifice of being a pilgrim (Basnyat, 2006). The Himalayan tradition of pilgrimage not only is linked to the physical act of visiting sacrosanct sites but also implies the mental and moral dimensions of a challenging journey, which are sought by devotees of such diverse religions as Hinduism, Buddhism, Bon, Islam, Christianity, Kirat Mundum, Jainism, and Sikhism (Bharadwaj, 1983; Zafren, 2007). Many Himalayan temples, shrines, and monasteries are held sacred by more than one religious group. One of the most famous examples is Mount Kailash in Tibet, which is visited and venerated by Hindus, Tibetan Buddhists, Bonpos, and Jains (Chakrabarty & Sadhukhan, 2020).

According to Himalayan customs, the deepest and holiest sanctuaries reveal themselves and their spiritual secrets only to those who have acquired a requisite level of religious training and enlightenment. Because Hinduism and Buddhism have common roots and share certain beliefs, Hindus and Buddhists share many sacred places in the Himalaya, although in some cases, only Hindus are permitted to enter a sacred shrine, such as the Pashupatinath Temple in Nepal. Himalayan pilgrims seek sacred places because they offer a pathway to personal salvation. Tibetans call these places *beyul*, which means 'treasure place'. The Sanskrit word *tirtha* (crossing place) also means 'holy place', and travelling to holy places in Sanskrit is *Tirtha yatra*. According to the Rigveda, an ancient Hindu text, the concept of tirtha is explained as 'a way or road' to 'truth, forgiveness, and kindness'. This suggests that the Hindu concept of sacred place, like that of Buddhists, ultimately rests in the inner transformational journey of the soul.

Himalayan pilgrimages embrace and intertwine the spirit of religion, tradition, and eco-sensitivity. Traditional pilgrimages were forces for social integration and cohesion. They became a way of preserving culture and providing a moral understanding of human existence and interaction with nature (S. Singh, 2006). *Tirtha yatra* usually describes transformational pilgrimage journeys that allow devotees to 'cross over' and connect with sacred forces (Lochtefeld, 2017). This type of pilgrimage occurs on special occasions, such as childbirth and recovering from sickness, as well as being a demonstration of gratitude after one's prayers have been answered (Bharati, 1963; Eck, 2012). *Tirtha yatras* are also undertaken to help deal with remorse or to seek forgiveness for wrongdoing (Bharadwaj, 1983; Lingat, 1973).

## Pilgrimage Destinations in the High Himalaya

There are thousands of pilgrimage destinations, sacred shrines, monasteries, and temples of many of faiths throughout the countries of the Himalaya, including Pakistan and Afghanistan. However, the following sections describe pilgrimage in several key localities and issues in India, Nepal, Bhutan, and China/Tibet.

### India

Throughout history, India has established a dense network of religious sites of different levels of importance, and it is believed that various types of pilgrimage have contributed to the preservation of Hinduism and the unification of the nation despite regional and cultural differences (Schwartzberg, 1978). Two of India's Himalayan states, Himachal Pradesh and Uttarakhand (formerly Uttaranchal), are among the most visited areas of the world for pilgrimage purposes (Price et al., 1997). Between these two states lies the Garhwal region (Uttaranchal), where Himalayan pilgrimages started and evolved (Singh, 1992). Pilgrimage tourism is a significant source of income in the Garhwal region, second only to gems and jewellery, and a major source of employment. The villages along the main pilgrimage trails are fully dependent on income from pilgrims and other tourists (Sati, 2013). This region has several world-famous pilgrimage centres for the devotees of different sects of Hinduism (Sati, 2015). Numerous shrines are located all through the Garhwal Himalaya. Among the highland pilgrimages, Yamnotri, Gangotri, Kedarnath, and Badrinath are the best known and together are called the Chota Char Dhams, or the four minor holy sites.

Yamnotri is located at an elevation of 3,291 metres and is the source of the Yamuna River in Uttarakhand (Figure 11.1). A motor road has been constructed from Uttarkashi up to Hanumanchatti, from which a 14-km trek on a rugged track is required to reach Yamnotri. Gangotri is at an elevation of 3,293 metres in Uttarakhand and is reachable by national highway. The trek from Gangotri temple

*Figure 11.1* Yamunotri in the Gharwal Himalaya

to the Gangotri glacier is 19 kilometres. Kedarnath lies at 3,553 metres in the Rudraprayag district and is known for its picturesque landscape (Figure 11.2). The Mandakini River originates from the Chaurabari glacier and flows alongside the Kedarnath temple. For half the year, the temple and its surroundings are covered with snow. A small village provides pilgrim services. Herders migrate with their animals to summer pastures in the surrounds of Kedarnath. During the same period, the *pandas* (religious teachers), shopkeepers, and small hotel owners migrate to Kedarnath to worship Lord Shiva and to operate their small seasonal businesses. Badrinath is located on the bank of the Vishnu Ganga (the Alaknanda River) at 3,100 metres.

These sites are visited each year by millions of Hindu pilgrims, who come from all over South Asia and other parts of the world (Table 11.1), as well as secular tourists. In the past, pilgrims were required to walk great distances, but with increased Chinese aggression in 1962, a road was constructed to the Badrinath site for military purposes and later used by pilgrims (Sati & Kumar, 2004). Today,

*Figure 11.2* Kedarnath pilgrimage centre

*Table 11.1* Number of pilgrims visiting Kedarnath and Badrinath Temples*

| Years | Kedarnath Pilgrimage | Badrinath Pilgrimage | Years | Kedarnath Pilgrimage | Badrinath Pilgrimage |
|---|---|---|---|---|---|
| 1990 | 117,774 | 362,757 | 2005 | 390,156 | 566,524 |
| 1991 | 118,750 | 355,772 | 2006 | 485,464 | 741,256 |
| 1992 | 141,704 | 412,597 | 2007 | 557,923 | 901,262 |
| 1993 | 118,659 | 476,523 | 2008 | 470,048 | 911,333 |
| 1994 | 104,639 | 347,415 | 2009 | 403,636 | 916,925 |
| 1995 | 105,160 | 461,435 | 2010 | 400,014 | 921,950 |
| 1996 | 105,693 | 465,992 | 2011 | 570,000 | 981,000 |
| 1997 | 60,500 | 361,313 | 2012 | 548,166 | 985,998 |
| 1998 | 82,000 | 340,510 | 2013 | 312,201 | 497,744 |
| 1999 | 80,090 | 340,100 | 2014 | 40,832 | 180,000 |
| 2000 | 215,270 | 735,200 | 2015 | 154,430 | 359,146 |
| 2001 | 119,980 | 422,647 | 2016 | 309,746 | 624,745 |
| 2002 | 169,217 | 448,517 | 2017 | 471,235 | 884,788 |
| 2003 | 280,243 | 580,913 | 2018 | 694,934 | 1,058,490 |
| 2004 | 274,489 | 493,914 | 2019 | 1,000,821 | 1,242,546 |

Source: Char Dham Yatra (2020)

* data were only available for two temples

Gangotri and Badrinath are well connected by road, but visiting Yamnotri and Kedarnath requires lengthy foot treks. Helicopter air services are now available at all these pilgrimage places to transport pilgrims and other tourists. Traditional pilgrimages embraced arduous treks, but modern-day pilgrims tend to want easier access, and non-pilgrimage tourism increases the need for transportation and infrastructure development.

Another famous place of worship and popular pilgrimage destination is the Amarnath Cave. The cave has been a pilgrimage destination for Hindus for centuries. Inside the cave, an ice stalagmite grows and decreases cyclically according to the phases of the moon and is worshipped as Shiva Linga. The cave is located in Jammu and Kashmir, 140 km east of Srinagar, at an altitude of 3,888 metres. Traditionally, pilgrims begin their journey on 1 July and reach their final destination in August after the snow melts. There are several alternative routes, but most pilgrims start in Pahalgam on a trek of approximately 43 km. The Amarnath Cave is visited by over 600,000 people each year.

## Nepal

The majority of Nepalese are Hindus, with a sizeable population of Buddhists and a very small percentage of Muslims, Kirat Mundhum, Prakritis, Bon, Jains, and Christians. Nepal is home to many sacred places, including temples and shrines that are important for devotees of several faiths (Bharadwaj, 1983). Muktinath is a famous temple complex dedicated to Vishnu and revered by Buddhists and Hindus, and Mu Gompa, one of the highest and remotest sites in Nepal on the border with Tibet, are two prominent examples. Muktinath is one of the world's highest temples (3,800 metres); it is also known as Saligrama, owing to the presence of the sacred black rocks known as saligrams. The smooth, polished stones are embedded in river deposits and display the spiralling pattern of fossilized ammonites. The wheel-like pattern is regarded by Hindus as an energy point and is of sacred ritual importance. Muktinath's physical setting high in the mountains at the head of a river is an auspicious location. Like many other sacrosanct localities, Muktinath contains a series of springs, which have been diverted into 108 fountains. There is also a natural gas fire located in a small cave, a large number of ancient Hindu temples, ritual bathing areas, and a sacred grove of ancient poplar trees. Tibetan Buddhists also revere Muktinath as a pilgrimage destination. They call it Chumig Gyatsa, meaning the place of 'a hundred odd springs'. For Tibetans, Muktinath is the abode of the Bodhisattva Avalokiteshvara, and was visited more than 1,200 years ago by the famous Buddhist saint Padmasambhava, who first introduced Buddhism to Tibet. Tibetans believe that the trees in the poplar grove derive from the walking sticks of 84 ancient Indian Buddhist magicians. They consider the saligrams to represent the Tibetan serpent deity, Gawo Jagpa. Moreover, they venerate the natural gas fires as the sexual union of tantric male and female deities. Alongside the Hindu temples are Buddhist monasteries filled with Tibetan icons; Buddhist and Hindu pilgrims share the same trails to Muktinath (Zurick & Pacheco, 2006).

Mu Gompa lies at an altitude of 4,000 metres and is the largest monastery in the Tsum Valley—a significant Himalayan pilgrimage destination. The temple is located in the upper Tsum Valley, which is part of the northernmost region of the Gorkha district, inside the Manaslu Conservation Area. It lies north of the Ganesh Himal and is surrounded on three sides by Tibet's Kyirong region. The Tsum Valley is also known as Beyul Kyimolung, or the valley of happiness. According to Buddhist tradition, *beyuls* are hidden/secret valleys only open to those with a pure mind and heart. The Mu Gompa monastery was established in 1921, and the nearby Rachen Nunnery was established in 1936 by Drupa Rinpoche, a Bhutanese lama who meditated in the caves along the rim of the valley. Previously closed to outsiders, this region opened for international visits in 2008.

The Kathmandu valley, a UNESCO World Heritage Site, is home to many significant temples. Kathmandu's Shiva temple complex at Pashupatinath is the most important Hindu site in Nepal (Sofield, 2001). The valley is also home to the 'Living Goddess', Kumari Devi, represented by a young girl who lives in Durbar Square (Tree, 2014). Nepal's sacred sites include natural features such as mountains and lakes, for example, Lake Gosaikunda in Langtang Valley. Lumbini, the birthplace of Lord Buddha and the country's busiest Buddhist pilgrimage site, lies in the lowlands just beyond the Himalayan foothills (Figure 11.3) (Nyaupane, 2009).

### Bhutan

Bhutan has an area of 47,000 square kilometres and a population of 2 million, making it one of the least densely populated parts of the Himalaya. Bhutan is

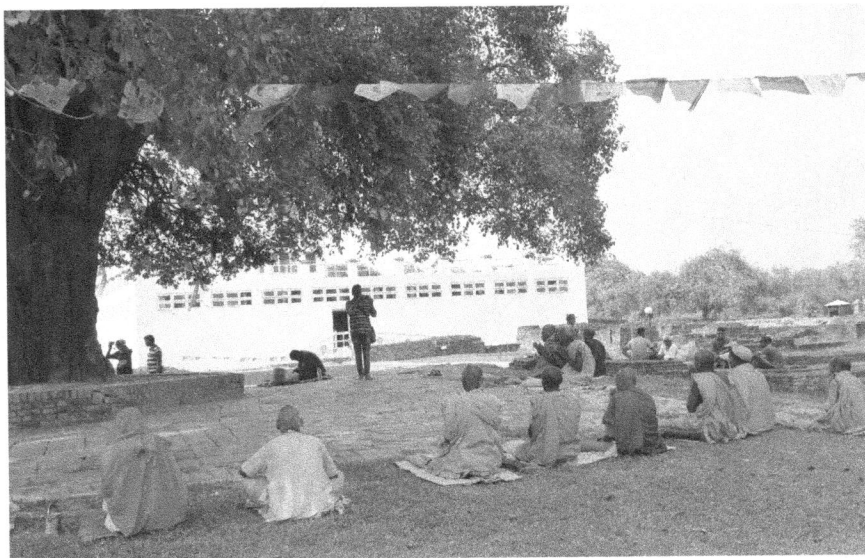

*Figure 11.3* Lumbini on the edge of Nepal's Himalaya

isolated geographically, bounded by dense tropical forests to the south and the Himalaya to the north, east, and west. The country's isolation and lack of colonial oversight have helped it preserve its highly distinctive culture and natural landscapes.

Bhutan possesses a rich tangible heritage of monuments and sites, the most prominent of which are the *dzongs*, or fortresses that serve religious, military, administrative, and social functions. In the past, dzongs were used as garrisons and armouries, and as trading posts between people of different districts. Presently, dzongs serve a dual purpose: religion and government, which in this kingdom are inseparable. Besides these functions, dzongs have taken on added significance as the main cultural tourist attractions in the country and every tour package includes requisite visits to at least a few dzongs. They are also venues for festivals that reflect the country's religious and cultural traditions, which are deeply rooted in, and inseparable from, the local form of Buddhism.

Taktsang Palphug Monastery, also known as Paro Taktsang or Tiger's Nest, is a Buddhist sacred site and temple complex in the Paro Valley, perched on a 900-metre cliff, 3,120 metres above sea level (Figure 11.4). According to legend, in the eighth century, Guru Padmasambhava flew to this place on the back of a flying tigress, landing on a cliff, which he then chose as the place to build a monastery. From that time onward, many famous Buddhist monks have visited the location. The current monastery was built in the seventeenth century. In 1998, it almost completely burned down but was restored in the early 2000s. The monastery is

*Figure 11.4* Taktsang Palphug Monastery (Tiger's Nest) in the Bhutanese Himalaya

one of Mahayana Buddhism's most sacred sites. It is a vital pilgrimage destination for domestic Bhutanese pilgrims and is included on nearly every foreign tourist's itinerary. From the Paro town, the monastery is accessible by foot or on horseback.

### China

One of the most popular religions in China is Chinese Buddhism (Tan, 2002; Yang, 2004). Tibet is home to dozens of sacred Buddhist pilgrimage sites, many of which also serve as world-renowned tourist attractions, namely the Potala Palace in Lhasa and the Jokhang Temple. Many temples, shrines, monasteries, and mountain peaks connect the Tibetans with their national and religious heritage, which attaches an additional layer of meaning to their pilgrimages. Many Tibetans find it difficult to separate leisure holidays from religious travel and combine motives and activities in both arenas (Yang, 2020).

Mount Kailash (also known as Mount Meru) is one such example of a place that connects Tibetans with their religio-cultural heritage and serves as a major destination for others. The mountain is 6,714 metres high and is in the Tibetan trans-Himalaya. It is the spiritual centre of Tibet and one of its holiest places (Chakrabarty & Sadhukhan, 2020). Every year, thousands of people from several religions undertake the set pilgrimage to Mount Kailash, believing that circling the mountain on foot is a sacred ritual that brings good fortune. The circumambulatory route around the mountain extends 52 km, and the trip lasts three days and four nights. It ends in Darchen at an altitude of approximately 4,600 metres, where the annual pilgrimage creates large crowds. For Hindus, the phallic shape of Mount Kailash signifies the deity Shiva, who is believed to reign over the entire Himalaya from his abode atop the summit. As such, the mountain is sometimes known as 'Shiva's Paradise'. Some pilgrims believe that the entire route should be completed in a single day, which is extremely difficult. Others perform slow rituals as they prostrate themselves on the ground and crawl to their final destination. These pilgrims take at least four weeks to complete the pilgrimage rituals. According to all the religions that worship the mountain, climbing it is a serious sin, and many believe that the stairs to Mount Kailash lead to heaven.

## Issues in Himalayan Pilgrimages

The extraordinary beauty of the Himalaya and the promise of salvation compel local pilgrims and those from outside to visit these mountains. The arduous nature of the journeys required to reach many of the pilgrimage sites only heightens the religious experience. The dangers inherent in snowbound passes and precipitous mountain trails are tempered by the practice of meditation and spiritual cleansing. Each year, millions of pilgrims visit Hindu temples and shrines located along well-established pilgrimage routes. Due to the growing number of pilgrims and the vulnerable montane environment, pilgrimage tourism in the High Himalaya faces a multitude of challenges and opportunities.

### *Pilgrimage and Accessibility*

In tourism development terms, inaccessibility is the primary characteristic of mountains (Price, 1995), due to their orographic topography and frequent natural hazards (e.g. landslides, earthquakes, avalanches, and floods). These conditions result in isolation, limited mobility, long distances, poor communications, and high transportation costs (Apollo, 2017b; Jodha, 2001). Many of the Himalayan pilgrimage sites are located at great elevations, often in places that are difficult to reach and which lack sufficient tourism infrastructure and services. Owing to conditions of inaccessibility and remoteness, Himalayan pilgrimage tourism is extremely sensitive to climate and weather conditions.

Simultaneously, isolation and inaccess help generate economic opportunities for the local people in providing accommodations and transportation, such as portering and yak or mule rides, which cannot easily be satisfied by outside interests. These opportunities help distribute economic benefits to villagers through the region (Sharma, 2000). Because inaccessibility necessitates the use of local resources, if managed inappropriately, pilgrimage tourism can severely degrade the natural environment, which in the High Himalaya, is the foundation of people's subsistence.

### Pilgrimage Impacts and Management

Pilgrimage is often lauded as an innocuous form of tourism, yet it rarely is, especially when, as noted earlier, it becomes mass tourism, exacting deep social and ecological impacts on the destination (Bleie, 2003; Sati, 2015). Tourism in the Himalaya is in danger of exhibiting all the detrimental effects of mass tourism and therefore needs more emphasis on conservation and promoting activities that are less harmful to the environment. High mountains have limited carrying capacities, and these thresholds should not be crossed. However, unfortunately, holy places in the Himalaya are often poorly regulated, or entirely unregulated, with over a million tourists visiting fragile environments, narrow trails, and small spaces every year.

The Himalayan region is ecologically fragile, tectonically and seismically active, geologically unstable, economically undeveloped, and geographically remote. Low carrying capacities and vulnerabilities to intense pressure characterize the ecologically fragile Himalaya. This includes vulnerability to natural resource degradation and depletion (Jodha, 2001). Steepness, high elevation, unstable geological movement, and extreme climatic conditions increase vulnerability to erosion, landslides, avalanches, floods, and loss of flora and fauna (Nepal & Chipeniuk, 2005).

According to Nyaupane and Chhetri (2009), the negative impacts of Himalayan pilgrimages typically include short-term peak demands exerting extreme stress on essential services, the accumulation of vast quantities of waste, high levels of pollution, and deforestation for cooking and heating water, and building lodging. Due to the periodic nature of traditional pilgrimage, these environmental impacts are likely to continue peaking and dissipating over time in and around the pilgrimage centres. Some pilgrimage destinations in the Himalaya attract tens of thousands of visitors over short summer periods. During these peak times, huge amounts of non-biodegradable waste (e.g. plastic and glass cups) are produced, water is polluted due to inadequate sewerage facilities, and trees and plants are harvested for firewood. Likewise, overcrowding and congestion are salient problems that contribute to compacted earth, which results in increased water runoff, floods, and mudslides. Unfortunately, these short-term problems accumulate over time and intensify with the constant influx of visitors. The negative effects on the environment are determined by two factors. First, the magnitude and pattern of visitation and the nature of the activities undertaken by visitors, which may include both religious and normative tourist behaviours that have certain direct impacts similar to those experienced in mass tourism destinations (Kaur, 1985; Singh, 2002). Second, tourism stimulates the growth of servicescapes (e.g. restaurants, souvenir shops, and travel agencies), and demand for these services leads to rapid urbanization (Mohanty, 1995; Rinschede, 1997) and gentrification, rendering the places unaffordable for local residents. The physical growth of tourism imposes further environmental pressures such as high stress on infrastructure by an increasing population, loss of forests to real estate development, depletion of water supplies, and increased traffic congestion.

On the other hand, initiatives such as the National Mission for Sustaining the Himalayan Ecosystem (NMSHE), are an important part of India's National Action Plan on Climate Change (NAPCC). The NMSHE is a multipronged, cross-cutting endeavour involving many sectors. It aims to foment management and policy measures for sustaining and safeguarding the Himalayan environment and providing mechanisms to assess its health in the long term. The NMSHE promotes sustainable tourism, adopting 'best practice' norms for infrastructure in mountain regions to avoid damaging sensitive ecosystems, despoiling landscapes, and exceeding tourism carrying capacities.

An additional consideration is that religious heritage sites in the Himalaya are increasingly becoming attractions for non-pilgrim tourists who visit for educational and leisure reasons, rather than strictly for religious motives. The caretakers of sacred places are tasked with managing tourist flows, in addition to protecting the place's religious values and spiritual sense of place (Olsen, 2006; Shackley, 2001). With the growth in additional tourist arrivals, many of the Himalayan sites suffer from noisy, insensitive tourists who disrupt those who have come to worship, in the process degrading the aesthetic qualities of a sacred place (Miles-Watson & Miles-Watson, 2011; Sati, 2018). This sense of place is critical in providing an atmosphere of worship and meditation for those who wish to commune with the divine. Caretakers, therefore, face significant challenges in maintaining a spiritual sense of place while satisfying the needs and expectations of both pilgrims and non-pilgrim tourists, and preserving the site's physical integrity. This is particularly acute when coupled with overcrowding, which can violate a site's sanctity, as in the case of Buddhist monasteries in Tibet and Nepal (Shackley, 1999). Overcrowding by pilgrims or other tourists has been seen to have a detrimental effect on both the built and natural environments in the Himalaya through vandalism/graffiti, theft, accidental damage, general wear and tear, microclimatic change, and litter. These impacts diminish the experience of religious visitors and others, and in some extreme cases can potentially lead to the site being unable to serve its original functions, becoming, in essence, a dedicated tourist attraction rather than a sacred site (Shackley, 2001).

Entrance fees can be a useful management tool to deter people who are not seriously interested in visiting, including some of the more leisure-oriented tourists. Visitor quotas have also been known to work, based upon maximum numbers of visitors admitted into an attraction or destination during a fixed period (Brunet et al., 2001; Puciato & Łoś, 2009). This may entail closing specific places or determining the maximum number of visitors allowed in a particular area. These instruments may retard overcrowding and, consequently, resource degradation. Although Bhutan has never had a visitor quota, the country's mandatory high daily tariff has functioned as a mechanism for restricting overtourism under the policy of 'high-value, low volume' tourism (Nyaupane & Timothy, 2010). This has led to Bhutan being able to maintain higher quality natural and cultural environments than many of its Himalayan neighbours that are open to mass tourism (Lundup, 2020; Nyaupane & Thapa, 2006).

Although tourism (including pilgrimage) to the high mountains might be a key socioeconomic development tool, it also can bring negative encounters between tourists and local residents (Apollo, 2015; Apollo et al., 2020b; Rybina & Lee, 2021). Himalayan mountain communities should use their limited resources precisely and purposefully to protect their cultures, social networks, and physical environments. Apparently, the inhabitants of higher altitudes have more patience and understanding of tourists than those in other more heavily touristified regions (Apollo et al., 2020b). Therefore, management plans are urgently needed in lower altitudes, and they need to address ways to ensure that communities are empowered in their tourism development decisions.

## *Pilgrimage and Safety*

As previously noted, mountains are particularly susceptible to environmental hazards. The Himalaya are regularly and severely affected by destructive natural processes. A rare weather system, last seen approximately 80 years earlier, converged over northwest India and western Nepal in June 2013, devastating communities in the region (Nirupama et al., 2015). This catastrophic event triggered widespread landslides, floods, and other disasters, killing hundreds of people, destroying communities and their livelihoods, and blocking access points, which led to thousands of pilgrims being stranded on a number of pilgrimage routes. This event caused many human casualties and irreparable damage to infrastructure and property within the Garhwal Himalaya (Mehta et al., 2017; Nirupama et al., 2015).

The Nepal earthquake (Gorkha earthquake) of 2015 is another example of the vulnerabilities of Himalayan pilgrimages. The Himalaya are seismically active and constantly in a state of motion. On April 25, 2015, a deep 7.8-magnitude earthquake struck the Gorkha District approximately 77 km northwest of Kathmandu. This extreme event killed thousands of people and physically destroyed several villages and large swaths of the capital city. Several aftershocks in different locations were also highly destructive. Although there were no reports specifically of pilgrims dying or pilgrimage routes being blocked, largely owing to the time of year the disaster occurred, the earthquake was particularly devastating to several pilgrimage destinations in the country. Many important sacred sites were levelled or damaged in numerous towns and villages. Several temples in Kathmandu's Darbur Square, a UNESCO World Heritage Site and one of Nepal's main pilgrimage centres and tourist attractions, collapsed. Approximately 1,000 temples, shrines, and stupas throughout the country were damaged or ruined (Bhagat et al., 2018).

Deadly bus and car crashes in the Himalaya frequently make the headlines. Vehicle accidents are common, given the narrow roads, poor road conditions, poor driving habits, often overcrowded buses, and the steep ravines that flank transportation corridors in many parts of the region. On July 16, 2017, 16 Hindu pilgrims were killed when a bus plunged off the road into a gorge in India's

Jammu and Kashmir (Agence France, 2017). On December 15, 2019, at least 14 Hindu pilgrims died in Nepal when their bus fell 70 metres off a mountainside on their return from visiting Kalinchowk Bhagawati, a popular Hindu temple located 4,000 metres above sea level (Al Jazeera, 2019).

Altitude sickness is a major safety concern for high-altitude pilgrims. Many pilgrimage destinations in the Himalaya are located well above 2,500 metres. When lowlanders visit high elevations, they experience the symptoms of hypoxia, or lack of adequate oxygen (see Pun, Bhandari and Basnyat chapter in this volume). These symptoms may be mild (e.g. headaches and sleepiness) or extremely severe and deadly, such as brain swelling or high-altitude cerebral oedema (Basnyat, 2014). Unfortunately, many pilgrims are unprepared for their high-elevation journeys, particularly the aged and those who prepare for spiritual encounters by fasting (Basnyat, 2006). In most cases, Himalayan pilgrimages occur outside urban areas with inadequate medical facilities, which results in frequent deaths and urgent medical evacuations (Koul et al., 2013).

Trekking accidents are often caused by a lack of information about the pilgrimage and the topography the trail traverses. Many pilgrims are caught unaware of the difficulties of the terrain or changeable weather conditions. Mass pilgrimage tourism has further accelerated the fragility and vulnerability of highland areas, and the carrying capacity of pilgrimage centres is low.

In certain regions, terrorist attacks have tainted the image of Himalayan pilgrimages. For example, the Amarnath Cave in Jammu and Kashmir has an unfortunate reputation for pilgrim deaths from terrorist attacks (in 2000, 2001, 2002, 2006, and 2017). Militants associated with the Lashkar-e-Taiba terrorist organization in Pakistan have been the source of most of these attacks on Amarnath pilgrims on several occasions over the past 20 years, stemming from the ongoing conflict between India and Pakistan in Kashmir (Tabasum et al., 2017). Political instability in Nepal, ethnic suppression in Bhutan, and territorial conflicts between India and China may also be sources of danger for pilgrims (Bhattarai et al., 2005; Bleie, 2003; Shrestha, 2018; Vogel & Field, 2020).

### Conflict and Communitas

More than any other sacred destinations in the world, the Himalaya are home to sites that are venerated by several religions. Christian, Muslim, and Jewish pilgrimages in the Mediterranean and Middle East are distinct from one another, and although some of their sacred spaces overlap, the majority of them have clearly drawn spatial, liturgical, and ceremonial boundaries. Nonetheless, where shared heritages do exist, conflict is redolent in places such as the Holy Land (Coleman, 2002; Ron & Timothy, 2019). Although there is some degree of dissonance at Himalayan pilgrimage localities that are shared by different faiths, or whose host community is of a different faith than that of the majority of pilgrims (see Nyaupane, 2009; Nyaupane et al., 2015), for the most part, the religions and their adherents get along with one another in these mountains, particularly those with a shared history (e.g. Hindus and Buddhists). In fact, the general lack of discord among religious sects in the Himalaya has been attributed to the role of shared pilgrimage spaces in promoting interfaith harmony and friendship (Bhat, 2013).

### Commercialization and the Pilgrimage as Commodity

Another unique perspective on Himalayan pilgrimages that is atypical of pilgrimages in places such as Europe, North America, East Asia, and the Middle East is that most Himalayan pilgrims are poor and less able to pay for pilgrimage holiday packages or even stay in hotels and lodges along the way. In this sense, they have a lower economic impact on the destination than pilgrims do in many other parts of the world, as they typically carry food with them and sleep in tents or under the stars (Bhat, 2013). Although originally, the majority of pilgrims in Mecca and Medina were poor and destitute, today most of them must spend huge sums of money on organized Hajj tour packages or risk not being able to attend (Qurashi, 2017). The same is true of much Christian and Jewish pilgrimage travel in Europe and the Levant, as well as many Hindu and Sikh pilgrimages in India (Jutla, 2016; Shinde, 2012). What were once outward demonstrates of humility are now consumer experiences.

In the postcolonial period (after 1947), the democratization of leisure, educational awareness, social mobility, paid holiday schemes, the development of transport networks, and ego enhancement contributed to the growth of mass tourism. The Himalayan region started to profit from tourism in the 1950s, which quickly grew into mass tourism in the fragile montane ecosystems (R.P.B. Singh, 2006), including 'exotic' pilgrimage destinations. This eventually led to the desacralization of shrines, pilgrimage routes, and religious practices, thereby changing the very nature of some Himalayan pilgrimages (Sharma, 2000). Some Himalayan localities have gone from a conservation and indigenous knowledge-based sustainability ethos to one of commercialism and mass consumerism. In addition to formal pilgrimages, which continue in the present day, many people travel to Himalayan sacred sites because these locations have become famous through marketing campaigns and heritage branding such as UNESCO World Heritage Sites. Many non-pilgrims also visit Himalayan sacred sites because of their perceived therapeutic environments and healing powers (Wei, 2010). Rishikesh, India, for example, is one of Hinduism's holiest pilgrimage destinations, but it is now also known informally as the 'yoga capital of the world', each year receiving thousands of secular tourists who come to participate in healing activities (e.g. yoga) and Ganges River rafting (Bowers & Cheer, 2017). Although, some High Himalayan pilgrimage places are seeing increased commercialization and overdevelopment, they have certainly not yet reached the level of commoditization that more accessible and mass pilgrimage destinations have reached.

## Conclusion

Pilgrimage tourism in the Himalaya has multiple socioeconomic and environmental impacts. It has a generally positive impact on economic development among the mountain populations, but as it now demonstrates many characteristics of mass tourism, it also has negative impacts on the delicate montane environment, including pollution/excessive waste and land degradation, just as other forms of tourism in the Himalaya do (e.g. trekking) (Nyaupane et al., 2014; Serenari et al., 2012). The fragile ecosystems of the highlands and alpine meadows, where most of these shrines are located, have largely been affected by pilgrims and other tourists as their numbers have swelled dramatically in recent years, especially with the Himalayan countries' policies of promoting pilgrimages as an economic driver. This has resulted in the over-exploitation of natural and cultural resources in many sacred locales.

It took only 20 years for the ultimate tourism thresholds in the Himalaya to be crossed and for traditional pilgrimage to bleed into a particular mix of religious tourism and secular pilgrimage. Contemporary pilgrimage tourism has caused overcrowding, vehicular and pedestrian congestion, environmental degradation, and the trespassing of tourists further into higher elevations. Regardless of the erosion of sacred places, changes in their unique attributes, and deterioration in the quality of the experience, the 'business of tourism' continues to grow. Today, these mountains host heterogeneous visitor groups of pilgrims and tourists with diverse travel motivations, which lie nebulously between the two extremes of reverent devotion and secular holiday-making (Olsen & Timothy, 2006).

The two most critical issues facing highland pilgrimages in the Himalaya are climate change and overtourism. These are mitigated for good or for bad by inaccessibility, safety and security, interfaith solidary or conflict, and the commoditization of the sacred for hedonic consumption. The highlands are highly vulnerable to climate change, which naturally affects pilgrimages and the environments that host them. Likewise, increased numbers of pilgrims and non-pilgrim tourists visiting the region have also created ecological and social concerns, particularly with regard to the extension of modern transportation deep into the region, which has encouraged greater numbers of pilgrims and other tourists and the social and environmental consequences that follow (Sati, 2013).

## References

Abbate, C.S., & Di Nuovo, S. (2013). Motivation and personality traits for choosing religious tourism: A research on the case of Medjugorje. *Current Issues in Tourism*, 16(5), 501–506.

Agence France (2017). Bus crash in northern India kills at least 16 Hindu pilgrims. Retrieved March 1, 2021, from www.theguardian.com/world/2017/jul/16/india-bus-crash-hindu-pilgrimage-jammu-kashmir-himalayan

Al Jazeera (2019). Many killed in Nepal pilgrimage bus crash. Retrieved February 2, 2021, from www.aljazeera.com/news/2019/12/15/many-killed-in-nepal-pilgrimage-bus-crash

Alley, K.D. (1998). Images of waste and purification on the banks of the Ganga. *City & Society*, 10(1), 167–182.

Amaro, S., Antunes, A., & Henriques, C. (2018). A closer look at Santiago de Compostela's pilgrims through the lens of motivations. *Tourism Management*, 64, 271–280.

Apollo, M. (2014). Climbing as a kind of human impact on the high mountain environment—based on the selected peaks of Seven Summits. *Journal of Selcuk University Natural and Applied Science*, 2, 1061–1071.

Apollo, M. (2015). The clash—social, environmental and economical changes in tourism destination areas caused by tourism: The case of Himalayan villages (India and Nepal). *Current Issues of Tourism Research*, 5(1), 6–19.

Apollo, M. (2016). Mountaineer's waste: Past, present and future. *Annals of Valahia University of Targoviste, Geographical Series*, 16(2), 13–32.

Apollo, M. (2017a). The good, the bad and the ugly: Three approaches to management of human waste in a high-mountain environment. *International Journal of Environmental Studies*, 74(1), 129–158.

Apollo, M. (2017b). The true accessibility of mountaineering: The case of the High Himalaya. *Journal of Outdoor Recreation and Tourism*, 17, 29–43.

Apollo, M. (2017c). The population of Himalayan regions—by the numbers: Past, present, and future. In R. Efe & M. Ozturk (Eds.), *Contemporary Studies in Environment and Tourism* (pp. 143–159). Newcastle upon Tyne: Cambridge Scholars Publishing.

Apollo, M., Andreychouk, V., Moolio, P., Wengel, Y., & Myga-Piątek, U. (2020b). Does the altitude of habitat influence residents' attitudes to guests? A new dimension in the residents' attitudes to tourism. *Journal of Outdoor Recreation and Tourism*, 31, 100312.

Apollo, M., Wengel, Y., Schänzel, H., & Musa, G. (2020a). Hinduism, ecological conservation, and public health: What are the health hazards for religious tourists at Hindu temples? *Religions*, 11(8), 416.

Barber, R. (1993). *Pilgrimages*. London: Boydell Press.

Basnyat, B. (2006). The pilgrim at high altitude. *High Altitude Medicine & Biology*, 7(3), 183–184.

Basnyat, B. (2014). High altitude pilgrimage medicine. *High Altitude Medicine & Biology*, 15(4), 434–439.

Bhagat, S., Samith Buddika, H.A.D., Kumar Adhikari, R., Shrestha, A., Bajracharya, S., Joshi, R., Singh, J., Maharjan, R., & Wijeyewickrema, A.C. (2018). Damage to cultural heritage structures and buildings due to the 2015 Nepal Gorkha earthquake. *Journal of Earthquake Engineering*, 22(10), 1861–1880.

Bharadwaj, S.M. (1983). *Hindu Places of Pilgrimages in India: A Study in Cultural Geography*. Berkeley: University of California Press.

Bharati, A. (1963). Pilgrimage in the Indian tradition. *History of Religions*, 3(1), 135–167.

Bhat, Z.A. (2013). Tourism industry and pilgrimage tourism in Jammu and Kashmir: Prospects and challenges. *International Journal of Research in Management & Technology*, 2, 105–113.

Bhattarai, K., Conway, D., & Shrestha, N. (2005). Tourism, terrorism and turmoil in Nepal. *Annals of Tourism Research*, 32(3), 669–688.

Bixby, B. (2006). Consuming simple gifts: Shakers, visitors, goods. In P. Scranton & J.F. Davidson (Eds.), *The Business of Tourism: Place, Faith and History* (pp. 85–109). Philadelphia: University of Pennsylvania Press.

Bleie, T. (2003). Pilgrim tourism in the Central Himalayas. *Mountain Research and Development*, 23(2), 177–184.

Bowers, H., & Cheer, J.M. (2017). Yoga tourism: Commodification and western embracement of eastern spiritual practice. *Tourism Management Perspectives*, 24, 208–216.

Brunet, S., Bauer, J., DeLacy, T., & Tshering, K. (2001). Tourism development in Bhutan: Tensions between tradition and modernity. *Journal of Sustainable Tourism*, 9(3), 243–263.

Chakrabarty, P., & Sadhukhan, S.K. (2020). Destination image for pilgrimage and tourism: A study in Mount Kailash region of Tibet. *Folia Geographica*, 62(2), 71–86.

Char Dham Yatra (2020). Sacred yatra. Retrieved November 20, 2020, from www.sacredyatra.com/

Chatterjee, S. (2007). Hindu pilgrimages. In A. Wilder-Smith, M. Shaw, & E. Schwartz (Eds.), *Travel Medicine: Tales Behind the Science* (pp. 269–279). Oxford: Elsevier.

Cohen, E.H. (2003). Tourism and religion: A case study—visiting students in Israeli Universities. *Journal of Travel Research*, 42, 36–47.

Coleman, S. (2002). Do you believe in pilgrimage? Communitas, contestation and beyond. *Anthropological Theory*, 2(3), 355–368.

Collins-Kreiner, N. (2016). Dark tourism as/is pilgrimage. *Current Issues in Tourism*, 19(12), 1185–1189.

Dasgupta, S., Mondal, K., & Basu, K. (2006). Dissemination of cultural heritage and impact of pilgrim tourism at Gangasagar Island. *Anthropologist*, 8(1), 11–15.

Digance, J. (2006). Religious and secular pilgrimage: Journeys redolent with meaning. In D.J. Timothy & D.H. Olsen (Eds.), *Tourism, Religion & Spiritual Journeys* (pp. 36–48). London: Routledge.

Dowson, R., Yaqub, J., & Raj, R. (Eds.). (2019). *Spiritual and Religious Tourism: Motivations and Management*. Wallingford: CABI.

Eck, D. (2012). *India: A Sacred Geography*. New York: Random House.

Gautret, P., & Steffen, R. (2016). Communicable diseases as health risks at mass gatherings other than Hajj: What is the evidence? *International Journal of Infectious Diseases*, 47, 46–52.

Gawlik, A. (2022). The internalization of tourism externalities: Methods and instruments. In M. Apollo & P. Moolio (Eds.), *Poverty and Development: Current Problems and Prospects*. Bristol: Channel View Publications.

Gladstone, D.L. (2005). *From Pilgrimage to Package Tour: Travel and Tourism in the Third World*. New York: Routledge.

Guha, S.B., & Gandhi, S. (1995). Evolution and growth of Hindu pilgrimage centres in greater Bombay. In D.P. Dubey (Ed.), *Sacred Places, Sacred Traditions, Society of Pilgrimage Studies* (pp. 179–189). Allahabad: Society of Pilgrimage Studies.

Jackowski, A., & Smith, V.L. (1992). Polish pilgrim-tourists. *Annals of Tourism Research*, 19, 92–106.

Jimura, T. (2016). World heritage site management: A case study of sacred sites and pilgrimage routes in the Kii mountain range, Japan. *Journal of Heritage Tourism*, 11(4), 382–394.

Jodha, N.S. (2001). *Life on the Edge: Sustaining Agriculture and Community Resources in Fragile Environments*. New Delhi: Oxford University Press.

Joseph, C.A., & Kavoori, A.P. (2001). Mediated resistance: Tourism and the host community. *Annals of Tourism Research*, 28(4), 998–1009.

Jutla, R.S. (2016). The evolution of the Golden Temple of Amritsar into a major Sikh pilgrimage center. *AIMS Geosciences*, 2(3), 259–272.

Karan, P.P. (1987). Population characteristics of the Himalayan region. *Mountain Research and Development*, 7(3), 271–274.

Kaur, J. (1985). *Himalayan Pilgrimages and the New Tourism*. New Delhi: Himalayan Books.

Koul, P.A., Khan, U.H., Hussain, T., Koul, A.N., Malik, S., Shah, S., Bazaz, S.R., Rashid, W., & Jan, R.A. (2013). High altitude pulmonary edema among "Amernath Yatris". *Lung India*, 30(3), 193–198.

Labor, H.L.A. (2018). The sacred in caves and mountains: Animist and Christian interface in the Philippines. In S. Yasuda, R. Raj, & K.A. Griffin (Eds.), *Religious Tourism in Asia: Tradition and Change* (pp. 49–57). Wallingford: CABI.

Lingat, R. (1973). *The Classical Law of India*. Berkeley: University of California Press.

Lochrie, S., Baxter, I.W.F., Collinson, E., Curran, R., Gannon, M.J., & Taheri, B. (2019). Self-expression and play: Can religious tourism be hedonistic? *Tourism Recreation Research*, 44(1), 2–16.

Lochtefeld, J. (2017). *Pilgrimage*. Oxford Bibliographies, Hinduism. Retrieved February 3, 2021, fromwww.oxfordbibliographies.com/view/document/obo-9780195399318/obo-9780195399318-0096.xml

Łoś, A. (2012). Wymiary efektywności i jej pomiar we współczesnej turystyce. In M. Morawski (Ed.), *Zarządzanie wiedzą w turystyce, a efektywność gospodarki turystycznej* (pp. 79–88). Wrocław: Akademia Wychowania Fizycznego we Wrocławiu.

Lundup, T. (2020). The sacred and the secular: Exploring mass tourism encounters at Lamayuru Monastery in Ladakh. *Himalaya*, 39(2), 119–130.

Mehta, M., Shukla, T., Bhambri, R., Gupta, A.K., & Dobhal, D.P. (2017). Terrain changes, caused by the 15–17 June 2013 heavy rainfall in the Garhwal Himalaya, India: A case study of Alaknanda and Mandakini basins. *Geomorphology*, 284, 53–71.

Miles-Watson, J., & Miles-Watson, S.B. (2011). Conflicts and connections in the landscape of the Manimahesh pilgrimage. *Tourism*, 59(3), 319–333.

Mohanty, M.P. (1995). Orissa's sacred city enforces strict laws to protect its pilgrims. *Hinduism Today*, Retrieved January 3, 2021, from www.hinduismtoday.com/modules/smartsection/item.php?itemid=3480

Nepal, S.K., & Chipeniuk, R. (2005). Mountain tourism: Towards a conceptual framework. *Tourism Geographies*, 7(3), 313–333.

Nirupama, N., Sharma, R., Verma, K., & Panigrahi, S. (2015). A multi-tier hazard-part I: Description of the event. *Natural Hazards*, 76(1), 259–269.

Nyaupane, G.P. (2009). Heritage complexity and tourism: The case of Lumbini, Nepal. *Journal* of Heritage Tourism, 4(2), 157–172.

Nyaupane, G.P., & Budruk, M. (2009). South Asian heritage tourism. In D.J. Timothy & G.P. Nyaupane (Eds.), *Cultural Heritage and Tourism in the Developing World: A Regional Perspective* (pp. 127–145). London: Routledge.

Nyaupane, G.P., & Chhetri, N. (2009). Vulnerability to climate change of nature-based tourism in the Nepalese Himalayas. *Tourism Geographies*, 11(1), 95–119.

Nyaupane, G.P., Lew, A.A., & Tatsugawa, K. (2014). Perceptions of trekking tourism and social and environmental change in Nepal's Himalayas. *Tourism Geographies*, 16(3), 415–437.

Nyaupane, G.P., & Thapa, B. (2006). Perceptions of environmental impacts of tourism: A case study at ACAP, Nepal. *The International Journal of Sustainable Development and World Ecology*, 13(1), 51–61.

Nyaupane, G.P., & Timothy, D.J. (2010). Power, regionalism and tourism policy in Bhutan. *Annals of Tourism Research*, 37(4), 969–988.

Nyaupane, G.P., Timothy, D.J., & Poudel, S. (2015). Understanding tourists in religious destinations: A social distance perspective. *Tourism Management*, 48, 343–353.

Olsen, D.H. (2006). Management issues for religious heritage attractions. In D.J. Timothy & D.H. Olsen (Eds.), *Tourism, Religion & Spiritual Journeys* (pp. 104–118). London: Routledge.

Olsen, D.H. (2021). Fan pilgrimage, religion, and spirituality. In D.H. Olsen & D.J. Timothy (Eds.), *Routledge Handbook of Religious and Spiritual Tourism*. London: Routledge.

Olsen, D.H., & Timothy, D.J. (2006). Tourism and religious journeys. In D.J. Timothy & D.H. Olsen (Eds.), *Tourism, Religion and Spiritual Journeys* (pp. 1–21). London: Routledge.

Olsen, D.H., & Timothy, D.J. (Eds.). (2021). *Routledge Handbook of Religious and Spiritual Tourism*. London: Routledge.

Olsen, D.H., & Trono, A. (Eds.). (2018). *Religious Pilgrimage Routes and Trails: Sustainable Development and Management*. Wallingford: CABI.

Price, M.F. (1995). Climate change in mountain regions: A marginal issue? *The Environmentalist*, 15(4), 272–280.

Price, M.F., Moss, L.A.G., & Williams, P.W. (1997). Tourism and amenity migration. In B. Messerli & J.D. Ives (Eds.), *Mountains of the World: A Global Priority* (pp. 249–280). London: Parthenon.

Puciato, D., & Łoś, A. (2009). Wykorzystanie opłat ekologicznych jako instrumentu zarządzania i finansowania pośredniej gospodarki turystycznej w obszarze ochrony środowiska. *Opole: Studia i Monografie Politechniki Opolskiej*, 8, 252.

Qurashi, J. (2017). Commodification of Islamic religious tourism: From spiritual to touristic experience. *International Journal of Religious Tourism and Pilgrimage*, 5(1), 89–104.

Raj, R., & Griffin, K.A. (Eds.). (2015). *Religious Tourism and Pilgrimage Management: An International Perspective*. Wallingford: CABI.

Rinschede, G. (1997). Pilgrimage studies at different levels. In R.H. Stoddard & A. Morinis (Eds.), *Sacred Places, Sacred Spaces: The Geography of Pilgrimages* (pp. 94–112). Baton Rouge: Louisiana State University.

Ron, A.S., & Timothy, D.J. (2019). *Contemporary Christian Travel: Pilgrimage, Practice and Place*. Bristol: Channel View Publications.

Rybina, L., & Lee, T.J. (2021). Traveller motivation and destination loyalty: Visiting sacred places in Central Asia. *Tourism and Hospitality*, 2(1), 1–14.

Sati, V.P. (2013). Tourism practices and approaches for its development in the Uttarakhand Himalaya, India. *Journal of Tourism Challenges and Trends*, 6(1), 97–112.

Sati, V.P. (2014). *Towards Sustainable Livelihoods and Ecosystems in Mountain Regions*. Cham: Springer.

Sati, V.P. (2015). Pilgrimage tourism in mountain regions: Socio—economic and environmental implications in the Garhwal Himalaya. *South Asian Journal of Tourism and Heritage*, 8(2), 164–182.

Sati, V.P. (2018). Carrying capacity analysis and destination development: A case study of Gangotri tourists/pilgrims' circuit in the Himalaya. *Asia Pacific Journal of Tourism Research*, 23(3), 312–322.

Sati, V.P., & Kumar, K. (2004). *Uttaranchal Dilemma of Plenties and Scarcities*. New Delhi: Mittal Publications.

Schwartzberg, J. (1978). *A Historical Atlas of South Asia*. Chicago: University of Chicago Press.

Serenari, C., Leung, Y.F., Attarian, A., & Franck, C. (2012). Understanding environmentally significant behavior among whitewater rafting and trekking guides in the Garhwal Himalaya, India. *Journal of Sustainable Tourism*, 20(5), 757–772.

Shackley, M. (1999). Managing the cultural impacts of religious tourism in the Himalayas, Tibet and Nepal. In M. Robinson & P. Boniface (Eds.), *Tourism and Cultural Conflicts* (pp. 95–112). Wallingford: CABI.

Shackley, M. (2001). *Managing Sacred Sites: Service Provision and Visitor Experience*. London: Continuum.

Sharma, P. (2000). *Tourism as Development: Case Studies From the Himalaya.* Kathmandu: Himal Books.

Shinde, K. (2003). Environmental crisis in God's abode: Managing religious tourism. In C. Fernandes, F. Mcgettigan, & J. Edwards (Eds.), *Religious Tourism and Pilgrimage— Conference Proceedings of the 1st Expert Meeting of ATLAS, Religious Tourism and Pilgrimage Research Group* (pp. 87–102). Leiria, Portugal: Tourism Board of Leiria/ Fatima.

Shinde, K. (2007). Pilgrimage and the environment challenges in a pilgrimage centre. *Current Issues in Tourism*, 10(4), 343–365.

Shinde, K. (2012). Policy, planning, and management for religious tourism in Indian pilgrimage sites. *Journal of Policy Research in Tourism, Leisure and Events*, 4(3), 277–301.

Shinde, K., & Olsen, D.H. (2020). The environmental impacts of religious tourism. In K.A. Shinde & D.H. Olsen (Eds.), *Religious Tourism and the Environment* (pp. 1–22). Wallingford: CABI.

Shrestha, R.K. (2018). Bhutan political crisis and Bhutanese refugees. *International Journal of Management Research*, 5(1), 1–4.

Shuo, Y.S., Ryan, C., & Liu, G. (2009). Taoism, temples and tourists: The case of Mazu pilgrimage tourism. *Tourism Management*, 30, 581–588.

Singh, R.P.B. (1997). Sacred space and pilgrimage in Hindu society: The case of Varanasi. In R.H. Stoddard & A. Morinis (Eds.), *Sacred Places, Sacred Spaces: The Geography of Pilgrimages* (pp. 191–208). Baton Rouge: Louisiana State University.

Singh, R.P.B. (2006). Pilgrimage in Hinduism: Historical context and modern perspectives. In D.J. Timothy & D.H. Olsen (Eds.), *Tourism, Religion and Spiritual Journeys* (pp. 220–236). London: Routledge.

Singh, R.P.B. (2013). *Hindu Tradition of Pilgrimage: Sacred Space and System.* New Delhi: Dev Publishers.

Singh, R.P.B., Rana, P.S., & Olsen, D.H. (2022). Environment as a sacred space: Religious and spiritual tourism and environmental concerns in Hinduism. In D.H. Olsen & D.J. Timothy (Eds.), *Routledge Handbook of Religious and Spiritual Tourism* (pp. 135–152). London: Routledge.

Singh, S. (2002). Managing the impacts of tourist and pilgrim mobility in the Indian Himalayas. *Revue De Geographie Alpine*, 90(1), 25–35.

Singh, S. (2004). Religion, heritage and travel: Case references from the Indian Himalayas. *Current Issues in Tourism*, 7(1), 44–65.

Singh, S. (2006). Tourism in the sacred Indian Himalayas: An incipient theology of tourism? *Asia Pacific Journal of Tourism Research*, 11(4), 375–389.

Singh, T.V. (1992). Development of tourism in the Himalayan environment: The problem of sustainability. *Industry and Environment*, 15(3/4), 22–27.

Sofield, T.H.B. (2001). Sustainability and pilgrimage tourism in the Kathmandu Valley of Nepal. In V.L. Smith & M. Brent (Eds.), *Hosts and Guests Revisited: Tourism Issues of the 21st Century* (pp. 257–271). New York: Cognizant.

Sridhar, S., Gautret, P., & Brouqui, P. (2015). A comprehensive review of the Kumbh Mela: Identifying risks for spread of infectious diseases. *Clinical Microbiology and Infection*, 21(2), 128–133.

Tabasum, R., Saha, A., Habib, U., Pal, J., & Singh, R. (2017). Political turmoil impacting pilgrimage tourism in Jammu & Kashmir: A geographical study. *International Journal of Economic Research*, 14(20), 607–616.

Tan, W. (2002). General study on Chinese lay Buddhist. *Social Science Front*, 5, 61–67.

Timothy, D.J. (2013). Religious views of the environment: Sanctification of nature and implications for tourism. In A. Holden & D. Fennell (Eds.), *The Routledge Handbook of Tourism and the Environment* (pp. 31–42). London: Routledge.

Timothy, D.J. (2021). Tourism and the multi-faith heritage of the Middle East and North Africa: A resource perspective. In C.M. Hall & S. Seyfi (Eds.), *Cultural and Heritage Tourism in the Middle East and North Africa: Complexities, Management and Practices* (pp. 34–53). London: Routledge.

Timothy, D.J., & Nyaupane, G.P. (Eds.). (2009). *Cultural Heritage and Tourism in the Developing World: A Regional Perspective*. London: Routledge.

Timothy, D.J., & Olsen, D.H. (Eds.). (2006). *Tourism, Religion and Spiritual Journeys*. London: Routledge.

Tree, I. (2014). A house for the living goddess: On the dual identity of the Kumari Chen in Kathmandu. *South Asia: Journal of South Asian Studies*, 37(1), 156–178.

Triantafillidou, A., Kortios, C., Chatzipanagiotou, K., & Vassilikopoulou, A. (2010). Pilgrimages: The "promised land" for travel agents? *International Journal of Contemporary Hospitality Management*, 22(3), 382–398.

UNWTO (2011). *Religious Tourism in Asia and the Pacific*. Madrid: UNWTO.

Verma, M., & Sarangi, P. (2019). Modeling attributes of religious tourism: A study of Kumbh Mela, India. *Journal of Convention & Event Tourism*, 20(4), 296–324.

Vogel, B., & Field, J. (2020). (Re)constructing borders through the governance of tourism and trade in Ladakh, India. *Political Geography*, 82, 102226.

Wang, W., Chen, J.S., & Huang, K. (2016). Religious tourist motivation in Buddhist Mountain: The case from China. *Asia Pacific Journal of Tourism Research*, 21(1), 57–72.

Wei, C.S. (2010). *Research on Buddhism from the Modern Perspective*. Beijing: The Orient Press.

Wright, K. (2008). Religious tourism: A new era, a dynamic industry. *Tourism-Review.com*. Retrieved December 3, 2020, from www.tourism-review.com/travel-tourismmagazine-religious-tourism-a-new-era-a-dynamicindustry-article695.

Yang, X.R. (2004). *Differentiation and Analyzing the Evolution of Buddhism Perspectives on Female Gender*. Unpublished doctoral dissertation, Sichuan University, Sichuan, China.

Yang, X.R. (2020). *Privileging Indigenous Voices: Narratives of Travel Experiences of Tibetans*. Unpublished doctoral dissertation, University of Waterloo, Canada.

Zafren, K. (2007). Pilgrimages in the high Himalayas. In A. Wilder-Smith, E. Schwartz, & M. Shaw (Eds.), *Travel Medicine: Tales behind the Science* (pp. 279–286). Amsterdam: Elsevier.

Zurick, D., & Pacheco, J. (2006). *Illustrated Atlas of the Himalaya*. Lexington: University Press of Kentucky.

# 12 Expanding the Boundaries of Local Cuisine

## The Context of Himalayan Community-Based Homestays

*Kishor Chitrakar, Neil Carr and Julia N. Albrecht*

## Introduction

This chapter explores an under-examined area within community tourism, namely the relationship between community-based homestays and local cuisine. Food is an indispensable part of human culture, the consumption of which is continually evolving (Cleave, 2013; Hjalager & Richards, 2002; Timothy, 2016). Many factors are responsible for creating the relevant dynamics, and tourism is prominent therein (Frisvoll et al., 2016). When people travel, they take their food choices, preferences, habits, and tastes with them, and they often bring back new culinary experiences and ingredients. The Himalayan region has been increasingly attracting the attention of adventure and cultural tourists, which exposes local food cultures to external audiences.

This chapter examines the influence of community-based homestay tourism (CBHT) on the local cuisine in the Himalaya with special reference to the Panauti Community Homestay (PCH) in Nepal. Community-based homestay tourism is an emerging type of tourism in the region that allows close interaction between hosts and guests. This interaction increases during food-related activities (Agyeiwaah, 2013). As a result, the host families' cuisine has become prone to being influenced by hosting guests. With only a few exceptions (e.g., Dutta, 2018; Giampiccoli & Kalis, 2012), there is relatively little research about the food aspects of community-based tourism. In particular, tourism's influence on local cuisine has not been examined well in the tourism literature. Thus, how community homestays bring about changes in the local food culture is not fully understood at this time. The first part of this chapter analyses the impact of homestays on regional cuisine in the Himalaya. Later, to provide a deeper understanding, a community-based homestay in Nepal is presented to illustrate how culinary changes occur because of homestay tourism. The evidence presented here may be relevant to other community-based tourism initiatives in the Himalaya and other mountain regions in the Global South.

## Cuisine Dynamics and Tourism

The term 'cuisine' is generally understood to refer to high-end gourmet food. In a broader sense, however, the term is used interchangeably to refer to food

DOI: 10.4324/9781003030126-15

culture, foodscapes, and foodways (Timothy, 2016). Timothy and Ron (2013) suggest that, from a heritage perspective, cuisine consists of tangible aspects such as ingredients and cooking utensils, as well as intangible elements such as tastes, smells, recipes, and traditions related to food, its preparation, and its consumption. Belasco (2008) posits that there are five elements of cuisine in any community: basic foods such as rice, a distinct manner of preparing food, unique ways of seasoning dishes, distinctive eating manners, and a specific food chain. In this chapter, the terms cuisine and local food culture are used to refer to food varieties, their tastes, and people's eating habits.

There is a growing body of literature on the relationships between tourism and food, which may indicate the significant influence of tourism on the local food culture. Tourism is a powerful agent that can bring about changes in the local cuisine (Frisvoll et al., 2016). Timothy and Ron (2013) demonstrate the multifaceted relationships between cuisine and tourism. Local food is often modified to meet the tastes, hygiene, or other requirements of tourists. Cohen and Avieli (2004) maintain that this transformation does not necessarily decrease the quality of local food but rather provides opportunities for creativity and innovation. Tourism can play a constructive role in local food culture in tourist destinations. While many residents imitate tourists' eating patterns, this demonstration effect may be a positive force for local food (Bélisle, 1983). On the basis of evidence from the Caribbean region, Bélisle suggests that when tourists desire to experience local food, destination residents may also start to consume more food products from their locality. This can benefit local food production and, thereby, minimise food importation, as well as develop a sense of pride in a community's culinary heritage. Tourism thus may play a vital role in expanding knowledge about the culture of a community's foodways (Kuhn et al., 2018). Dougherty et al. (2013) suggest that local food-based tourism increases social and human capital, which may outweigh environmental costs, such as carbon emissions created by tourism. Tourism's influence on local food culture is more significant in community-based tourism where experiencing regional food is a major attraction for tourists (Chitrakar et al., 2020).

## Himalayan Cuisine

The Hindu Kush Himalayan region covers the territories of eight Asian countries in Central and South Asia, extending from Afghanistan in the west to Myanmar in the east (Rasul et al., 2018). The region is comprised of the High Himalaya, midland valleys, and vast river valleys. Because of its distinctive physical, climatic, and cultural environment, the Himalayan region has a unique culinary heritage. The High Himalaya have limited plant and animal products compared to the fertile river basins. Despite, or perhaps even because of, this, the region has inimitable food cultures that have evolved on the basis of the abundance or scarcity of resources, as well as rich indigenous knowledge. The Himalayan people's knowledge of local plants and animals has evolved over thousands of years. Based on agroclimatic conditions, geographical isolation, and socio-economic and cultural factors, different communities and ethnic groups have developed their own unique food cultures (Tamang, 2016; Timothy, 2021). Examples include the use of fermented soya products, called *kinima* in the Eastern Himalaya (Tamang, 2001), and dried foods, known as *badi* in Central Himalayan villages in Garhwal, India (Gusain, 2017). Gusain (2017) maintains that traditional Himalayan gastronomic knowledge, such as fermenting and drying, to preserve food is crucial for food security and reducing the consumption of unhealthy fast and processed food products.

The ethnicities in the Himalaya have distinctive food knowledge and gastronomic practices. For instance, pastoral communities share knowledge of more than 32 local ingredients and recipes for fermented and non-fermented foods (Kala, 2021). Likewise, much of Himalayan cuisine is a fusion between Tibeto-Mongolian foodways in the north and Indo-Aryan dishes in the south, with numerous variations among ethnic groups per their sociocultural characteristics (Tamang, 2009). The region boasts more than 100 types of fermented foods, 300 types of non-fermented dishes, and more than 50 varieties of ethnic alcoholic drinks (Tamang, 2009). Because of the need to preserve food in a harsh climate, fermented foods and drinks occupy a significant place in traditional Himalayan cuisine (Joshi et al., 2015). Examples of fermented and dried staples include *gundruk* and *sinki* (dried vegetables, which can be used for soups, curries, or pickles), *dachi* (a milk-based curry), and *chhurpi* (a hardened cheese). *Dheroh* (maize or millet porridge) and buttermilk are examples of popular non-fermented foods. *Chyyang* (mildly sweet, acidic beer) and *raksi* (distilled hard liquor) are the best-known examples of regional fermented alcoholic drinks (Tamang, 2001). Rice and maize are used in daily meals in the Eastern Himalaya, while wheat and barley are more common in the Western Himalaya (Tamang, 2009). Himalayan food culture also includes several types of millets and wild plants with rich nutritional value. However, there has been a significant decline in native crops in the region, so there is an urgent need to create awareness about, and a market for, them. There is also a need to address the issue of declining traditional food knowledge through proper documentation and training (Kala, 2021).

Despite a strong history, local food culture in the Himalaya has changed considerably in recent years. External socio-economic factors, such as road construction and improved access, education, and media and communication development are largely responsible for this (Rasul et al., 2018). Grocke (2016) believes that the development of roads, in particular, has brought notable changes to the middle and lower mountains, where local grains are being replaced by fast food and industrial processed products. Parts of the Western Himalayan regions, such as Kumaon and Garhwal, India, are witnessing changes to their traditional food cultures due to socio-economic changes, including urbanisation (Pant, 2016). Many unique local food cultures that are based on traditional subsistence farming are in peril, which is contributing to a significant decrease in the local crop diversity. There also appears to be a decline in the consumption of traditional wild plants (Boesi, 2014), which is threatening the sustainable use of native vegetation—one of the main manifestations of indigenous knowledge.

In contrast, Kikon (2015) argues that increased accessibility and globalisation have enabled indigenous cuisines to find a place on the nation's food table in India. One example is *akhuni*, a fermented soybean dish popular in the Eastern Himalaya. *Akhuni* has made a name for itself in the Indian capital New Delhi after a long struggle and negotiations with mainstream Indian palates, which tend to prefer masalas (Kikon, 2015). Peaty (2009) believes that tourism can be an avenue for preserving Himalayan foodways. The region is home to many ethnic food production enterprises that could be potential tourist attractions. These include *chhurpi* production in Sikkim, *mohi* production in Nepal, and *chhyang* production in Ladakh (Tamang, 2009). The rich culinary heritage of the Himalaya offers significant potential for tourism development in the region.

## Tourism and Change in Himalayan Food Culture

Authentic experiences are an important theme in discussions of tourism and cultural change. Although the concept of authenticity is hotly debated, it is intrinsic to tourism. Many people travel in search of authentic experiences to escape from their complicated lives in a modern world surrounded by much 'inauthenticity' (Martin, 2014). Some studies on food culture support Cohen's (1979) seminal work, which argues that authenticity is a highly subjective phenomenon. Yang and Wall (2009) argue that both domestic and international, organised, and independent tourists have their own perceptions of what is authentic. At a homestay in western Nepal, domestic tourists consider a familiar rice meal with chicken to be authentic local fare, whereas locals consider crab and snail to be more 'local' (Nepal Rastra Bank, 2015). Strychacz (2010, p. 157) posits that, although Western adventure tourists may also seek authentic food experiences, they tend to maintain a "flexible positional superiority" regarding their food preferences over traditional foods in the destination. Among adventure tourists in the Sagarmatha National Park, Nepal, Strychacz (2010) found that although the tourists are quite accepting of certain sociocultural differences, their preference for Western-style food is deeply embedded in their travel behaviours.

Tourism in the Himalayan region has been influential in effecting changes in local food cultures. In the past, montane adventure tourists had to deal with food scarcity and lack of variety. However, nowadays tourists in many parts of the Himalaya find extended menus that include foods that are more aligned with Western tastes (Nyaupane et al., 2014; Strychacz, 2010). This includes many examples of fast food in the Mount Everest region (Ramsay, 2002). Traditional food items are increasingly replaced by imported industrial products. For instance, in many locales, *chhyang* and *raksi* have been replaced by mass-produced and imported beer and other alcoholic products (Fisher, 1990; Nyaupane et al., 2014). Likewise, with the increase in income from tourism and exposure to imported food, many local people prefer to consume expensive imported foods, such as rice, which are replacing native crops, such as buckwheat (Fisher, 1990). Nepal (2016) points out the increasing number of international chains, such as Starbucks and Irish pubs, in the tourist areas of the Sagarmatha National Park. These were originally meant to offer food options for international tourists but have significantly influenced the food habits of the local residents as well. Adhikari and Kandel (2017) reported similar findings, noting that as tourism has grown in the Nepali Himalaya, authentic local cuisines have begun to disappear in some areas. Swenson (2001) notes similar conditions in Tibet, where local Tibetan gastronomy has largely been supplanted by mainstream Han Chinese cuisine. River rafting and camping on the Ganges River in Uttarakhand, India, appear to be shifting agricultural practices from subsistence traditions to commercial farming, causing a loss of traditional horticultural and animal husbandry methods (Farooque et al., 2008a). Farooque et al. (2008b) note how these practices also have resulted in the increased consumption of pre-packaged and junk food among tourists and local inhabitants.

Despite these changes and the influential power of tourism, many studies reveal that local communities can manage tourism's impacts rather than silently and passively accepting them by becoming active players in resident-tourist interactions (Zhang et al., 2017). Oakes (1999) argues that despite a dominant national culture and pressure from globalisation, local communities can in fact find dynamic ways to maintain their identities, and tourism can be an avenue for this. To address this issue in the context of food, Strychacz (2010) suggests that local and imported cuisines can share the same menu spaces, encouraging tourists to try local gastronomy. Interestingly, hybrid foods are developing, such as the yak burger, which uses local ingredients but may appeal to the tastes of international tourists. Importantly, these studies recognise local communities not only as silent receivers of impacts but also as being capable of deliberately choosing to adapt tourism to their local traditions.

## Community-Based Homestay Tourism and Local Cuisine

Though food consumption is the core of any travel experience, there is a dearth of studies that examine the influence of food culture on tourist experiences and, conversely, how tourism influences local food cultures (Long, 2004). Most literature focuses on established 'foodie' destinations such as Italy and France, and the findings from the mainstream research may not be entirely applicable to destinations with less-familiar food cultures (Peštek & Činjarević, 2014). Furthermore, there is only a small number of academic works on the topic of community homestays and local food culture (e.g., Giampiccoli & Kalis, 2012; Mnguni & Giampiccoli, 2016).

Traditional destination food experiences play a significant role in homestay-based tourism. Giampiccoli and Kalis' (2012) research in rural South Africa investigated how local food products can positively contribute to community-based tourism development. Indigenous food there is recognised as a cultural resource that can be tapped for tourism development. Mnguni and Giampiccoli (2016) propose that community-based food tourism, in which local indigenous food is a central attraction, can become the foundation of rural tourism development.

Food experiences in community-based homestay tourism can contribute to and detract from visitor satisfaction. In their study of homestay operations in

Malaysia, Aziz et al. (2014) found that local food contributed significantly to tourists' satisfaction. Tasting traditional food and observing its preparation wielded a significant appeal to tourists. A similar study carried out on homestay operations in Malaysia demonstrates that local cuisine, customary farming activities such as monkey coconut harvesting, and traditional cooking demonstrations were some of the most highly appreciated activities among tourists (Jamaludin et al., 2012). At a homestay in Nepal, local cuisine is one of the top three determinants of guests' satisfaction with their homestay experience (Biswakarma, 2015). Jamaludin et al. (2012) suggest that destinations could create unique food items in each village so that visitors can associate food and place in their memories.

Although there are several positive links between local cuisine and visitor satisfaction, some studies reveal that homestay food can negatively impact guests' experiences. Unhygienic conditions (Aziz et al., 2014; Mnguni & Giampiccoli, 2016), a lack of basic infrastructure such as electricity and clean water (Kunjuraman & Hussin, 2017), poor food quality, and lack of information about traditional food (Ogucha et al., 2015) all can detract from visitors' satisfaction with community homestay experiences.

Although food experiences can be part of the core appeal of homestay tourism, as noted earlier, there is a lack of knowledge about how tourism influences local food cultures. By extension, few studies examine the relationships between homestay tourism and food. Likewise, although there have been many studies on host-guest interaction, little is known about the impact of this on local food culture. This chapter aims to help fill these gaps and explain how tourism can influence the local cuisine of the host community in the context of Himalayan homestay tourism.

### Tourism and Changes in Local Cuisine: Community Homestays in the Himalaya

This section explores how tourism can bring about changes in local cuisine, in the context of community-based homestay in the Himalaya. Information on community-run homestays in India, China, Bhutan, and Nepal form the basis of the analysis.

Homestays usually take the form of rooms in people's homes that are let out to tourists for a nightly fee, although in some localities, homestays are built nearby but separately for privacy. In either case, homestays are characterised in part by the close interactions between tourists and the community members who host them. By utilising homestay lodging, tourists become more immersed in local cultures as they share the lives of their hosts and participate in discussions, family activities, farm chores, and meals. The close interactions between hosts and guests in homestay tourism have the potential to impact the destination culture, including food. Dong (2020) notes two different types of cultural changes relevant in the context of homestays: acculturation and cultural drift. The former entails the subsuming of cultural groups and their practices into dominant cultures and the simultaneous loss of the distinct cultural identity of the subsumed group. The latter regards the phenotypic behaviour of hosts and guests during their interaction. The host community is more likely to receive long-term impacts than the visitors. However, such changes may prevail only in the presence of guests, and once they depart, the host communities may return to normal life again. Dong (2020) suggests that tourism is not the only factor driving food changes in homestays; these changes are also driven by globalisation, such as media, technological development, and food availability.

Two kinds of homestays operate in the Himalayan region: private and community-based. Private homestays have more flexibility in food preparation and serving, and concern is given to the needs and preferences of guests. Some homestays in China that work with the model of homestays developed in Australia and New Zealand where international students stay with host families for a specific term require both hosts and guests to understand and adapt to each other's food cultures (Lee et al., 2018). Tourists and host families can solve cultural food differences by understanding one another, which can make the homestay experience an opportunity for learning.

Experiencing local life and interacting closely with hosts and guests are salient features of CBHT and part of the reason tourists choose this vacation option (Jamal et al., 2011; Kontogeorgopoulos et al., 2015). Local food and the foodways associated with it are a core part of the overall experience of homestay tourism and is a vital element of the immersive cultural experience. Guests typically spend a significant amount of time in food-related activities, including harvesting, cooking, eating, and visiting farms and markets. In CBHT, host families are instructed to use local food cultures insofar as it is practical (MoTCA, 2010). Government homestay regulations in India, Nepal, and Bhutan focus on requirements for safe and enjoyable food experiences for visitors. This usually requires some modifications to local food traditions. For instance, some basic continental/Western meals are recommended, especially for breakfast. Homestay training manuals developed by non-government organisations (NGOs) provide some flexibility to host families regarding the timing of meals, selection of ingredients, and synchronisation of food tastes to make the comestibles enjoyable for visitors. As part of the culinary cultural experience, local eating manners, traditional kitchens, dining room ambience, and customary utensils are encouraged as much as possible.

Current research on community homestays in the Himalaya suggests that, although tourism requires the hosts to modify their native food cultures, tourism activity can boost the preservation, promotion, and rejuvenation of local ethnic cuisine. Locally produced ethnic foods are prioritised in the region's community homestays (Peaty, 2009; Walter, 2020). Kulshrestha and Kulshrestha (2019) found that homestay tourism in Uttarakhand, India, supports local food production and has helped make local cuisine popular among tourists. Similar findings were reported by Lama (2013) in homestays in western Nepal, where imported packaged foods were discouraged. Similarly, Anand et al. (2012) note that homestays in Korzok, India, promote traditional Ladakhi cuisine, such as warm noodle soup (*thukpa*), butter tea, and rice meals with local flavours. Furthermore, Dahal et al. (2020) found that in a community-based homestay in southern Nepal, tourism has aroused interest among local youth in learning traditional culinary skills.

However, some studies highlight the need for maintaining a balance between the essence of native cuisines and the desires of tourists. Walter (2020) argues that visitors in community homestays are not looking for entirely authentic local food culture, but rather are content with the wider menu consisting of popular food from other parts of the country. Dutta (2018) advocates a balanced approach that addresses the needs and desires of tourists on the one hand, and the continuation of indigenous cuisines on the other. Similarly, Dong (2020) observes that homestay host families in Briddim, Nepal, imitate the food cultures of tourists such as using cake in birthday celebrations. However, they continue to use traditional foods and beverages alongside these adaptations.

Homestay tourism's impact on native food culture is more nuanced in destinations where domestic tourists are the main market. Although authentic gastronomy is the major motivation factor for domestic tourists, they also look for familiar food experiences. Taking the case of India's Eastern Himalayan region, Dutta (2018) suggests that for domestic tourists, the addition of everyday breakfast items for tourists from the mainstream areas of India alongside indigenous food from the locality is warranted. A similar case was found at the Dalla Homestay in Nepal. There, domestic tourists demanded familiar food, such as rice with chicken curry, which is favoured by mainstream Nepalese people, despite the fact that locality has other authentic local cuisines (Nepal Rastra Bank, 2015). Walter (2020) found that visitors in another homestay in Nepal were happy with the gustatory options that combine local ethnic food with imported food items and drinks from other parts of the country.

As previously noted, studies on CBHTs in the Himalaya provide an overview of tourism's impact on local culinary cultures as regards food services for tourists. However, there is still a dearth of studies dedicated exclusively to tourism's impact on the host community's food culture. To gain insight into this phenomenon, the case of a CBHT in central Nepal is presented in the next section. To provide a greater understanding of the influence of homestay tourism on local cuisine, an in-depth analysis of the Panauti Community Homestay (PCH) is provided. In-depth interviews with 49 members from 17 households and other relevant stakeholders form the empirical basis of this case.

## Panauti Community Homestay (PCH) and Changes in Local Food Culture

Panauti is a small town located just outside the Kathmandu Valley. Panauti is a proposed UNESCO World Heritage Site with diverse tangible and intangible cultural heritage. The Newar ethnic group comprises the majority of the town's population, who have a dominant influence on the cultural life, including cuisine. Tourism in Panauti is still relatively small in scale, compared to the nearby established tourist destinations of Bhaktapur and Dhulikhel. Panauti is one of the few urban areas of Nepal where community-based homestays are more successful than commercial tourist accommodations. Established in 2014, the PCH is the most popular accommodation in Panauti among international tourists. In 2018, the PCH was the €225,000 winner of the Top Ten Sustainable Tourism Start-Ups Worldwide under Booking.com's Booster program. Among guests, food-related activities are the most enjoyed experiences at the PCH. Apart from enjoying local meals, cooking classes and visiting farms are also part of the homestay package.

The local food culture in Panauti has changed drastically in the last couple of decades. First, Panauti is going through a nutritional transition in which dietary patterns are shifting from agricultural staple foods towards industrially processed foods. Eating rice meals (*dalbhat*) twice a day has become an established habit for the locals. This has replaced several native food items such as *dhindo* (corn porridge) and *chiura* (beaten and flattened roasted rice). Second, several historical food taboos have been abandoned, and a culture of eating out, once considered unsuitable for 'respectable' people, is now widely accepted. Third, several influences from cuisines other than those of Panauti are entering the foodscape of the town. The dynamics of local cuisines are mirrored by changes in feast patterns. As Liechty (2001) points out, the traditional system of *bhoj*[1] has been largely replaced by a modern buffet or catering system. As a result, traditional local dishes are rarely available in this new feasting trend. Indian food dominates these parties, followed by Chinese cuisine to some extent. Economic, social, technological, infrastructure, and media influences can be linked to these changes, while tourism is an emerging factor (Chitrakar et al., 2021).

### The PCH and Cuisine Change

This section focuses on the role played by the community homestay operation on the dynamics of local cuisine of the host community in Panauti, Nepal. Hosting an international guest, who may be more used to staying in tourist-standard accommodations, requires great care. Food is a core of the homestay experience, which also requires host families to modify their daily food in order to create enjoyable culinary experiences for the guests. This modification, which was adopted for tourists, has had far-reaching consequences for host families' food culture in several ways.

First, the PCH has added new varieties of food to the host community's local food culture. Though Panauti has a rich variety of food and drinks, the richest culinary experiences can normally only be enjoyed during the festivals and celebrations related to lifecycle events, such as weddings. Homestay tourism, however, has enriched the food experience throughout the year for guests, as well as the hosts by bringing those elements of culinary heritage into daily food preparation. According to the homestay rules, the host families are supposed to provide different food items for each meal. So, if guests are given a rice meal during lunch, the hosts usually make other dishes for dinner. For one-night guests, a typical meal of rice, roti, and curry is served for lunch and dinner, whereas breakfast is normally a mixed platter of local ethnic food items, such as *wocha* (a Newari pizza-like pancake made from lentils), Indian dishes like *puri-tarkari* (fried chapattis and spicy curry), and Western food preferences like fruits, yogurt, and pancakes.

The longer the length of stay, the easier it is for host families to offer traditional meals that differ from everyday rice meals. Alongside this, new creations are added to the local food repertoire. Each homestay family has self-invented recipes by changing ingredients, cooking methods, and presentation style. Creolisation is another dimension that is enriching the local food culture. One of the most popular creolised items is chow-chow pizza or noodle pizza, which is a fusion of local Newari, Thai, and Italian cuisines. A pizza lookalike cooked in a Newari way with a base of Thai-style instant noodle, chow-chow pizza has become the second iconic food item in the study area after traditional *yomari*. This was initially an experimental dish of a homestay operator's husband, who was a chef in a tourist restaurant in Kathmandu. Now, the item has become a favourite among all homestay families and is identified as part of their emerging local food culture.

The second way in which host families' foodscapes have changed is the revitalisation of local food heritage: an apparent positive outcome of tourism in the PCH (Chitrakar et al., 2020). This has included arranging the traditional feast pattern *bhoj* for guests who stay longer. Similarly, homestay operators have provided festive and seasonal culinary specialities, particularly those from Newari cuisine, for the guests. Old gustatory practices such as using traditional utensils and providing traditional dishes have also been revived through tourism. The host families are proud to observe and be a part of this revitalisation. Young family members (in their early twenties) are particularly enthusiastic about the cultural resurrection capacity of tourism and keen to be involved in it. Furthermore, homestay operators in Panauti aspire to create a more authentic foodscape in the PCH in the future. The establishment of ethnic restaurants, such as the Sunrise Cafe, which offer Tamang[2] food and drinks, is also one of the revitalising results of homestay tourism. This revitalising role is distinctive when set against a backdrop of rapidly declining traditional gastronomy elsewhere.

Thirdly, the PCH has helped preserve and promote local gastronomic knowledge through vertical and horizontal transmission. Enthusiasm has been generated among young homestay operators regarding traditional gastronomic heritages, such as *bhoj* and festive cuisines. In addition, the PCH has helped to expand local gastronomic knowledge beyond ethnic boundaries. The homestay business has prompted the sharing of local gastronomic skills among community members (Chitrakar et al., 2021), which has enhanced the richness and sustainability of local cuisine. Homestay community members share their ethnic cuisine knowledge with members of other ethnic groups, which has resulted in new food items being added to each household. For instance, the Brahmin community has learned how to make Newari ethnic food specialities like the steamed, sweet winter bread *yomari* and *wocha*, whereas members of the Newari community have learned how to make *selroti*, a sweet frybread, from Brahmin families. These patterns affirm the finding of Ishak et al. (2013), who assert that the interaction between different local cultures may inspire each of them to make changes in their own food traditions. The majority of homestay operators expressed a desire to learn about ethnic food items from other ethnicities to provide a wider range of food choices for guests as well as to satisfy their own culinary curiosity.

The fourth change relates to healthier and more home-based cooking. The experience of running homestays has inspired home cooking and healthy food habits in the PCH (Chitrakar et al., 2020). Catering to international guests has boosted host families' confidence in their cooking skills, which has created more interest in home cooking. Some of the homestay operators described how they have learned to prepare food items that are served in restaurants. Besides, cooking for international guests has made them aware of the superiority of homemade over commercial food, in terms of both quality and hygiene. Cooking at home is favoured because it is more cost-effective and accessible to all members of the family. Consequently, some family members observed that there had been a decrease in eating out in restaurants or buying takeaway food. Alongside the encouragement to participate more in home cooking, tourism has also inspired healthier food habits. One of the most noteworthy impacts of tourism was the acceptance of a more balanced diet. Typically, a Nepalese rice meal would comprise a large amount of rice with only a few side dishes of lentil soup and curry. However, homestay families have started to use larger portions of vegetables, meat, and fruits, with a relatively small amount of rice. This change was not made only for the guests but also occurs when the families dine by themselves. Also as a consequence of host-guest interaction, the homestay families have increased the use of safe drinking water in their daily lives.

Finally, using local cuisine in the homestay has helped strengthen the destination identity. The local food experience, particularly Newari cuisine, has been associated with the Panauti area as a whole. In particular, homestay families have utilised *yomari* as an iconic local food. Though the sweet dish is usually made during a festival in December, it has become a signature item in homestay families as a unique part of the identity of Panauti. The delicacy is believed to have originated in Panauti, and the local people are becoming increasingly proud of this particular culinary inheritance. As a result, the management committee of the homestays has encouraged all member families to serve *yomari* more often to their guests. Following the steps of the PCH, a large effigy of the uniquely shaped *yomari* has been installed at the entrance of Panauti municipality, and local restaurants now feature *yomari* as a centrepiece of their menus.

## Conclusion

This chapter emphasises the importance of traditional foodways as an important part of the Himalayan cultural heritage. In particular, it examines the role of community-based homestay tourism in the evolution of local cuisine and culinary traditions. In the context of the rapidly changing cuisine in this region, tourism has emerged as an influential factor in determining the nature of local foods. Tourism is a double-edged sword for local food culture in the Himalaya. It draws on, and can reinforce, authenticity in local cuisines, while at the same time, it demands a familiarity with tourists' tastes and preferences. Frequently, in areas of high tourist visitation, local ethnic cuisines are dominated by mainstream national and international foods. However, community-based homestay tourism in the region has shown a significant concern for maintaining the localness of the food offered to tourists. Experiencing authentic local flavours is one of the core attractions of community homestays, making it necessary to raise the profile of local food traditions. Although host communities tend to accept some familiar food cultures from tourists and other external influences, there is still space to maintain local cultural identities through culinary means. Community homestays have provided a platform for local ethnic cuisines that was lacking in commercial tourist food and beverage services in the region.

Alongside strengthening ethnic cuisines, community homestay tourism in some parts of the Himalaya has helped enrich local food cultures, in terms of both its quantity and quality. It has broadened local foodscapes by embracing influences from both domestic and international kitchens. In addition, new food items have been created through the process of creolisation. Community homestays have also emphasised the use of local indigenous plants and traditional food production activities, which had been gradually replaced by imported industrial products. Homestay projects have also encouraged the improvement of food hygiene and safety.

Community-based homestay projects show a greater capacity to deal with impacts from tourism demand, to adapt and contextualise according to the local environment, compared to other mainstream tourist accommodation or private homestays. Because eating local cuisine is an essential part of the community-based homestay immersive experience, emphasis should be placed on the maximum use of local cuisines. Thus, local food specialities become popular among people outside the locality. In some cases, cultural pride and revitalisation of culinary heritages are boosted by homestay projects. Although the creolisation process accepts external influences, it favours the use of local ingredients and cooking methods, highlighting how the shifting nature of local cuisine is influenced externally but controlled internally. Overall, tourism in community homestays enables local cuisines to adapt to tourist demands and external influences while ensuring the continuity of local gastronomic legacies. The new way of celebrating a birthday in one of these community homestay mirrors such a phenomenon where traditional food culture continues while, at the same time, influences from tourist food cultures are accepted.

The case of the PCH provides insight into the discussion of tourism's impact on local food culture. This example shows that community homestays have broadened the spectrum of local cuisines in Panauti, Nepal. Tourism has not just played a destructive or constructive role with regard to local food culture. Rather, it has been an integral component of the dynamism of local foodscapes. On the one hand, PCH has strengthened customary food heritage, while on the other hand it has added new dimensions to local cuisines. Rather than keeping local foodways frozen in time as historic artefacts, the homestay operation has encouraged operators to accept influences from different cuisines, which enrich the local culinary heritage and makes it more vibrant. In line with Robertson (2012) and Timothy

(2016), the case of Panauti shows that local food culture can react to globalisation as a synthesis of globalisation and localisation, in the form of 'glocalisation'. This process of glocalisation can lead to creolisation (Mak et al., 2012) and the diversification of local food (Richards, 2012).

The influence of hosting guests in the homestay is not limited only to guests' meals. Rather, it also exerts long-lasting impacts on different aspects of the host community's food culture in daily life, such as food preferences and eating habits. The community homestay project places the host community in a powerful position while negotiating with guests in terms of food service. Rather than accepting influences directly from tourists' cuisines, which are predominantly European, host families have made local cuisine vibrant by accepting influences from different national and international gastronomies, as well as revitalising some traditional gastronomic practices. Further, the PCH case shows that by using local food specialties, community homestays can play a role in building a local identity, cultural solidarity, and continuity of local food knowledge.

This chapter examines the influence of community-based homestay tourism on the local food cultures of the Himalaya. Local food culture in the region is evolving and tourism has emerged as an influential factor in these changes. The influence of community homestays in terms of local cuisine is not limited to food services for tourists. It also encompasses the everyday food culture of host communities. We have demonstrated how community homestays can expand the boundaries of local foodscapes by preserving and revitalising indigenous foodways while at the same time accepting new influences from other food cultures. The major challenge is to create a balance between authentic local food and guest food preferences. Alongside formal training in food and beverage services, local food knowledge has to be considered. For example, community tourism initiatives like homestays can play a pivotal role in documenting culinary knowledge in written form and providing workshops to ensure its survival and intergenerational transfer. In addition, homestay operators must maintain their localness or uniqueness in their cuisines. A bottom-up approach to homestay management enables host communities to prioritise traditional food specialities in tourist food services. To conclude, community-based homestay tourism can play an important role in connecting the past with the present and future of traditional foods, which can help extend the boundaries and vitality of local cuisines.

## Notes

1 Traditional feast pattern in which guests sits cross-legged in rows and food is served by the hosts
2 An ethnic group belonging to the hilly areas of central Nepal.

## References

Adhikari, L., & Kandel, P. (2017). *Biodiversity and Sustainable Tourism.* ICIMOD. Retrieved from www.icimod.org/article/biodiversity-and-sustainable-tourism/

Agyeiwaah, E. (2013). International tourists' motivations for choosing homestay in the Kumasi Metropolis of Ghana. *Anatolia*, 24(3), 405–409.

Anand, A., Chandan, P., & Singh, R. (2012). Homestays at Korzok: Supplementing rural livelihoods and supporting green tourism in the Indian Himalayas. *Mountain Research and Development*, 32(2), 126–136.

Aziz, N.I.A., Hassan, F., & Jaafar, M. (2014). Exploring tourist experiences in Kampung Beng Homestay Programme. *Asia-Pacific Journal of Innovation in Hospitality and Tourism*, 3(1), 1–20.

Belasco, W.J. (2008). *Food: The Key Concepts*. Oxford: Berg.

Bélisle, F.J. (1983). Tourism and food production in the Caribbean. *Annals of Tourism Research*, 10(4), 497–513.

Biswakarma, G. (2015). On the dimensionality of measuring tourist satisfaction towards homestay. *International Journal of Hospitality and Tourism Systems*, 8(2), 51–63.

Boesi, A. (2014). Traditional knowledge of wild food plants in a few Tibetan communities. *Journal of Ethnobiology and Ethnomedicine*, 10(1), 1–19.

Chitrakar, K., Albrecht, J.N., & Carr, N. (2021). Community-based homestay tourism and social inclusion. In S. Walia (Ed.), *The Routledge Handbook of Community Based Tourism Management: Concepts, Issues & Implications* (pp. 238–248). London: Routledge.

Chitrakar, K., Carr, N., & Albrecht, J.N. (2020). Community-based homestay tourism and its influence on indigenous gastronomic heritage. *Journal of Gastronomy and Tourism*, 4(2), 81–96.

Cleave, P. (2013). The evolving relationship between food and tourism. In C.M. Hall & S. Gössling (Eds.), *Sustainable Culinary Systems: Local Foods, Innovation, Tourism and Hospitality* (pp. 156–168). London: Routledge.

Cohen, E. (1979). A phenomenology of tourist experiences. *Sociology*, 13(2), 179–201.

Cohen, E., & Avieli, N. (2004). Food in tourism. *Annals of Tourism Research*, 31(4), 755–778.

Dahal, B., K C, A., & Sapkota, R.P. (2020). Environmental impacts of community-based home stay ecotourism in Nepal. *The Gaze: Journal of Tourism and Hospitality*, 11(1), 60–80.

Dong, T.B. (2020). Cultural tourism: An ethnographic study of home stay in Briddim Village, Nepal. *The Gaze: Journal of Tourism and Hospitality*, 11(1), 10–36.

Dougherty, M.L., Brown, L.E., & Green, G.P. (2013). The social architecture of local food tourism: Challenges and opportunities for community economic development. *Journal of Rural Social Sciences*, 28(2), 1–27.

Dutta, M. (2018). Gastronomic behaviour in community based tourism: Ethnic food & beverages of North Eastern India. *Amity Research Journal of Tourism, Aviation and Hospitality*, 3(1), 6–25.

Farooque, N.A., Budal, T.K., & Maikhuri, R.K. (2008a). Environmental and socio-cultural impacts of river rafting and camping on Ganga in Uttarakhand Himalaya. *Current Science*, 94(5), 587–594.

Farooque, N.A., Nehal, N., Budal, T K., & Maikhuri, R.K. (2008b). Cultural and social impact analysis of adventure tourism in Himalayan River Ganga in India. *Indian Journal of Youth Affairs*, 12, 104–111.

Fisher, J.F. (1990). *Sherpas: Reflections on change in Himalayan Nepal*. Berkeley: University of California Press.

Frisvoll, S., Forbord, M., & Blekesaune, A. (2016). An empirical investigation of tourists' consumption of local food in rural tourism. *Scandinavian Journal of Hospitality and Tourism*, 16(1), 76–93.

Giampiccoli, A., & Kalis, J.H. (2012). Tourism, food, and culture: Community-based tourism, local food, and community development in Mpondoland. *Culture, Agriculture, Food and Environment*, 34(2), 101–123.

Grocke, M.U. (2016). *On the Road to Better Health? Impacts of New Market Access on Food Security, Nutrition, and Well-Being in Nepal, Himalaya*. Unpublished doctoral dissertation, University of Montana, Missoula.

Gusain, P. (2017). A traditional preserved food Badi from ethnic village Devrara, Garhwal Region. *Trends in Biosciences*, 10(3), 1068–1070.

Hjalager, A.-M., & Richards, G. (Eds.). (2002). *Tourism and Gastronomy*. London: Routledge.

Ishak, N., Zahari, M.S.M, & Othman, Z. (2013). Influence of acculturation on foodways among ethnic groups and common acceptable food. *Procedia—Social and Behavioral Sciences*, 105, 438–444.

Jamal, S.A., Othman, N., & Muhammad, N.M.N. (2011). Tourist perceived value in a community-based homestay visit: An investigation into the functional and experiential aspect of value. *Journal of Vacation Marketing*, 17(1), 5–15.

Jamaludin, M., Othman, N., & Awang, A.R. (2012). Community based homestay programme: A personal experience. *Procedia—Social and Behavioral Sciences*, 42, 451–459.

Joshi, P., Sharma, N., Roy, M.L., Kharbikar, H.L., Chandra, N., & Sanwal, R. (2015). Traditional food practices in North Western Himalayan region: Case of Uttarakhand. *Journal of Agricultural Engineering and Food Technology*, 2(3), 170–174.

Kala, C.P. (2021). Ethnic food knowledge of highland pastoral communities in the Himalayas and prospects for its sustainability. *International Journal of Gastronomy and Food Science*, 23, 1–9.

Kikon, D. (2015). Fermenting modernity: Putting *akhuni* on the nation's table in India. *South Asia: Journal of South Asian Studies*, 38(2), 320–335.

Kontogeorgopoulos, N., Churyen, A., & Duangsaeng, V. (2015). Homestay tourism and the commercialization of the rural home in Thailand. *Asia Pacific Journal of Tourism Research*, 20(1), 29–50.

Kuhn, E., Haselmair, R., Pirker, H., & Vogl, C.R. (2018). The role of ethnic tourism in the food knowledge tradition of Tyrolean migrants in Treze Tílias, SC, Brazil. *Journal of Ethnobiology and Ethnomedicine*, 14(1), 14–26.

Kulshrestha, S., & Kulshrestha, R. (2019). The emerging importance of "homestays" in the Indian hospitality sector. *Worldwide Hospitality and Tourism Themes*, 11(4), 458–466.

Kunjuraman, V., & Hussin, R. (2017). Challenges of community-based homestay programme in Sabah, Malaysia: Hopeful or hopeless? *Tourism Management Perspectives*, 21, 1–9.

Lama, M. (2013). *Community Homestay Programmes as a Form of Sustainable Tourism Development in Nepal*. Kokkola: Centria University of Applied Sciences.

Lee, J.M., Contento, I., & Gray, H.L. (2018). Change in food consumption and food choice determinants among East Asian international students in New York. *Journal of Hunger & Environmental Nutrition*, 15(3), 1–24.

Liechty, M. (2001). Consumer transgressions: Notes on the history of restaurants and prostitution in Kathmandu. *Studies in Nepali History and Society*, 6(1), 57–101.

Long, L.M. (Ed.). (2004). *Culinary Tourism*. Lexington: University Press of Kentucky.

Mak, A.H.N., Lumbers, M., & Eves, A. (2012). Globalisation and food consumption in tourism. *Annals of Tourism Research*, 39(1), 171–196.

Martin, M.S. (2014). Authenticity, tourism, and Cajun cuisine in Lafayette, Louisiana. In R. Cobb (Ed.), *Paradox of Authenticity in a Globalized World* (pp. 13–22). New York: Palgrave Macmillan.

Mnguni, E.M., & Giampiccoli, A. (2016). Community-based tourism and food: Towards a relationship framework. *African Journal of Hospitality, Tourism and Leisure*, 5(1), 1–12.

MoTCA (2010). *Home-stay Sanchalan Karyabidhi, 2067 [Homestay Operations Procedures, 2010]*. Kathmandu: Ministry of Tourism and Civil Aviation.

Nepal Rastra Bank (2015). *A Study on Dallagaon Homestay and Its Sustainability*. Nepalgunj: Banking Development and Research Unit.

Nepal, S.K. (2016). Tourism and change in Nepal's Mt Everest region. In H. Richins & J.S. Hull (Eds.), *Mountain Tourism: Experiences, Communities, Environments and Sustainable Futures* (pp. 270–279). Wallingford: CABI.

Nyaupane, G.P., Lew, A.A., & Tatsugawa, K. (2014). Perceptions of trekking tourism and social and environmental change in Nepal's Himalayas. *Tourism Geographies*, 16(3), 415–437.

Oakes, T. (1999). Eating the food of the ancestors: Place, tradition, and tourism in a Chinese frontier river town. *Ecumene*, 6(2), 123–145.

Ogucha, E.B., Riungu, G.K., Kiama, F.K., & Mukolwe, E. (2015). The influence of homestay facilities on tourist satisfaction in the Lake Victoria Kenya Tourism Circuit. *Journal of Ecotourism*, 14(2–3), 278–287.

Pant, R. (2016). The disappearing rainbow colours: From the central Himalayan platter. *TerraGreen*, 8(11), 30–33.

Peaty, D. (2009). Community-based tourism in the Indian Himalaya: Homestays and lodges. *Journal of Ritsumeikan Social Sciences and Humanities*, 2, 25–44.

Peštek, A., & Činjarević, M. (2014). Tourist perceived image of local cuisine: The case of Bosnian food culture. *British Food Journal*, 116(11), 1821–1838.

Ramsay, C.R. (2002). *Sir Edmund Hillary & the People of Everest*. Kansas City: Andrews McMeel.

Rasul, G., Hussain, A., Mahapatra, B., & Dangol, N. (2018). Food and nutrition security in the Hindu Kush Himalayan region. *Journal of the Science of Food and Agriculture*, 98(2), 429–438.

Richards, G. (2012). An overview of food and tourism trends and policies. OECD, Food and the Tourism Experience: The OECD-Korea Workshop, OECD Publishing, Paris.

Robertson, R. (2012). Globalisation or glocalisation? *Journal of International Communication*, 18(2), 191–208.

Strychacz, T. (2010). Yak burgers and black tea: Consumption, deprivation, and the literature of Himalayan adventure travel. *Food and Foodways*, 18(3), 145–167.

Swenson, K. (2001). Culture & thought—Fine dining: Tables with a view—High in the Himalayas, Chez Tashi's makes a mean yak burger. *Asian Wall Street Journal*. Retrieved January 3, 2021, from https://search-proquest-com.ezproxy.otago.ac.nz/docview/31540 8374?accountid=14700

Tamang, D.D. (2016). Culinary tradition and indigenous knowledge in the Nepal Himalayas. In J. Xing & P. Ng (Eds.), *Indigenous Culture, Education and Globalization: Critical Perspectives from Asia* (pp. 27–66). Berlin: Springer-Verlag.

Tamang, J.P. (2001). Food culture in the eastern Himalayas. *Himalayan and Central Asian Studies*, 5(3–4), 107–118.

Tamang, J.P. (2009). *Himalayan Fermented Foods: Microbiology, Nutrition, and Ethnic Values*. Boca Raton: CRC Press.

Timothy, D.J. (2016). Heritage cuisines, foodways and culinary traditions. In D.J. Timothy (Ed.), *Heritage Cuisines: Traditions, Identities and Tourism* (pp. 1–24). London: Routledge.

Timothy, D.J. (2021). Heritage and tourism: Alternative perspectives from South Asia. *South Asian Journal of Tourism and Hospitality*, 1(1), 35–57.

Timothy, D.J., & Ron, A.S. (2013). Understanding heritage cuisines and tourism: Identity, image, authenticity, and change. *Journal of Heritage Tourism*, 8(2/3), 99–104.

Walter, P. (2020). Community-based ecotourism projects as living museums. *Journal of Ecotourism*, 19(3), 233–247.

Yang, L., & Wall, G. (2009). Minorities and tourism: Community perspectives from Yunnan, China. *Journal of Tourism and Cultural Change*, 7(2), 77–98.

Zhang, J., Xu, H., & Xing, W. (2017). The host—guest interactions in ethnic tourism, Lijiang, China. *Current Issues in Tourism*, 20(7), 724–739.

# 13 Tea Heritage and Tourism as Sustainable Development in the Eastern Himalaya

*Lee Jolliffe*

## Introduction

This chapter investigates the potential for merging tea heritage and tourism for sustainable development in the Eastern Himalaya. This region, extending from eastern Nepal across northeastern India, Bhutan, the Tibet Autonomous Region, Yunnan (China) and northern Myanmar, is a biodiversity hotspot with notable cultural diversity as well. The climate is influenced by the South Asian monsoon from June to September. The region is an important tourist destination with potential for growth and development. It is, however, vulnerable to environmental (Nyaupane & Thapa, 2006) and social change issues, including natural disasters (Kala, 2014), population pressures (Nepal, 2000) and geopolitical conflicts (Nyaupane et al., 2014). Tourism in the region is also subject to the effects of climate change (Nyaupane & Chhetri, 2009; Pandey & Bardsley, 2015). However, throughout the region, the impacts of poverty, population growth, environmental deterioration and politics are intertwined (Nepal, 2000).

In this chapter, the emergence of tea-related tourism is examined within the context of the varied agricultural and culinary heritage traditions of the area. Tea tourism has been defined as tourism motivated by an interest in the history, traditions and consumption of tea (Jolliffe, 2007, 2016; Jolliffe & Aslam, 2009). This tourism niche is of particular relevance to destinations producing tea, as some tea farms and gardens use tourism to diversify their operations, with the goal of positively impacting local livelihoods (Cheng et al., 2010, 2012). Studies in China have identified the benefits of tea tourism for tea farmers (e.g. Chee-Beng & Yuling, 2010) as an approach to sustainable livelihoods (Su et al., 2019).

Some tea-producing areas of the Eastern Himalaya, with their landscapes of rolling cultivated tea or forest grown tea, tea factory productions, small tea holders and tea shops or tea bungalows, are therefore potential locations for the development of tea-related tourism. Tea tourism is a growing niche segment, especially in the Darjeeling and Doors regions of the state of West Bengal, India (Bhutia, 2014). In nearby Assam, tea tourism on a small scale could be employed for sustainable development with the goal of improving local livelihoods within a changing climatic situation (Biggs et al., 2018). A sustainable livelihoods approach can be taken to integrate tea and tourism for the benefit of local communities as

DOI: 10.4324/9781003030126-16

has been advocated in China (Su et al., 2019). The wide-ranging tea culture of the Eastern Himalaya is therefore relevant for developing heritage and culinary focused community-based tourism.

Tea is an agricultural product that is consumed as a beverage brewed from the tea leaf. The origins of the beverage are in mountainous regions, with the varieties of *camellia sinesis var. sinesi*s and *cameilia sinesis var. assamica* having originated in China and India, respectively. Historically, tea is a widely traded agricultural product that has had historical influences. This includes, for example, ancient trade, such as the Tea Horse Road running primarily between China and Tibet, and colonization involving the establishment of tea gardens and the migration of workers (e.g. the introduction of the *camellia sinesis var. sinesis* variety from China into India by the British). From a contemporary perspective, tea and its associated trade in the region are influenced by workers movements (e.g. the 2017 closure of tea gardens in Darjeeling), economics (e.g. global tea markets) and political relationships (e.g. international diplomacy). Moreover, climate change is affecting tea-growing areas with more severe cold weather, which impacts production (Pandey & Bardsley, 2015).

Therefore, tea heritage tourism is considered here within the broad context of various circumstances in the region that have led to the development of differing traditions and practices surrounding the tea beverage. This relates to agricultural production and consumption as a beverage and a part of the region's heritage cuisines. Today, tea, if merged with tourism, may offer opportunities for sustainable development in the Eastern Himalaya that are suitable for cultivating tea and/or exhibiting unique tea traditions that will appeal to tourists.

Common issues in the contemporary evolution of tea-related forms of tourism in the Himalaya will thus be considered in this chapter, in particular the ability of niche tourisms such as homestay tourism and small-scale tea tours, to contribute to sustainable development and improved livelihoods for residents. Examples of issues related to the implementation of tea tourism from various locations in the region accompany the discussion. This provides a conceptual overview of the current state of tea-related heritage tourism in relation to sustainable development in the Eastern Himalaya and a foundation for future studies on the topic.

The subject of tea heritage and tourism and its relation to sustainable development is investigated within this chapter from a number of perspectives. First, the overall agricultural heritage background to tea and sustainable livelihoods is reviewed, including its historical and geographical aspects and an identification of where it is currently grown in the region. Second, the culture of tea is profiled, in terms of forms of consumption, which is an important form of intangible cultural heritage. Tea is investigated as a traditional Himalayan beverage that can be experienced. Third, tea tourism is examined as a form of sustainable development. The challenges and issues of the current hospitality and tea industries in the region and their relation to local communities are reviewed. Types of tea tourism in the Eastern Himalaya are identified, and the contribution of tea tourism to improving livelihoods of tea-producing communities is examined. Fourth, empirical examples of tea-related tourism and sustainable development from Nepal, India

(Darjeeling and Sikkim) and Yunnan (China) are presented, identifying issues for future consideration.

## Tea as Agriculture in the Himalaya

Agriculture, including farming and herding, has long been part of the cultural practices of the Himalaya that has sustained life there for thousands of years (Knörzer, 2000). The unique climates of the Himalaya, rainfall measures, topographies and soil types have determined much of what can be grown in the region, and agricultural products and practices have adapted to these distinctive environmental situations over many millennia. The mountainous environments of the Eastern Himalaya were eventually established as key locations for the development of tea-based agriculture. "The complex geo-environmental and agro-climatic conditions in the region lend to the tea grown in the area a distinct quality" (Khawas, 2002, p. 1). Historically and contemporarily, one of the most important agricultural products grown in the Eastern Himalaya is tea.

Set amidst the world's tallest mountain range, the Himalaya, the Eastern Himalaya of India and surrounding countries (i.e. Nepal, Bhutan, China and Myanmar) are an important tea growing region, producing mostly black teas, although some growers are experimenting with green tea. This includes the significant production of a number of tea gardens in Darjeeling and one in Sikkim, both in India, and in areas of Nepal. These areas have a similar climate and thus their teas are on occasion blended together. Teas grown in this region have a distinct high-grown Himalayan character as opposed to those grown in other mountainous regions such as in Nuwara Eliya, Sri Lanka and Nantou, Taiwan.

Table 13.1 identifies the major tea-growing areas of the Eastern Himalaya in terms of locations and types of teas produced. In addition, geographical indications (GIs) received for the use of terminology and protection of its value are noted with the years this designation was received. Nepal tea has received its GI and is grown mainly in areas bordering Darjeeling. Major tea production in the region also occurs in Darjeeling, where 87 tea gardens have been recognized with the geographical indicator (GI) for Darjeeling tea. Darjeeling is also seeing

*Table 13.1* Major Tea-Producing Regions of the Eastern Himalaya

| Country | Region | Production/Recognition With Dates |
|---------|--------|-----------------------------------|
| Nepal | Jhapa, Ilam, Panchthar, Dhankuta and Terhathum Districts | Nepal, CTC, Orthodox and special tea types (including green and oolong) GI Nepal Tea (2018) |
| India | Darjeeling, Kalimpong, West Bengal | Darjeeling, black, green, oolong and white teas GI Darjeeling Tea (2004) |
| China | Yunnan's 639 towns in 11 prefectures and cities, including Pu'er and Dali | Pu'er, black and green teas. GI Pu'er Tea (2008) |

a growing number of smallholder tea farms. In Sikkim, there is one large govern-ment-owned tea garden established in 1969 and now operating as a cooperative producing a high-quality tea marketed as Temi Tea. Major tea production occurs in Yunnan where pu'er, black and green teas are produced. In other areas of the Eastern Himalaya, there is limited tea production. Bhutan, for example, report-edly only has one tea garden area in production as a cooperative of about 10 acres (Beckwith, 2018).

The region's tea trade was historically connected by the Tea Horse Trade Route. Here, the production of brick tea (compressed bricks of black tea) was traded from Yunnan to Tibet on this route with side routes to Sichuan and India. Today in Yun-nan, this tea is still produced, and the fermented tea, pu'er (also known as pu-erh, pureh, pu-er, pu-eh and puer), is both experienced for its health benefits and col-lected for its economic value. Pu'er tea has also received a GI.

## Tea as Culture in the Eastern Himalaya

In the Himalaya, tea has become a salient part of the living culture. Tea grow-ing, making and drinking have a role to play in everyday life, special events, social gatherings and religious celebrations. In the Himalaya, the culture of tea drinking permeates well beyond the main areas of production mentioned earlier. Wilkes (1968) indicates that the high altitude and severe weather of the Eastern Himalaya and Tibet have been conducive to developing beverages that are both high in nutritional value and served warm. In Tibet, Bhutan and India, the day begins with tea, but its origins and preparation may differ according to location (Wilkes, 1968). Tea is not produced in Tibet but is consumed in the form of butter tea, known as Tibetan Tea. This beverage is made from tea leaves and yak butter and is the traditional drink both in Tibet and in the Tibetan communities of India and Nepal. Yunnan, where the brick tea used by Tibetans to prepare butter tea is produced, has a traditional pu'er tea culture.

In India, brewed tea is drunk without additions, as with Darjeeling tea, or with the addition of milk and spices as in Chai tea. In Bhutan, tea may be prepared with either the Tibetan or the Indian methods (Wilkes, 1968). Tibetan tea has a direct historic link through the Tea Horse Trade Road to other tea-producing areas of the Eastern Himalaya. The origins of using brick tea to prepare butter tea dates back to the ancient Tea Horse Trade Road, when tea from Yunnan, China, was transported to Tibet. Although the tea beverage was introduced from China, its preparation in this region is distinctively Tibetan (Wilkes, 1968). Tea is also grown in Myanmar where it has made its way into local cuisine through Laphet, a traditional ethnic food consisting of fermented tea leaves (Han & Aye, 2015).

Tea as an element of culture is often offered as a sign of hospitality and is embedded within hospitality services for visitors. For example, many trekkers in Nepal stop at small wayside tea shops which, beyond serving tea, also provide food and on occasion accommodation (Nepal, 2000). However, the concentra-tion of these tea shops in some areas has also brought localized problems, such as environmental, health and sanitization degradation. The hospitality industry in the region was also impacted by British colonization in the nineteenth century, as many of the mountainous areas served as hill stations for British colonial officials and workers, such as in Darjeeling and Sikkim. These locations were desirable destinations during the hot seasons in the lowlands and were developed with transportation, accommodations and food services for colonial visitors. This setup has transitioned into much of today's domestic tourism (Sacareau, 2007).

## Tea and Tourism in the Eastern Himalaya

Agriculture-based tourism (i.e. agritourism) has seen a rapid growth throughout the world in recent years (Barbieri, 2019). Although it has long been part of rural tourism in the Himalaya, agritourism has taken a strong foothold in the region during the past 20 years (Bhatta et al., 2019; Das, 2019; Khanal & Shrestha, 2019; Mehta & Yadav, 2015). Likewise, with the specialization of certain tourism experiences and tourist markets, tea-based tourism is becoming increasingly popular in major tea growing regions, such as China and Sri Lanka (Aslam & Jolliffe, 2015; Jolliffe, 2007; Jolliffe & Aslam, 2009; Huang, 2006), as well as in the Eastern Himalaya.

Linking tea agriculture and processing to tourism in the Himalaya is an increasingly important manifestation of rural tourism, agritourism and culinary tourism. Consumers can visit tea farms, gardens and estates; experience the growing procedure; observe tea processing; and consume the product in situ—just as agritourists do in other farming and plantation contexts (Pezzi et al., 2019; Sznajder et al., 2009). In the Western Himalaya, Himachal Pradesh where some tea is also produced, agritourism is seen as a means of agricultural diversification (Singh et al., 2017). While the government has been slow to support this endeavour, Singh, Raman and Hansra note that farmers with small holdings can potentially generate more income for their families through tea tourism. In Nepal, the prospects for agritourism have been noted as the integration of diversified resources, landscapes, biodiversity, cultural heritages and unique traditions (Thapa, 2013).

Tea grown in the Eastern Himalaya is a key agricultural product. The rural tea landscapes provide a setting for the development of tea-related tourism that has the potential to contribute to sustainable development by improving the livelihoods of local tea workers and communities. The region's food and beverage culture is relevant for tourism development (Tamang, 2001), especially the potential for 'ethnic food tourism', as culinary and beverage traditions vary throughout the region.

Tea tourism in the Eastern Himalaya tends to be small scale due to a number of key factors. The tea-growing areas are often difficult to reach owing to poor road conditions and their distance from airports and other regional gateways. Many areas also lack infrastructure, such as accommodations, resulting in limited capacity to receive large numbers of visitors. The interest in dedicated tea tourism is limited to a small niche market comprised of a few domestic tourists, international tea enthusiasts and global tea industry representatives who often visit as tea buyers in season. There is also 'accidental tea tourism', which occurs when the main purpose of visiting (mainly for domestic tourists) is to enjoy nature, fresh air, tea landscapes and fresh local teas and cuisine. The visitor guest book at the Glenburn Tea Estate in Darjeeling examined by the author reveals very few comments that would indicate that visitors saw themselves as 'tea tourists'. Accounts of travelling for tea often are found in tea company blogs, some of which host tea tours for international visitors to experience tea culture and production.

As noted previously, tea-related tourism in the region has strong connections to other types of tourism including agritourism, heritage tourism and culinary tourism (Jolliffe, 2007). Tea tourism itself can take different forms that offer visitors opportunities to experience tea in different ways (Jolliffe, 2016). This includes stays at colonial tea estates and tea garden/factory tours (Darjeeling and Sikkim); homestays in tea communities (Nepal and Darjeeling); tea tours (Nepal and Darjeeling); hands-on processing, tasting and blending (all tea-producing locations); tea house treks (Nepal); and nature tourism in ancient tea forests (Yunnan).

In terms of heritage tourism and culinary tourism, the heritage food and drink traditions of an area are a resource for tourism (Jolliffe, 2003; Timothy, 2016, 2021). This is the case in the Eastern Himalaya where strong tea traditions are part of the culinary heritage that visitors can experience. Besides gaining an in-depth understanding of tea by visiting tea gardens and factories, tourists can experience diverse Himalayan tea traditions. In particular, Darjeeling and Sikkim play on the charms of English afternoon tea, albeit adapted to the Indian postcolonial context (Skinner, 2019). For example, the all-inclusive rate to stay at the Glenburn Tea Estate luxury property in Darjeeling includes bed tea (tea delivered to the room early in the morning) and afternoon tea. The Glenburn Tea Estate has developed tourism as a diversification tool beyond solely tea production, employing 50 workers in this endeavour. Four guest rooms were opened in 2002 in the renovated original planters bungalow, and four additional guest rooms were added in another building in 2008 (Koehler, 2015). This is an example of tea-related tourism contributing to both employment and human resource development in Darjeeling (Bhutia, 2014).

Tea-based tourism in Yunnan, China, focuses on the rich pu'er tea culture of the area, dating back several thousand years, including links to the ancient Tea Horse Trade Route. The importance of the ancient tea trade as manifested in this region has made this part of China an important heritage destination, especially as regards tea-based cultural tourism (Ru & Xu, 2001).

## Regional Examples

### Nepal

Nepal is a small land-locked country bordering India and China. The 2015 constitution guarantees inclusive socioeconomic and political development, building an egalitarian and pluralistic society and eliminating all forms of discrimination. Having achieved success with the millennium development goals (MDGs), the implementation of the sustainable development goals (SDGs) began in 2016 with the aim of becoming a middle-income country by 2030. Despite the environmental challenges of natural disasters and climate change, Nepal is making progress towards achieving the SDGs in poverty reduction, social and geographic inclusion and gender parity (Government of Nepal, 2017).

Mountains and hills cover 83% of the country's land mass with the remaining 17% in the Terai Plains (Government of Nepal and International Trade Center, 2017). Most of the population, including that of the designated tea-growing areas, is rural. The tea industry is made up of thousands of small-scale farmers (Mohan, 2016) and some larger tea estates (Table 13.2). The main tea cultivation area in eastern Nepal borders the Himalaya and Darjeeling in India. Tea is Nepal's third leading agricultural export (Nepal Tea and Coffee Development Board, n.d.). Tea planting dates back to the late 1800s with recent tea industry developments in the districts of Jhapa, Ilam, Panchthar, Dhankuta and Terhathum, designated by the government as tea zones in 1982 and the not-for-profit Tea and Coffee Development Board being established in 1993.

The country ranks 19th in global tea production (Nepal Tea and Coffee Development Board, n.d.). The types of tea produced include CTC, Orthodox and special types (including green and oolong). Orthodox tea is harvested seasonally during four periods: spring flush (February to April), summer flush (May and June), monsoon flush (after June) and autumn flush, resulting in teas with different qualities and characteristics. Orthodox tea is picked by hand using the traditional one-bud two-leaf principle resulting in specialty tea production, whereas CTC tea is picked by machine, allowing for mass production. A collective trademark for Nepal Orthodox tea (implemented in 2018) aims to raise the profile of Nepalese tea. To use the logo, producers must meet government guidelines for quality standards, workers' employment security and sensitivity to environmental protection (The Himalayan Times, 2018).

Tea is part of the culture of Nepal with a high domestic consumption rate and small tea shops being staging areas for daily social interactions. The Nepal Tourism Policy of 2009 identified the sector as being important for economic and social development (Government of Nepal, 2016). In part, the motivation for developing tea tourism was to increase revenue and expand employment opportunities to improve living standards. A report noted, "Tea tourism which involves tours to tea-growing estates and experiencing the process of local tea leaf production is becoming quite popular among western tourists in search of exotic experiences"

*Table 13.2* Producers in the Tea Industry in Nepal

| *Type* | *Number* | *Notes* |
|---|---|---|
| Tea estates | 142 | Represent 47% of planting area |
| Tea processing factories | 54 | Orthodox 19, CTC 35 |
| Tea cooperatives | 53 | As of 2011 |
| Registered small holders | 9941 | Represent 53% of planting area |

Source: adapted from Nepal Tea and Coffee Development Board (n.d.).

(Nepal Ministry of Tourism, 2012, n.p.). The tea value chain in Nepal is predominately made up of small holders (Mohan, 2016). While the concept of upgrading livelihoods could apply to introducing community-based tea tourism endeavours into the equation, Mohan (2016) observed that these efforts do not always result in increased income, poverty reduction, enhanced livelihoods and encouragement of gender equality.

Nepal's diverse cultural and natural resources (including its mountains) have led to developing adventure-based tourism through trekking, often involving stopping at small tea shops that serve tea and food and occasionally provide accommodation. For example, in the Annapurna region, over 1,000 locally owned lodges/tea shops provide jobs for locals (Nyaupane et al., 2006). However, this concentration of tea shops has raised environmental, health and sanitization concerns (Nepal, 2000). The significant role of tea in Nepal has been noted in the government export development plan for the tea sector (Government of Nepal and International Trade Centre, 2017). This report calls for a study in tea-producing areas to assess the potential for ecotourism in tea estates and for small-scale producers to augment their incomes through tea tourism activities.

Tea-related tourism in Nepal beyond trekkers visiting tea shops appears to be limited in part by the lack of infrastructure (i.e. transportation, accommodations and related services), which deters some independent travellers and buyers visiting tea gardens, as well as visits being included in guided tours of the tea areas such as Illam. Nonetheless, as tea-based tourism has grown over the last 40 years in neighbouring Darjeeling, this could also be possible in Nepal (Thapa, 2004).

### Darjeeling and Sikkim, India

Darjeeling is both a town and a partially autonomous district of the Indian state of West Bengal. It is located in the foothills of the Lesser Himalaya at an elevation of 6,700 ft (2,042.2 metres) and borders Nepal, Bhutan and Sikkim. It is noted for its tea industry, views of Kangchenjunga, the world's third highest mountain and also for the famed UNESCO World Heritage Site, the Darjeeling Himalayan Railway. The nearby Indian state of Sikkim borders Bhutan, Nepal and Tibet.

The distinctive Darjeeling black tea, now recognized with an international trademark, is currently produced on 87 recognized tea gardens and a number of small holdings. It is ranked globally among the most popular black teas (Table 13.3). Due to Darjeeling's unique agroclimatic conditions, the tea has a distinctive

*Table 13.3* Tea Producers in Darjeeling

| Type | Number | Notes |
|------|--------|-------|
| Tea estates | 87 | Cultivate about 19,000 hectares (i.e. 46,950 acres) |
| Tea processing factories | 87 + | On each tea estate and small holders establishing small factories |
| Registered small holders* | 880 | Cultivate about 192 hectares of land |

Source*: Dutta (2017).

natural flavour recognized by a geographical indicator (GI). The region is famous for four tea seasons: the First Flush in spring, the Second Flush in early summer, the Monsoon Flush in the rainy season and the fall Autumnal teas. In recent years, the tea industry here has faced competition from other areas of India and from neighbouring Nepal. The industry has also been affected by worker absenteeism on tea estates, workers' demands for higher wages, worker strikes and most recently by the 2017 Gorkhaland movement, which seeks an independent Gorkha state. Although much tea is produced on large estates, there are a growing number of small producers and a few cooperatives.

A few tea cooperatives have been established, for example Potong Tea Garden; once a colonial tea plantation, it is now collectively run by workers along democratic lines. This new model is expected to benefit the estate's 350 employees and their families. While the workers now own 51% of the gardens, they also benefit from partnering with an existing company, Tea Promoters of India, a network of organic growers and a non-governmental organization, Equal Exchange, which works globally to promote authentic fair trade and green partnerships (Equal Exchange Coop, n.d.). Potong Tea Garden has been abandoned and closed from time to time, leading to employment insecurity. At another abandoned tea garden, locals started to tend the tea bushes from the former Mineral Springs and have now formed a cooperative, called *Sanjukta Vikash*, made up of around 400 small-farmer families (Heinze, 2017).

At Makalberi Tea Estate, an innovative homestay programme was established to augment the livelihoods of those living on the estate. Upon establishing the programme in 2005, 44 families on the tea estate were involved (Koehler, 2015). It was part of the vision of Rajh Banjaree, a well-known Darjeeling tea planter, who advocated an organic agricultural revolution through cooperatives for Darjeeling (Vater, 2018). The emergence of small tea farmers, who both produce tea and provide limited visitor services, such as a café and homestay, is also a notable development in tea tourism offerings in the region (Sharma, 2018).

Darjeeling became a popular tourist destination as early as 1860. During colonial times from the early nineteenth century, the British used this location as a hill station for leisure. They also established intensive tea gardens there from around the 1830s, and tea production flourished. Most workers for the tea gardens were migrants from Nepal. Tea estates were initially established as communities with workers being provided on-site housing. Later, schools and medical facilities were added.

Tourism and the tea industry are noted as the most significant contributors to the economy of Darjeeling. It is reported to be the only location in eastern India receiving large numbers of foreign visitors. A study of sustainable tourism development in Darjeeling identified unplanned tourism growth in some areas such as Darjeeling town and lack of infrastructure as barriers to further developing tourism (Bhutia, 2015). The tourist experience here may involve staying at tea estates, touring tea gardens, observing growing and production processes, experiencing tea-related hospitality by consuming the beverage and participating in afternoon tea, a meal tradition adapted from British times (Skinner, 2019) and shopping for tea as a souvenir. This is an important element of the area's heritage tourism, for Darjeeling promotes its tea estates as a reflection of its British colonial past.

With a few estates being converted to cooperatives, some others are adding accommodation and visitor services to create tea resorts alongside tea production. However, the policy of the West Bengal state government limits the amount of land within tea estates that can be used for tourism to generate additional revenue (The Economic Times, 2013). Some tea estates have converted historic manager bungalows into visitor lodging, such as at the Ging and Singtom Tea Estates, while others have augmented this approach with purpose-built accommodations.

In 2018, the author observed at the Glenburn Tea Estate that accommodation and meals (bed tea, breakfast, lunch, afternoon tea and dinner) are provided in the dining room of the original manager's bungalow and in a newly build bungalow with a full range of activities. This included tours of the gardens and tea factory, visits to the local school and hiking and picnicking in the tea gardens. Food menus use local produce from the estate and combine Western and Indian dishes prepared by a chef/innkeeper. An on-site shop sells a range of tea-related gifts and textiles and teas that are sought out by tourists as souvenirs (Jolliffe, 2007). The meal traditions have been adapted to reflect both colonial (Western) and local (Indian) tastes; bed tea (tea brought by a servant at an early hour) is a typical colonial service (Skinner, 2019). Since the road to Glenburn is difficult, the estate employs drivers to pick up guests and bring them to Glenburn from the closest airport (Bagdogra Airport).

It has been previously noted that tea tourists have a range of motivations (Jolliffe, 2007). In 2018, the author's examination of the Glenburn Tea Estate guest books for an annual period revealed few dedicated tea tourists, with only a few tea buyers from international firms reporting that they came specifically for the tea. It was observed that visitors to Glenburn appreciate the peaceful rural setting and tea landscape. However, Glenburn also hosts groups such as those of World Tea Tours, whose participants could be considered dedicated and highly motivated tea tourists (Jolliffe, 2007).

A proposal to create a tourism experience combining a ride on the UNESCO-designated Darjeeling Historic Railway (DHR) with stops at a number of tea gardens has raised concerns over the integrity of the DHR and the possible visitor impacts on tea gardens in terms of damaging tea leaves, plants and soil quality (Singh, 2018). It has also been noted that only the tea gardens on the proposed five train stops would benefit from this initiative.

In the nearby state of Sikkim, the Temi Tea Garden was established in 1969. It is the state's only tea garden producing high-quality organic tea, sought out by the European market. A resort, Cherry Resort, is also located within the tea garden. In addition, it is possible to stay in the historic and restored Temi Tea Garden Bungalow. A stay in a tea bungalow is a typical activity as part of tea tourism in nearby Darjeeling (Bhutia, 2014). A public-private partnership to promote tea tourism at Temi and the use of the restored tea bungalow with four guest rooms was recently announced (Sikkim Express, 2018). At the Temi Tea Garden, visitors are charged a nominal fee for tours of the facility, and there is a small shop outlet operated by a workers' cooperative. Sikkim is an organic state, and thus the tea garden's production of organic tea fits in with state policy. With limited arable land in Sikkim but bountiful natural beauty, tourism has been identified as an economic growth segment engine (Choudhury, 2001) and a source of alternate livelihoods. With only one tea garden in the state, the tea-related tourism potential based on visiting and staying at tea gardens is limited. However, tea traditions can be an integral part of hospitality provision there. Tourism in Sikkim is recognized as being at a developmental stage; most visitors are domestic with relatively few international tourists spending time in Sikkim (Joshi & Dhyani, 2009).

### Yunnan, China

Yunnan is a mountainous southwestern province of China, bordering Sichuan (China), Tibet, Myanmar, Laos, Cambodia and Vietnam. The ancient Tea Horse Trade Road to Tibet with side routes to Sichuan and India starts in Yunnan (Forbes & Henley, 2011). This is China's most diverse province, both culturally and in terms of biodiversity. The Old Town of Lijiang, designated a World Heritage Site in 1997, has a history dating back more than a thousand years, as it was once a confluence for trade along the Tea Horse Road. The road is also referred to as the 'Tea Caravan Route' due to caravans of horses loaded with tea and other goods along the trade route (du Cros, 2007).

The tea produced in Yunnan, both historically and today, is a black fermented tea known as 'pu'er'. It is typically pressed into bricks of different shapes, as it was in ancient times so it could be transported. This tea is cultivated in forests, rather than in the manicured tea gardens of other locations. Pu'er tea received its GI designation in 2008. Dedicated tea tourists are motivated to visit the area covered by the designation and to travel to see the old tea trees in the forest. The tea produced in cakes has also become quite collectable and is a local souvenir for visitors to the area.

Tourism is well developed in Yunnan. Numbers of tours focusing on different elements of tea culture, as well as of the ancient Tea Horse Road, are available for both domestic and international tourists. These tours visit local tea farms where participants can learn about tea growing and processing. Domestic tourists are interested in relaxing, tea tasting and viewing the tea landscapes, while international tourists want to learn more about the tea gardens and tea processing (Griggs, 2014). Most tourism in Yunnan, however, is domestic rather than international (Nyaupane et al., 2006). It is notable that the focus of tea tourism in Yunnan is on the distinctive pu'er tea culture of the province. There are a few tours specifically for tea professionals, with the annual tour of Seven Cups Fine Chinese Tea being a prime example.

Some adventurers, such as Canadian Jeff Fuchs, have travelled to Yunnan to trace some of the paths and remains of the Tea Horse Road (Fuchs, 2008). The best known pu'er comes from a group of mountains near Xishuangbanna in the very south of the province. This location is known for ethnic tourism, in which tea is a dominant beverage and its production plays an important part in local culture (Yang & Wall, 2008). It has become a popular tourist destination, and it is possible for persistent tourists to experience some locations of the original Tea Horse Road, such as at Lijiang described earlier. The overall potential of the route for developing tourism has been examined (du Cros, 2007). The Tea Horse Road is increasingly being promoted for tourism, especially for the expanding Chinese domestic tourism market (Forbes & Henley, 2011).

## Comparing the Examples

Identifying the relationship of tea to sustainable development in the Eastern Himalaya has included the nature of tea production and culture, as well as related forms of tea tourism that have potential for the improvement of livelihoods. In this section, the strengths, weaknesses, opportunities and threats for developing such a form of tourism as reflected by the case studies are identified and discussed.

From the review, above it is evident that the rich tea traditions of the region provide ample resources for developing tea-related tourism, albeit at a small scale. As indicated by a SWOT analysis (Table 13.4), while the tea tourism sector here has strengths and opportunities, ahead there are also weaknesses and threats that the various destinations must address.

There are cases documented in this chapter where tea tourism has the potential to support sustainable development with prominent examples from Nepal and India's Sikkim and Darjeeling areas. Particular types of endeavours may have a positive impact, such as homestays and educational tours and heritage tea garden stays, which might contribute to local employment, human resource development

*Table 13.4* SWOT Analysis of Tea Heritage Tourism in the Eastern Himalaya

| Strengths | Weaknesses |
|---|---|
| • Distinctive tea production<br>• Authentic, unique tea history<br>• Heritage beverage and cuisine<br>• Tea garden estates into resorts<br>• Homestay and community-based tourism (Nepal, Darjeeling, Yunnan)<br>• GI designations for tea (Nepal tea, Darjeeling tea, Pu'er tea) | • Access (transport, poor road conditions) and cost<br>• Lack of infrastructure (accommodation and food service)<br>• Hospitality skills<br>• Volatile political situations<br>• Limited government policies for implementation of tea tourism |
| Opportunities | Threats |
| • Involvement of small producers and cooperatives in tea-related tourism<br>• Develop tea tourism based on ancient trade routes (Tea Horse Road)<br>• Develop methods to ensure tea tourism benefits local livelihoods<br>• Planning for heritage tea garden stays, employment beyond tea, heritage preservation | • Tea estate worker issues, labour shortages, absenteeism<br>• Commodification of tea culture<br>• Geopolitical situations<br>• Guarding against becoming too developed and popular (overtourism)<br>• Impacts of climate change on tourism seasons and resources<br>• Natural disasters |

and preservation of the built heritage. The recognition and protection of regional teas offered by GI designation is notable, as it protects the authenticity of the products and can be beneficial to tourism development. Linking various locations in the Eastern Himalaya to the ancient Tea Horse Road provides opportunities to be explored for communities and regions to join together to develop contemporary, yet authentic, tourism routes.

Government policies that recognize the potential of tea tourism in some cases (e.g. Nepal) and limiting development in others (e.g. Darjeeling) need to be structured to reflect a sustainable livelihoods approach to developing tea-related tourism products and experiences. Visitors to tea-producing regions of the Himalaya, in particular tea buyers and educators, can play an important role in promoting the potential for tea-related tourism to support sustainable development.

Of particular note is the notion that the homestay movement could play an important role in creating a sustainable form of tea tourism. This could have an effect on improving livelihoods within tea settlements while preserving and conserving built and living heritage, as well as the sociocultural and natural environment. Such is the case illustrated by the Makaiberi Tea Estate Homestay Program in Darjeeling, India.

Examples of the most significant individual strengths, weaknesses, opportunities and threats for developing tea tourism in the region are found in Table 13.5. This shows some common threads in terms of strengths. In all of the specific regions profiled, tea has received a GI designation. Common weaknesses include poor transportation infrastructure (in two of the cases) with roads dating from

*Table 13.5* Top Strengths, Weaknesses, Opportunities and Threats From Eastern Himalayan cases

| Case | Strength | Weaknesses | Opportunities | Threats |
|------|----------|-----------|---------------|---------|
| Nepal | GI designation for tea | Less developed destination Poor road infrastructure | Improve livelihoods through tea tourism Develop domestic tea tourism | Climate change Natural disasters |
| Darjeeling | Well-developed tea destination GI designation for tea | Poor road infrastructure Accommodation extremes, budget to luxury | Improve livelihoods through tea tourism Develop human resources through tea tourism | Climate change Seasonality Political disturbances |
| Yunnan | Popular tea destination GI designation for tea | High levels of visitation | Improve livelihoods through tea tourism | Climate change Seasonality Overtourism |

colonial times. In all cases considered here, there are opportunities for improving livelihoods through tea tourism. However, appropriate forms of tourism need to be developed that will be created by locals and will deliver benefits in terms of improving their standards of living. In particular, where tea-producing areas are within reach of significant population centres and markets, the advantages of developing tea tourism as a domestic tourism product should be considered.

Common threats are found in the environmental arena and include climate change, which is affecting tea production and landscapes, as well as natural disasters such as earthquakes, which have affected some parts of the region. Political disturbances have also impacted a number of the areas discussed. The most recent event, the Ghorka strike in Darjeeling, closed down all 87 registered tea gardens for most of the 2017 production season.

## Conclusion

This chapter has provided a preliminary review of the potential for tea as an important element of the agricultural heritage of the Himalaya. Tea-based heritage tourism, reflecting tea heritage, traditions and production, is closely related to sustainable development with the goal of improving local livelihoods in the Eastern Himalaya. As tea is essentially an agricultural product, this review may have some applicability to looking at the connection of other types of produce to tourism and the issues that are likely to be encountered in developing such experiences.

In the case of tea-based tourism and sustainable development, this chapter has introduced some of the issues and challenges facing the development of a viable and sustainable (in terms of local populations) tea tourism sector. The continued impact of environmental factors, including climate change coupled with geopolitical uncertainties in the region and outmoded models of tea production (e.g. tea estates with a high level of absenteeism), will effect challenges going forward. It may be that the small tea holders in the region, less encumbered with the production issues of the past, may be able to reinvent tea tourism at a small scale in support of sustainable local livelihoods. This potential is reflected by the emergence of worker and cooperative run homestays in tea communities.

## References

Aslam, M.S.M., & Jolliffe, L. (2015). Repurposing colonial tea heritage through historic lodging. *Journal of Heritage Tourism*, 10(2), 111–128.

Barbieri, C. (2019). Agritourism research: A perspective article. *Tourism Review*, 75(1), 149–152.

Beckwith, S. (2018). Bhutan's only green tea. *In Pursuit of Tea*, August 23. Retrieved November 14, 2019, from www.inpursuitoftea.com/tea-blog/bhutan-green-tea

Bhatta, K., Itagaki, K., & Ohe, Y. (2019). Determinant factors of farmers' willingness to start agritourism in rural Nepal. *Open Agriculture*, 4(1), 431–445.

Bhutia, S. (2014). The role of tourism for human resource development in Darjeeling district of West Bengal, India. *Journal of Tourism and Hospitality Management*, 2(1), 113–128.

Bhutia, S. (2015). Sustainable tourism development in Darjeeling hills of West Bengal, India: Issues and challenges. *Global Journal of Human-Social Science Research*, 15(3), 1–10.

Biggs, E.M., Gupta, N., Saikia, S.D., & Duncan, J.M. (2018). The tea landscape of Assam: Multi-stakeholder insights into sustainable livelihoods under a changing climate. *Environmental Science & Policy*, 82, 9–18.

Chee-Beng, T., & Yuling, D. (2010). The promotion of tea in South China: Re-inventing tradition in an old industry. *Food and Foodways*, 18(3), 121–144.

Cheng, S., Hu, J., Fox, D., & Zhang, Y. (2012). Tea tourism development in Xinyang, China: Stakeholders' view. *Tourism Management Perspectives*, 2, 28–34.

Cheng, S., Xu, F., Zhang, J., & Zhang, Y. (2010). Tourists' attitudes toward tea tourism: A case study in Xinyang, China. *Journal of Travel & Tourism Marketing*, 27(2), 211–220.

Choudhury, M. (2001). Community development and tourism: The Sikkim experience in the Eastern Himalayas. Paper presented at the World Mountain Symposium, Interlaken, Switzerland.

Das, S. (2019). Towards the development of sustainable tourism in Sikkim, India: Issues and challenges. *International Journal of Research in Social Sciences*, 9(2), 575–592.

du Cros, H. (2007). China's Tea and Horse Trade Route and its potential for tourism. In L. Jolliffe (Ed.), *Tea and Tourism: Tourists, Traditions and Transformations* (pp. 167–179). Clevedon: Channel View Publications.

Dutta, I. (2017). Darjeeling small tea growers seek relief, too. *The Hindu*, December 9. Retrieved December 10, 2019, from www.thehindu.com/business/Industry/darjeeling-small-tea-growers-seek-relief-too/article21381676.ece

The Economic Times (2013). Mamata Banerjee government's new tea tourism policy not practical. Retrieved October 14, 2019, from https://economictimes.indiatimes.com/industry/services/travel/mamata-banerjee-governments-new-tea-tourism-policy-not-practical-dta/articleshow/24426046.cms?from=mdr

Equal Exchange Coop (n.d.). Potong tea garden. Retrieved August 14, 2019, from https://equalexchange.coop/our-partners/farmer-partners/potong-tea-garden

Forbes, A., & Henley, D. (2011). *China's Ancient Tea Horse Road*. Chiang Mai, Thailand: Cognoscenti Books.

Fuchs, J. (2008). The tea horse road. *The Silk Road*, 6(1), 63–71.

Government of Nepal (2016). Nepal tourism statistics. Retrieved November 10, 2019, from www.tourism.gov.np/downloadfile/Nepal_Tourism_Statistics_2015_forwebsite_edited_1486627947.pdf

Government of Nepal and International Trade Centre (2017). Nepal national sector export strategy tea, 2017–2012. Retrieved November 11, 2019, from www.intracen.org/uploadedFiles/intracenorg/Content/Redesign/Audience/policy_makers/Nepal%20Tea%206-2_web.pdf

Griggs, M.S. (2014). Tourists now taking time to see all the tea in China. *Smithsonian Magazine*. Retrieved September 30, 2019, from www.smithsonianmag.com/smart-news/touring-all-tea-china-180951836/#MQQw3r3XME3pXYeb.99

Han, T., & Aye, K.N. (2015). The legend of laphet: A Myanmar fermented tea leaf. *Journal of Ethnic Foods*, 2(4), 173–178.

Heinze, K. (2017). Tea promoters India: Organic future for Darjeeling. Retrieved September 30, 2019, from https://organic-market.info/news-in-brief-and-reports-article/tea-promoters-india-organic-future-for-darjeeling.html

The Himalayan Times (2018). Nepal tea gets international trademark after 154 tears. Retrieved October 11, 2019, from https://thehimalayantimes.com/nepal/nepali-tea-gets-intl-trademark-after-154-years/

Huang, R. (2006). Linking tea and tourism: The experience of Hunan province (China). *China Tourism Research*, 2(4), 579–593.

Jolliffe, L. (2003). The lure of tea: History, traditions and attractions. In C.M. Hall, L. Sharples, R. Mitchell, N. Macionis, & B. Cambourne (Eds.), *Food Tourism around the World: Development, Management and Markets* (pp. 121–136). Oxford: Butterworth-Heinemann.

Jolliffe, L. (Ed.). (2007). *Tea and Tourism: Tourists, Traditions and Transformations*. Clevedon: Channel View Publications.

Jolliffe, L. (2016). Coffee and tea tourism. In J. Jafari & H. Xiao (Eds.), *Encyclopedia of Tourism* (p. 158). New York: Springer.

Jolliffe, L., & Aslam, M.S. (2009). Tea heritage tourism: Evidence from Sri Lanka. *Journal of Heritage Tourism*, 4(4), 331–344.

Joshi, R., & Dhyani, P.P. (2009). Environmental sustainability and tourism—implications of trend synergies of tourism in Sikkim Himalaya. *Current Science*, 97(1), 33–41.

Kala, C.P. (2014). Deluge, disaster and development in Uttarakhand Himalayan region of India: Challenges and lessons for disaster management. *International Journal of Disaster Risk Reduction*, 8, 143–152.

Khanal, S., & Shrestha, M. (2019). Agro-tourism: Prospects, importance, destinations and challenges in Nepal. *Archives of Agriculture and Environmental Science*, 4(4), 464–471.

Khawas, V. (2002). *Environment and Rural Development in Darjeeling Himalaya: Issues and Concerns*. Ahmedabad: Centre for Environmental Planning and Technology.

Knörzer, K.H. (2000). 3000 years of agriculture in a valley of the High Himalayas. *Vegetation History and Archaeobotany*, 9(4), 219–222.

Koehler, J. (2015). *Darjeeling: The Colorful History and Precarious Fate of the World's Greatest Tea*. New York: Bloomsbury.

Mehta, S., & Yadav, K.K. (2015). Rural tourism as the river of regional development: A case of north eastern states, India. *Indian Journal of Regional Science*, 47(2), 22–28.

Mohan, S. (2016). Institutional change in value chains: Evidence from tea in Nepal. *World Development*, 78, 52–65.

Nepal Ministry of Tourism (2012). Campaign to popularize tea in Nepal. Retrieved September 12, 2019, from www.explorehimalaya.com/2012/06/03/campaign-to-popularize-tea-tourism-in-nepal/

Nepal, S.K. (2000). Tourism in protected areas: The Nepalese Himalaya. *Annals of Tourism Research*, 27(3), 661–681.

Nepal Tea and Coffee Development Board (n.d.). About NTCDB. Retrieved September 12, 2019, from www.teacoffee.gov.np/

Nyaupane, G.P., & Chhetri, N. (2009). Vulnerability to climate change of nature-based tourism in the Nepalese Himalayas. *Tourism Geographies*, 11(1), 95–119.

Nyaupane, G.P., Lew, A.A., & Tatsugawa, K. (2014). Perceptions of trekking tourism and social and environmental change in Nepal's Himalayas. *Tourism Geographies*, 16(3), 415–437.

Nyaupane, G.P., Morais, D.B., & Dowler, L. (2006). The role of community involvement and number/type of visitors on tourism impacts: A controlled comparison of Annapurna, Nepal and Northwest Yunnan, China. *Tourism Management*, 27(6), 1373–1385.

Nyaupane, G.P., & Thapa, B. (2006). Perceptions of environmental impacts of tourism: A case study at ACAP, Nepal. *The International Journal of Sustainable Development and World Ecology*, 13(1), 51–61.

Pandey, R., & Bardsley, D.K. (2015). Social-ecological vulnerability to climate change in the Nepali Himalaya. *Applied Geography*, 64, 74–86.

Pezzi, M.G., Faggian, A., & Reid, N. (2019). *Agritourism, Wine Tourism, and Craft Beer Tourism: Local Responses to Peripherality Through Tourism Niches*. London: Routledge.

Ru, B.L., & Xu, Y. (2001). An initial inquiry into the development of tea culture tourism in Yunnan. *Journal of Yunnan Normal University*, 33(4), 61–65.

Sacareau, I. (2007). Himalayan hill stations from British Raj to Indian Tourism. *European Bulletin of Himalayan Research*, 31, 30–47.

Sharma, A. (2018). Small tea maker makes it big in Darjeeling. *Nikkei Asian Review*. Retrieved October 3, 2019, from https://asia.nikkei.com/Life-Arts/Life/Small-tea-maker-makes-it-big-in-Darjeeling

Sikkim Express (2018). Temi to open for tea tourism. Retrieved November 11, 2019, from www.sikkimexpress.com/NewsDetails?ContentID=9916

Singh, G. (2018). Misgivings abound over the proposed toy train-tea tourism initiative in Darjeeling. *The Hindu Business Line*. Retrieved July 3, 2019, from www.thehindubusinessline.com/blink/know/in-the-loophole/article24407540.ece

Singh, S., Raman, N.L.M., & Hansra, B.S. (2017). Perspectives of agritourism in Himachal Pradesh: A new dimension in hill agricultural diversification. *Journal of Community Mobilization and Sustainable Development*, 12(2), 207–215.

Skinner, J. (2019). *Afternoon Tea: A History*. Lanham, MD: Rowman & Littlefield.

Su, M.M., Wall, G., & Wang, Y. (2019). Integrating tea and tourism: A sustainable livelihoods approach. *Journal of Sustainable Tourism*, 27(10), 1591–1608.

Sznajder, M., Przezbórska, L., & Scrimgeour, F. (2009). *Agritourism*. Wallingford: CABI.

Tamang, J.P. (2001). Food culture in the Eastern Himalayas. *Journal of Himalayan Research and Cultural Foundation*, 5(3–4), 107–118.

Thapa, D. (2013). *Development of Agrotourism Nepal, an Alternative to Emigration for Nepalese Youths. Case: BishramBatika Restaurant*. Unpublished thesis, Centria University of Applied Science.

Thapa, Y. (2004). Commodity case study—Tea. In *The Implications of WTO Membership on the Nepalese Agriculture*. Kathmandu: FAO, UNDP and Ministry of Agriculture and Cooperatives.

Timothy, D.J. (Ed.). (2016). *Heritage Cuisines: Traditions, Identities and Tourism*. London: Routledge.

Timothy, D.J. (2021). *Cultural Heritage and Tourism: An Introduction* (2nd ed.). Bristol: Channel View Publications.

Vater, T. (2018). Darjeeling's tea king wants an organic agricultural revolution. Raj wants a cooperative revolution. *Nikkei Asian Review*. Retrieved November 12, 2019, from https://asia.nikkei.com/Editor-s-Picks/Tea-Leaves/Darjeeling-s-tea-king-wants-an-organic-agricultural-revolution

Wilkes, H.G. (1968). Interesting beverages of the Eastern Himalayas. *Economic Botany*, 22(4), 347–353.

Yang, L., & Wall, G. (2008). Ethnic tourism and entrepreneurship: Xishuangbanna, Yunnan, China. *Tourism Geographies*, 10(4), 522–544.

# Part 4

# Potential Solutions

# 14 Corporate Social Responsibility for Sustainable Mountain Tourism in the Himalayan Region

*Rojan Baniya and Brijesh Thapa*

## Introduction

People have always been fascinated and inspired by the peace, heritage, and religious rites of the Himalaya. Historically, travel to the Himalaya was mostly in the form of pilgrimage for Hindu and Buddhist devotees (Grötzbach, 1985) and later evolved into recreation for restorative purposes, especially during the 19th-century British colonial era in South Asia (Sacareau, 2009). For example, over the course of half a century, several summer resorts were created in Shimla (1819), Mussoorie (1827), Darjeeling (1835), and Nainital (1839) (Grötzbach, 1994). In the late 1830s, these regions, better known as hill stations, drew a variety of clientele from the British aristocracy and middle class to more affluent native tourists seeking an escape from the heat and summer monsoons (Grötzbach, 1985). However, with India's independence in 1947, the region saw a dramatic fall in British tourist arrivals, although tourism began to recover after a few years when the Indian urban middle class frequented the hill stations as vacation destinations (Kennedy, 1996).

The popularity of the Himalayan hill stations grew with the momentous global exposure of the ascent of Mt. Everest by Edmund Hillary and Tenzing Norgay on May 29, 1953. This global exposure initiated a major tourist influx to the Himalaya during the 1950s (Walder, 2000), but transportation services were still limited (Grötzbach, 1994). However, following the Indian-Chinese border war in 1962, around 10,000 km of roads were built for military purposes and subsequently opened for tourists in the Indian Himalaya (Grötzbach, 1994). This led to a significant surge of tourists in the region. Similarly, in 1955, Nepal's government welcomed foreigners and granted tourist visas, which was unusual because the country had been closed off to outsiders until that time (Nepal, 2000). Since then, numerous mountaineering expeditions and millions of tourists have continued to flock to the mountains for clean air, diverse landscapes, rich biodiversity, and unique cultures (Baral et al., 2017; Charters & Saxon, 2007). Currently (until the blow of the COVID-19 pandemic in 2020), the Himalayan region continues to witness major inflows of visitors as adventure activities, such as trekking, mountain climbing, bungee jumping, sightseeing, and winter sports, have grown and created significant positive and negative impacts to the local economy, environment, and society.

DOI: 10.4324/9781003030126-18

Increased visitor flows have led to economic benefits as tourism has improved local livelihoods and prosperity (Slusariuc & Bîcă, 2015). More specifically, the economic diversification from agro-pastoralism to tourism-based activities has enormously benefitted the Himalayan regions of Nepal, India, Pakistan, Afghanistan, China, and Bhutan. Mountain-based tourism has diversified local economies, generated employment opportunities, and established an essential source of income that significantly contributes to the development of the countries in the region (Gioli et al., 2019). In the small kingdom of Bhutan, travel was permitted in 1974 with the primary goal of foreign exchange generation (Dorji, 2001). Though still somewhat restricted, the number of tourists has surged exponentially from 7,000 in 1999 to more than 274,097 in 2018 and created an annual revenue stream of US$85.41 million (Tourism Council of Bhutan, 2018). Tourism is also a key industry for Nepal, most notably mountain-based tourism. In 2019, nearly 1.2 million tourists visited Nepal for trekking, mountaineering, pilgrimage, and other reasons (Ministry of Culture, Tourism and Civil Aviation, 2020). Other specific destinations in the Himalaya such as Himachal Pradesh, Uttarakhand, Sikkim, Arunachal, and Meghalaya in India and Gilgit-Baltistan in Pakistan have also benefitted economically from mountain-based tourism (Gupta, 2015; us Saqib et al., 2019).

While the economic importance of mountain-based tourism is apparent, the industry still faces some major challenges, such as extreme seasonality, lack of suitable infrastructure and planning, and negative environmental impacts (Gaur & Kotru, 2018; Geneletti & Dawa, 2009; Nyaupane & Thapa, 2004, 2006; Upadhayaya, 2015). Increased tourism has compromised the fragile sustainability of mountain destinations. This is evident from many incidents of opposition by the local population over the intensity of tourism's resource utilization (Naitthani & Kainthola, 2015), over-exploitation, and construction of second homes (Tiwari & Joshi, 2016). Overall, tourism has contributed to extensive deforestation, unchecked pollution, and inadequate solid waste management (Gaur & Kotru, 2018; Upadhayaya, 2015). Additional impacts are related to the local infrastructure, such as substituting conventional eco-friendly and aesthetically pleasing architecture with inappropriate, unappealing, and unsafe construction, inadequately designed roads, and concomitant infrastructure (Gaur & Kotru, 2018; Nepal, 2011; Rai, 2017). Sociocultural impacts are also of major concerns and include a loss of authenticity and identity within communities, inter-generational conflicts, commercialization of art and local heritage, and social tensions (Nepal, 2002, 2011; Rai, 2017; Soni & Hussain, 2016) (see Figure 14.1).

Growing adverse impacts have led to a sense of urgency to protect the region from over-exploitation and to conserve cultural and environmental resources (Mutana & Mukwada, 2018). Comparatively, this is even more pressing and critical due to the Himalayan range's vulnerable landscapes, geography, lifestyles, and ecosystems. Given the projected visitor growth in the greater Himalayan region, the negative impacts will need to be mitigated along with programs and activities to address socio-economic issues and environmental effects.

| POSITIVE IMPACTS | POSITIVE IMPACTS | POSITIVE IMPACTS |
|---|---|---|
| • Contribute to environmental protection and conservation<br>• Raise awareness about environmental value<br>• Fund conservation of natural areas<br>• Enhance economic status of the environment in the region | • Promotion of local culture and traditions<br>• Foster pride in local tradition<br>• Avoid urban relocation by creating local jobs<br>• Supportive force for peace | • Diversify local economy<br>• Job Creation<br>• Promotion of local entrepreneurship<br>• Capital for Conservation<br>• Women entrepreneurship and jobs<br>• Increased standard of living |
| NEGATIVE IMPACTS | NEGATIVE IMPACTS | NEGATIVE IMPACTS |
| • Exploitation of environmental resources<br>• Generation of environmental pollution<br>• Pressure of seasonal traffic and congestion<br>• Degradation of trails and pathways<br>• Effect of unregulated construction and unplanned structures | • Commercialization of tradition, arts, and culture<br>• Altercation of local culture and lifestyle<br>• Conflicts between benefitting and non-benefitting from tourism<br>• Increased workload on mountain women | • Seasonality of Employment and Business<br>• Lack of Retention of Benefits<br>• Over-reliance on Tourism<br>• Inflation of goods, services, and properties<br>• Leakage of revenue |
| **Environmental Impacts** | **Socio-cultural Impacts** | **Economic Impacts** |

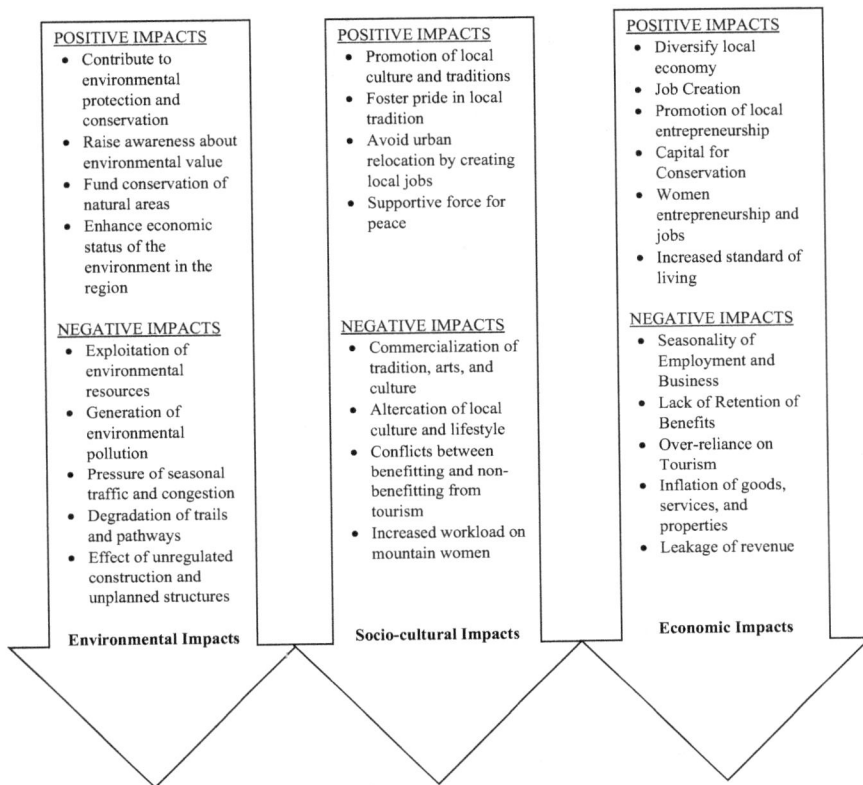

*Figure 14.1* Impacts of Tourism in the Himalayan Region

Sustainable mountain tourism in the Himalaya is frequently recognized as a panacea for the region's current socioeconomic problems and future economic advancement. The linkages between the economic, environmental, and socio-cultural aspects of tourism need to be further established to create sustainable development opportunities. This may be an effective mechanism to enhance local livelihoods and protect the region's natural and cultural heritage. This chapter outlines the impacts of tourism in the Himalayan region along with a response strategy, namely Corporate Social Responsibility (CSR), to promote sustainable mountain tourism in this part of the world.

## Tourism's Impacts in the Himalayan Region

The Himalayan region has its own unique set of challenges and associated impacts directly and indirectly through tourism. The communities are deeply reliant on available natural resources and products, such as agriculture, forestry, and livestock. The complex, diverse, vulnerable, and extreme ecological and social conditions distinctive to the region are somewhat incompatible with the region's increasing tourism demand. Furthermore, highland inhabitants are usually less affluent than the people who reside in lowland areas, even within the same countries; in reality, most mountainous areas of the region are poverty-ridden (Godde et al., 2000; Gioli et al., 2019; Messerli & Ives, 1997). The situation is exacerbated by difficulties in access and a tough climate, which makes infrastructure development extremely challenging (Singh & Mishra, 2004). Additionally, the Himalayan nations are susceptible to political instability, often due to the proximity and divisive nature of their international borders (Nepal & Chipeniuk, 2005) and conflicts over natural resources. Hence, despite rich resources, abundant biological reserves and cultural assets, the region remains underdeveloped.

Tourism provides an alternative livelihood and is often over-relied upon for its economic development potential, job opportunities, and income generation (Geneletti & Dawa, 2009). However, inflated prices, revenue leakage, and the seasonality of employment and benefits remain some of the region's key issues (Rai, 2017).

Increased tourism has also exacted severe impacts on the region's environment and society. For instance, it has led to extensive deforestation, flooding in the lowlands, pollution from vehicles, and waste left behind by trekkers. All of these outcomes have threatened the region's rich biodiversity to the extent of extinction of some species (Singh, 2006). The environmental impacts are more pronounced in the Himalayan region due to extreme tourism seasonality over short summer seasons, absence of appropriate infrastructure and planning, and interference with fragile ecosystems and protected areas (Buckley et al., 2000; Geneletti & Dawa, 2009). Furthermore, the region's high-altitude ecosystems are delicate, irresilient and vulnerable to human interference.

With respect to sociocultural impacts, changes in the local community have been experienced due to host-tourist interactions with exposure to a visible and robust Western lifestyle (Kunwar, 2012), which have been known to alter local cultures and lifestyles. Similarly, the commoditization of cultural artifacts, performances, and ceremonies that are deliberately altered to fit tourists' tastes and desires has been evident (Rai, 2017). Although women are more gainfully employed in tourism today, they are still typically subject to inequality and lack access to credit (Sherpa, 2012).

In summary, sustainability concerns and obstacles in the Himalaya are amplified because of the region's specific characteristics, including poverty and underdevelopment; difficulty in earning livelihoods; a general absence of adequate infrastructure and transportation; a dearth of sound planning; and a lack of enforceable tourism, environmental, and social policies (Geneletti & Dawa, 2009; Singh, 2002). The Himalayan countries are for the most part developing and riddled with political, social, and economic problems. Many of the region's problems derive from a general lack of social responsibility within the tourism sector, and thus a paradigm shift with respect to tourism planning and development is direly needed.

CSR offers a potential avenue to address and mitigate negative impacts, as well as formulate sustainable tourism enterprises and development.

## CSR and Sustainable Mountain Tourism in the Himalayan Region

In the 1990s, sustainable tourism gained prominence as a potential answer to the challenges posed by the scale, scope, and consequences of tourism development (Sharpley, 2010). According to UNEP and UNWTO (2005, pp. 11–12), sustainable tourism is "tourism that takes full account of its current and future economic, social and environmental impacts, addressing the needs of visitors, the industry, the environment, and host communities". This definition emphasizes that people should be responsible, respectful, and desirous to preserve the balance of economic, environmental, and sociocultural dimensions (Patterson, 2015). Essentially, sustainable tourism assures good ecological stewardship with the optimal utilization of environmental resources, while simultaneously valuing the preservation and promotion of host communities' sociocultural characteristics and ensuring their long-term economic benefits (Dodds & Butler, 2010).

To achieve sustainable tourism goals, all stakeholders (e.g. tour operators, travel agencies, transportation companies, tourism businesses, governments, community members, NGOs, and tourists) need to work toward common goals (Kasim, 2006). In this pursuit, tourism businesses can play especially significant and positive roles (Behringer & Szegedi, 2016) and participate through CSR engagement. According to the World Business Council for Sustainable Development (1999, p. 3), CSR is a "commitment by business to behave ethically and contribute to sustainable economic development while improving the quality of life of the workforce and their families as well as of the local community and society at large". Hence, CSR in tourism is an evolving business practice that incorporates sustainable development into its operating model to mitigate adverse impacts.

CSR emphasizes the economic, social, and environmental dimensions of sustainability (Panapanaan et al., 2003; Panwar et al., 2006). Martínez et al. (2013, p. 372) explain that good economic stewardship requires safeguarding long-term economic accomplishments to enable all stakeholders to benefit socioeconomically. The social dimension contributes to multicultural tolerance and acceptance, and preserving and fortifying traditional values and cultural assets. The environmental dimension emphasizes using natural assets in a way that protects and conserves them. CSR aims to achieve economic, social, and environmental sustainability by balancing the needs and benefits of these dimensions (Alvarado-Herrera et al., 2017). CSR encourages businesses to achieve financial success while following ethical values and valuing individuals, communities, and natural environments (Manente et al., 2014) to uphold the principles of sustainability (Martínez & Del Bosque, 2013).

The adoption of CSR principles by individual businesses, corporations, and destinations has yielded numerous business, most notably an increase and retention of consumer trust (Font & Lynes, 2018; Martínez & Del Bosque, 2013).

Consumers choose companies that take responsibility for their actions, namely those with positive CSR reputations (Eraqi, 2010). Besides maximizing profits, CSR influences tourism companies to value their staff, local communities, the environments where they operate, and other stakeholders (Inoue & Lee, 2011). Most importantly, implementing CSR leads to environmental improvements, including energy conservation, green production/services, support for non-profit environmental organizations, and pollution reduction. It also enhances destination inhabitants' quality of life and safeguards employees' well-being (Font et al., 2012). In reality, the focus and actions of tourism companies that practice CSR have made tourism more sustainable (Camilleri, 2017; Paskova & Zelenka, 2019), which is sorely needed in the Himalayan region.

## Tourism and CSR in the Himalayan Nations

Numerous institutions and governments have promoted the sustainable development of mountain tourism globally (Köhler et al., 2015), but there is lack of perspectives on CSR and sustainability in the Himalayan region. This section attempts to fill a gap in the literature on our understanding of CSR in the countries of the region: Bhutan, Pakistan, China, India, and Nepal. Although Afghanistan is within the broader Himalayan region and there are nascent efforts at CSR there (see Azizi & Jamali, 2016), the current political situation in that country is inconducive to CSR's widespread application. As well, there is a dearth of information about the CSR paradigm in Afghanistan's embryonic tourism sector, if it exists at all.

### *Bhutan*

Situated in the Eastern Himalaya, Bhutan is a relatively new entrant to global tourism, embracing it as late as the 1970s (Dorji, 2001; Nepal & Karst, 2017). Since transitioning towards global tourism, Bhutan has adopted a controlled model based on "high-value, low-impact" (officially, "high-value, low-volume") tourism aimed at protecting its environment and its unique culture from the potential negative influences of unrestricted tourism (Nyaupane & Timothy, 2010; Teoh, 2012). Under this model, the government appointed a single authorized body to be responsible for all tourism-related services and to control the types and quantity of tourism (Tourism Council of Bhutan, 2010). Independent travel is not allowed; non-Indian tourists must buy tour packages which include ground transportation, lodging, meals, and guide services, and they are required to enter or leave the country (at least one way) via the national airline, Drukair, with the cost of the air ticket not included in the package prices. For their tour packages, tourists are required to spend a minimum amount of money each day—$250 during high season, $200 during low season, although more upmarket experiences are available for higher daily tariffs. Although there are no visa quotas, the high cost of visiting the country and the policy against independent travel naturally limit the number of people who visit each year (Kingdom of Bhutan, 2021; Nyaupane & Timothy,

2010), although the policy does not apply to Indians and several other nationalities from the SAARC countries, which may pose some threats to over-visitation (Nyaupane & Timothy, 2010).

Owing to persistent pressure from the private sector, in the 1990s, the government permitted increased entrepreneurial activity and engagement of Bhutanese tour operators, which continues even today (Kingdom of Bhutan, 2021). Tourism stakeholders are committed to maintaining the visual and cultural aesthetics of the country. In fact, they initially coerced the government to revise its ambitious plan to target 250,000 annual tourist arrivals with no daily tariff to 100,000 with a minimum daily tariff of US$200–250 per day, depending on the season (Schroeder & Sproule-Jones, 2012).

The kingdom's unique Gross National Happiness (GNH) index has guided the tourism industry's sustainability effort to prioritize cultural and environmental interests over economic gains (Nepal & Karst, 2017). In recent years, Bhutan has witnessed a growing trend towards nature-based, community-based, and ecotourism initiatives to disperse the economic benefits to rural regions (Rinzin et al., 2007). In general, the country's unique tourism policies have encouraged sustainable tourism by minimizing its environmental and cultural impacts (Nyaupane & Timothy, 2010).

Despite the positive role of tourism in Bhutan's economy and social well-being, the adoption of CSR is still in a nascent state as it lacks a strategic focus and wider stakeholder engagement as it is customarily linked to philanthropy, such as endowments to education, sports, and hermetic causes (BCCI, 2013). With a lack of regulatory reforms to incentivize CSR, the business community is generally ambivalent about CSR implementation and instead favours the national mandate of GNH, which also supports traits such as ethics and holistic vision (Sebastian, 2015). In 2014, the Company Bill was enacted to inch closer toward CSR; it required businesses to formulate CSR policies to encourage compliant behaviour (Benn et al., 2014). Collectively, sustainable tourism in Bhutan seems to require a government mandate, as tourism businesses' voluntary contributions via CSR appears to be limited.

### *Pakistan*

Located in South Asia, Pakistan welcomes trekkers, climbers, hikers, and mountaineers in the Karakoram, Hindu Kush, Himalaya, and Pamir mountain ranges in the country's northern regions (Manzoor et al., 2019). In 2017, Pakistan witnessed record growth in international tourism with 1.75 million visitors and US$19.4 billion in revenue (WTTC, 2017). Pakistan's tourism has mainly focused on income generation and the associated economic benefits, with inadequate emphasis on the environmental costs (Shujahi & Hussain, 2016). Specifically, the principles of sustainability have not received adequate consideration from the tourism industry, as the country was ranked 121st out of 140 economies in 2019 in the Travel and Tourism Competitiveness Index (TTCI), which takes into account certain elements of sustainable tourism (World Economic Forum, 2019). While there have been several legislative actions (e.g., Pakistan Tourist Guides Act, 1976; Travel Agencies Act, 1976; Pakistan Hotels & Restaurants Act, 1976) to promote standardized sustainable practices among tourism businesses, implementation has been difficult (Arshad et al., 2018).

CSR emerged only a decade ago and was mostly a topic of media discussion (Ehsan et al., 2018). It was promoted largely by multinational companies at the forefront of the implementation process via dedicated departments (Khan et al., 2018). While some small and medium enterprises (SMEs) have formulated CSR policies, most remain unaware about its implementation and lack social responsibility (Raza & Majid, 2016). Generally, CSR has been associated with philanthropic investments in the health and education sectors, but recent trends have shown broader contributions towards environmental, community, and social issues (Khan et al., 2018). Such charitable endeavours are historically considered outcomes of cultural, religious, and family traditions (Sajjad & Eweje, 2014). Nevertheless, Pakistan has an inconsistent CSR implementation process due to a lack of uniformity in laws (Ahmed & Ahmad, 2011).

Recently, the government published voluntary guidelines for CSR via the Securities and Exchange Commission of Pakistan (SECP) to encourage businesses to practice socially responsible conduct (Lone et al., 2016). The guidelines recommend that businesses formulate a CSR policy to be incorporated by their boards and subsequently report related activities to reflect their commitment. However, the extent of these CSR disclosures seems to vary considerably across sectors, thereby reducing the effectiveness of the SECP.

## *India*

India hosted over 10 million foreign tourists in 2017 (Ministry of Tourism, Government of India, 2018). The country's Himalayan states have significant tourism potential due to their unique scenery and natural landscapes, opulent biodiversity, and cultural multiplicity. As a result, these states have attracted sizeable volumes of tourists and have transitioned from being traditional pilgrimage destinations to centres of mass tourism. Consequently, the mountain states have experienced levels of tourism growth and diversification that have transformed tourism into one of the fastest-growing economic sectors in the country (Gaur & Kotru, 2018).

Along with this growth have come impacts on climate, water security, and livelihoods (Wang et al., 2019) and several other negative impacts. These include the substitution of customary eco-friendly and aesthetically pleasing architecture with inappropriate, unappealing, and dangerous structures; poorly designed roads and associated infrastructure; insufficient solid waste management; increased air pollution; degradation of watersheds and water sources; and the loss of natural resources, biodiversity, and ecosystem services (Gaur & Kotru, 2018; Singh et al., 2009). The deficiency in environmental compliance and the very nominal investments made to promote sustainable tourism have adversely affected the sustainable development outlook in the region. Consequently, the tourism industry has observed a significant decrease in environmental sustainability rankings (from 41 in 2007 to 134 in 2017) in the Travel and Tourism Competitiveness Index provided by the World Economic Forum (Dahiya & Batra, 2018).

CSR in India has been synonymous with philanthropy where most large corporations contribute to social welfare with charitable donations (Gautam & Singh, 2010). This is attributed to the Gandhian philosophy of community development, which inspired charitable initiatives to support the underprivileged and establish educational institutions (Chaudhri & Wang, 2007). Recently, CSR engagement has been driven either by the desire to increase shareholder value or by the idea that it is the "right thing to do" (Sharma, 2018, p. 165). Hence, CSR has likely been executed on unplanned acts of charity rather than long-term investments due to the lack of a holistic CSR philosophy being thoroughly ingrained in businesses' missions or strategies.

Recently, the government has taken a grassroots approach to put more emphasis on the private sector's role in community development (Sharma, 2018). Through the Company Act 2013, it mandated that Indian companies with an annual return of about US$160 million or a net profit of more than US$830,000 during any of the previous three fiscal years must contribute 2% of their profits to CSR activities (Arul, 2014). This law institutionalized CSR and directed a combined total of over US$2 billion annually towards CSR undertakings (Baroth & Mathur, 2019; Sharma, 2018). This also marked a shift from philanthropic CSR to strategic investments in social projects as companies were encouraged to collaborate with local NGOs as viable implementation partners (Ray & Raju, 2014). Nevertheless, investments still energized businesses more towards philanthropy than other relevant aspects of CSR responsibilities (Deodhar, 2016).

### China

The tourism industry in China has flourished since the reforms of 1978. The annual contribution of the national tourism industry is about 9.13 trillion yuan (US$1.40 trillion), which translates into about 11.04% of the total GDP, and accounts for 10.28% of the total employment (Jiang et al., 2018). Tourism in the mountain regions of China, including the Himalayan reaches of the Tibetan Plateau, has increased in recent years and developed as a strategic principal sector for local economic progress (Zhang et al., 2017). But this growth has also led to major negative impacts on the fragile mountain ecosystems (Zhang et al., 2017).

CSR involves direct participation of the government (Tang, 2012), which has led to comprehensive programmes run by state-owned corporations to establish laws and regulations, guidelines, standards, and organizations. These include the Chinese Company Law in 2006, CSR principles for central-government controlled companies, Labor Contract law in 2008, and the Taida Environmental Index in the Shenzhen Stock Exchange (Lin, 2010). The most crucial policy has been the state-owned Asset Supervision and Administration Commission initiative, which involved the issuance of the "Guidelines on Fulfilling Social Responsibility by Central Enterprises". These guidelines were established by law in 2008 and indicated that CSR is politically accepted and endorsed in the state (Tang, 2012). The government-led China National Tourism Administration (CNTA) advocated for the incorporation of sustainable tourism development principles for all projects and ultimately became the standard (Sofield & Li, 2011).

Nevertheless, a CSR philosophy has not been widely adopted by Chinese corporations. Chinese managers tend to prioritize profits and business survival over social responsibility and their shareholders' interests atop any other business objectives (Shafer et al., 2007). Similarly, around 70% of managers in one study anticipated that CSR would lead to higher costs, and the brand reputation was considered the main driver to be socially responsible (Kolk et al., 2010). Hence, the private sector has been driven overwhelmingly by increased profit motives, and the expense of a reduced sense of responsibility towards society and the environment. One of the reasons for this might be that the public feels that state organizations and companies should play the role of CSR providers (Zhang et al., 2017). Overall, there appears to be inadequate mandates and facilitators for private companies to engage in CSR without any justification to do so of their own volition.

## Nepal

Nepal is a landlocked country in South Asia where mountain tourism began in 1949 as the country began opening up to foreign tourists (Sharma, 2009). It is home to eight 8,000-metre or higher mountains including Mount Everest, making it a primary destination for trekkers, mountaineers, and other tourists (Salisbury & Hawley, 2011). At the macro-level, tourism is the country's principal source of foreign exchange; at the micro-level, it is widely recognized as the most promising livelihood opportunity for mountain communities (Kruk & Banskota, 2007). In this milieu, tourism is Nepal's mainstay industry, and mountain tourism constitutes a significant part of it.

Mountain tourism lacks stringent regulations, which has enabled virtually uncontrolled development and led to major environmental (increased extraction of fuelwood and timber, increased loss of tree and shrub cover, pollution), socio-economic (local inflation of the costs of goods and services), and sociocultural problems (alienation of locals and ethnic prejudices) (Nepal, 2002, 2011; Upadhayaya, 2015). Following early signs of problems, several famous protected areas were created to preserve the environment and promote sustainable tourism in the Nepali Himalaya (Sacareau, 2009). The development of national parks, wildlife preserves, buffer zones, conservation areas, and cultural heritage has helped protect wildlife and historic sites and enhance ecotourism opportunities that have significantly influenced sustainability (Baral et al., 2012).

The tenets of CSR have been practiced less formally by businesses for the past century, but mostly in a philanthropic sense, such as the construction of religious inns, contributions to educational institutions, and donations for temples (Adhikari et al., 2016). Such undertakings were ad hoc acts of altruism that lacked both strategic and systematic approaches. This situation has prevailed in most private and public companies, where "profit-orientation" and "building family brand image" were primary concerns (Adhikari, 2012, p. 646). This is largely due to the fact that most Nepali businesses are small or mid-sized and family-owned, driven by a traditional management approach and the owners' convictions (Welzel, 2006). Over time, businesses have started to focus on the economic aspects of CSR, which is a shift from the previous charity and donation approach (Adhikari et al., 2016). Tourism SMEs are increasingly aware of CSR and are willing to participate, but most participation is still philanthropic in nature, driven by the owner's judgement and undertaken mostly for marketing purposes (Baniya et al., 2019). The government has addressed this in the form of the Industrial Enterprise Act 2016, which mandates specific industries to participate in CSR based on their fixed capital investment and annual turnover. However, there is a lack of strict monitoring and oversight of the law, making it largely ineffective.

Nonetheless, the tourism industry has also presented some best practices of CSR to address social, economic, and environmental challenges. For example, Three Sisters Adventure trekking company has rallied against gender inequality by engaging women to lead other women on treks. Similarly, Bakery Café, a popular restaurant chain employs deaf employees to mitigate discrimination against people with disabilities. Likewise, many SMEs have dedicated some of their resources to philanthropic CSR activities in their vicinities such as for building schools, donating clothing to students, donating money to local monasteries, and building trails. Overall, these initiatives are laudable, but they still remain scattered and lack the strength to address the broader sustainability issues in the region.

## CSR for Sustainable Mountain Tourism: A Way Forward

CSR actions can play a crucial role in sustainable mountain tourism in the Himalayan region. Sustainable mountain tourism emphasizes local ecological, social, and economic conditions to achieve a balanced use of natural resources, preservation and promotion of socio-culture, and long-term economic gains. Tourism businesses in the Himalaya have a significant role to play in establishing sustainable mountain tourism through their CSR engagements. However, businesses need to consider several factors in implementing CSR. First, CSR should be integrated with core business strategies. The strategies, business models, and practices of tourism enterprises can uphold sustainable development. Tourism businesses should incorporate social and environmental objectives into their core business strategies "to do well while doing good" (Spiller, 2000, p. 149). This would place CSR at the center of decision-making and guide business operations.

A second consideration in implementing CSR is formulating its principles within a sustainable development framework. The UN's 2030 Agenda for Sustainable Development sets the pattern for all states to follow with its 17 Sustainable Development Goals (SDGs). Tourism can play a pivotal part in attaining these goals (Adie et al., in press; Tham et al., 2020; Timothy, 2021) by encouraging tourism enterprises—accommodations, transportation, tour operators, and other travel intermediaries—that practice CSR to align their strategies with the SDGs.

For example, SDG 15 is to safeguard, restore, and promote sustainable use of terrestrial ecosystems and end biodiversity loss. This could be a key strategy for tourism businesses in the region. Under this initiative, enterprises might consider contributing to wildlife and ecosystems protection, pollution reduction, better waste treatment systems, buying green and locally-produced products, and sharing green information with customers.

Third, businesses can address local sustainability issues more succinctly. Himalayan tourism is unique in terms of its scope, opportunities, and challenges. Accordingly, business CSR initiatives should account for the context in which the business operates and address local environmental, social, and economic challenges. For instance, environmental hazards due to fire, deforestation, and the use of plastics are highly prominent in the Himalaya, so the business should consider the practices that help reduce environmental hazards.

A fourth viable action is promoting and supporting the government's role in CSR implementation. Governments at every level should collaboratively play a principal role in promoting, monitoring, and controlling CSR, as governments tend to be the primary drivers of sustainable development. With appropriate policies, governments can be catalysts in implementing CSR via tourism. Five thematic strategies would be well suited for the Himalayan region:

1. Raise awareness and build capacities for CSR within the tourism business community
2. Improve disclosure and transparency within the tourism business community
3. Facilitate socially responsible investments for tourism businesses
4. Lead by example with socially-accountable practices that can foster CSR within the tourism business community
5. Establish policies that incentivize CSR activities.

Fifth, enterprises can promote stakeholder engagement in implementing CSR. Tourism in the region is a network of complex phenomena that entails intricate interactions between various stakeholders from customers, tour operators, accommodation providers, porters, large hotels, government departments, and special interest groups. CSR requires stakeholder engagement and includes identifying common problems and interests, consulting with one another, communicating openly, dialoguing, and exchanging ideas. The importance of collective interactive engagement cannot be understated, as every stakeholder is unlikely to possess all the knowledge and resources needed to achieve their goals individually (Timothy, 1998). For example, proper waste management is a prominent concern in the region. Hotels can incorporate waste management as CSR in their operations and engage with stakeholders along with local and national hotel associations. They can also educate and train their customers through brochures, websites, menus, and employees. In addition, they can learn proper waste management techniques from local NGOs/INGOs and the government. Thus, there is a great deal of potential for stakeholder engagement to manage waste effectively.

A sixth way of adopting CSR in business is through sustainable supply chain management. The principles of sustainability should be incorporated into existing supply chain management to consider the environmental, social, and economic impacts of business activities. Sustainable supply chain management may be particularly beneficial for tourism businesses as a result of their close connections and influence on suppliers. Such actions incorporate sustainable transport development and sustainable use of resources; reduces, minimizes, and prevents pollution and waste; safeguards plants, animals, ecosystems, and protected areas (biodiversity); and preserves landscapes and natural heritage. Socio-economic and cultural perspectives include actions that contribute to the economic development and well-being of local communities, safeguarding cultural identities, respecting human rights—local communities' and indigenous people's rights (Tourism Operators Initiative & Center for Environmental Leadership in Business, 2003).

Finally, tourism businesses can adopt CSR certification to monitor, measure, and report. Demand for greater accountability, transparency, and credibility from the business community can be met based on their performance. Independent institutions and governments can play a monitoring role and use existing certifications to facilitate the entire process. For example, the Global Sustainability Tourism Council (GSTC) provides certification for hotels, tour operators, and destinations. Accordingly, GSTC performance indicators can be utilized by tourism enterprises in the region in the following ways:

1.  *Demonstrate effective sustainability management*—Sustainability management entails three distinct undertakings by tourism enterprises for CSR. First, tourism enterprises should incorporate sustainable tourism policies formulated by the local government and SDGs in their strategy and actions. For example, enterprises' strategies and actions may include managing visitor volumes and activities, following legal compliance, promoting staff engagement, providing superior customer experience, respecting and promoting local culture, ensuring effective waste management, erecting buildings and infrastructure, utilizing local produce, and ensuring decent work life and wages. Second, tourism enterprises should be ready to invest resources to effectively address the region's economic, sociocultural, and environmental issues. Third, tourism enterprises should follow robust periodic monitoring and reporting of their compliance with regard to all three sustainability dimensions.

2.  *Maximize social and economic benefits to the local communities and minimize negative impacts*—Tourism enterprises can actively support local infrastructure and social development initiatives, provide employment and growth opportunities for residents, purchase local goods and services, and support local entrepreneurs in their sustenance and development. Likewise, they can employ zero tolerance to exploitation and harassment, provide equal opportunities to their employees for growth, and promote local access tourist sites.

3.  *Maximize benefits to cultural heritage and minimize negative impacts*—Tourism enterprises must demonstrate awareness of, and compliance with, local

cultural sites and Himalayan indigenous communities. They must play an active role in protecting the region's culture and heritage. Additionally, they should promote local culture and heritage in the best possible way, according to the values and norms of the local communities.

4.  *Maximize benefits to the environment and minimize adverse effects, by*

    •   Conserving resources—For tourism enterprises, this entails purchasing environmentally friendly products, consumables and goods with minimal waste, and ensuring water and energy conservation.
    •   Reducing pollution—Tourism enterprises can transform themselves into carbon zero institutions, utilize less polluting transportation, use alternative cleaner transport, minimize wastewater and solid waste production, avoid harmful substances, and minimize other pollutions such as light and noise.
    •   Protecting biodiversity, ecosystems, and landscapes—Tourism enterprises should support and contribute to biodiversity conservation with minimal disturbances to natural eco-systems, and rehabilitate or provide compensation if any impacts are created to the ecosystem. Furthermore, enterprises should create and follow guidelines for visiting natural areas and engage in appropriate wildlife interaction.

## Conclusion

CSR practices by tourism companies in the Himalayan nations would demonstrate a philanthropic ethos. However, CSR practices have not yet been widely adopted by tourism businesses and are therefore ineffective in contributing to sustainable mountain tourism. For effective implementation, tourism businesses should promote stakeholder participation, establish sustainable supply chain management, and adopt a certification process. Also, the government should exhibit a decisive role in promoting and facilitating CSR activities among tourism businesses to contribute to sustainable tourism in the Himalaya.

CSR can be a critical tool for harnessing tourism's positive benefits in the Himalayan region and addressing its adverse effects. CSR would be a suitable guiding business philosophy for tourism enterprises, as it is in many other industries where economic, social, and environmental concerns are integrated in their missions, strategies, and operations commensurate with their scope and size. The primary sustainability goals for tourism enterprises are to manage the region's existing environmental, sociocultural, and economic assets to satisfy the needs of current stakeholders and ensure protection for future generations.

CSR can play an influential role in sustainable mountain tourism in the Himalaya if it is guided by clear visions of SDGs within a triple-bottom-line framework, endorsed by local authorities in developing sustainable tourism policies, and owned and implemented by the region's tourism enterprises. CSR can help Himalayan tourism enterprises become increasingly more responsible for addressing the region's economic, sociocultural, and environmental issues in relation to their size and scope. Through corporate social responsibility, companies of every size can play an active role in balancing the triple-bottom-line dimensions in all decisions and actions undertaken. Whether by compliance with rules and regulations or by their own volition, tourism enterprises in the Himalaya can fulfil their responsibility as good corporate citizens by adopting sustainability principles in implementing their commercial activities.

# References

Adhikari, D.R. (2012). Status of corporate social responsibility in selected Nepalese companies. *Corporate Governance*, 12(5), 642–655.

Adhikari, D.R., Gautam, D.K., & Chaudhari, M.K. (2016). Corporate social responsibility domains and related activities in Nepalese companies. *International Journal of Law and Management*, 58(6), 673–684.

Adie, B.A., Amore, A., & Hall, C.M. (in press). Just because it seems impossible, doesn't mean we shouldn't at least try: The need for longitudinal perspectives on tourism partnerships and the SDGs. *Journal of Sustainable Tourism*.

Ahmed, A., & Ahmad, I. (2011). *Corporate Conscience: CSR in Pakistan—A Study*. Bangalore: Sustainable Development Policy Institute.

Alvarado-Herrera, A., Bigne, E., Aldas-Manzano, J., & Curras-Perez, R. (2017). A scale for measuring consumer perceptions of corporate social responsibility following the sustainable development paradigm. *Journal of Business Ethics*, 140(2), 243–262.

Arshad, M.I., Iqbal, M.A., & Shahbaz, M. (2018). Pakistan tourism industry and challenges: A review. *Asia Pacific Journal of Tourism Research*, 23(2), 121–132.

Arul, P.G. (2014). Policy implication of Indian companies act 2013 on corporate social responsibilities (CSR) and its impact on domestic company and MNCs spending. *SJCC Management Research Review*, 4(1), 1–21.

Azizi, S., & Jamali, D. (2016). CSR in Afghanistan: A global CSR agenda in areas of limited statehood. *South Asian Journal of Global Business Research*, 5(2), 165–189.

Baniya, R., Thapa, B., & Kim, M.S. (2019). Corporate social responsibility among travel and tour operators in Nepal. *Sustainability*, 11(10), 2771.

Baral, N., Hazen, H., & Thapa, B. (2017). Visitor perceptions of world heritage value at Sagarmatha (Mt. Everest) National Park, Nepal. *Journal of Sustainable Tourism*, 25(10), 1494–1512.

Baral, N., Stern, M.J., & Hammett, A.L. (2012). Developing a scale for evaluating ecotourism by visitors: A study in the Annapurna conservation area, Nepal. *Journal of Sustainable Tourism*, 20(7), 975–989.

Baroth, A., & Mathur, V.B. (2019). Wildlife conservation through corporate social responsibility initiatives in India. *Current Science*, 117(3), 405.

Behringer, K., & Szegedi, K. (2016). The role of CSR in achieving sustainable development-theoretical approach. *European Scientific Journal*, 12(22).

Benn, S., Edwards, M., & Williams, T. (2014). *Organizational Change for Corporate Sustainability*. London: Routledge.

BCCI (2013). *Corporate Social Responsibility in Bhutan: General Perception and Some Related Problems*. Thimphu: BCCI.

Buckley, R.C., Pickering, C.M., & Warnken, J. (2000). Environmental management for Alpine tourism and resorts. In P.M. Godde, M.F. Price, & F.M. Zimmermann (Eds.), *Tourism and Development in Mountain Regions* (pp. 27–45). Wallingford: CABI.

Camilleri, M.A. (2017). *Corporate Sustainability, Social Responsibility and Environmental Management*. Cham, Switzerland: Springer.

Charters, T., & Saxon, E. (2007). *Tourism and Mountains: A Practical Guide to Managing the Environmental and Social Impacts of Mountain Tours*. Washington, DC: UNEP.

Chaudhri, V., & Wang, J. (2007). Communicating corporate social responsibility on the internet: A case study of the top 100 information technology companies in India. *Management Communication Quarterly*, 21(2), 232–247.

Dahiya, K.S., & Batra, D.K. (2018). India—Sustainability and the tourism rankings. *African Journal of Hospitality, Tourism and Leisure*, 7(3), 1–20.

Deodhar, S.Y. (2016). Trapping India's CSR in a legal net: Will the mandatory trusteeship contribute to triple bottom line? *Vikalpa*, 41(4), 267–274.

Dodds, R., & Butler, R. (2010). Barriers to implementing sustainable tourism policy in mass tourism destinations. *Tourismos*, 5, 35–53.

Dorji, T. (2001). Sustainability of tourism in Bhutan. *Journal of Bhutan Studies*, 3(1), 84–104.

Ehsan, S., Nazir, M., Nurunnabi, M., Raza Khan, Q., Tahir, S., & Ahmed, I. (2018). A multimethod approach to assess and measure corporate social responsibility disclosure and practices in a developing economy. *Sustainability*, 10(8), 2955.

Eraqi, M.I. (2010). Social responsibility as an innovative approach for enhancing competitiveness of tourism business sector in Egypt. *Tourism Analysis*, 15(1), 45–55.

Font, X., & Lynes, J. (2018). Corporate social responsibility in tourism and hospitality. *Journal of Sustainable Tourism*, 26(7), 1027–1042.

Font, X., Walmsley, A., Cogotti, S., McCombes, L., & Häusler, N. (2012). Corporate social responsibility: The disclosure—performance gap. *Tourism Management*, 33(6), 1544–1553.

Gaur, V.S., & Kotru, R. (2018). *Report of Working Group II: Sustainable Tourism in the Indian Himalayan Region*. New Delhi: NITI Aayog.

Gautam, R., & Singh, A. (2010). Corporate social responsibility practices in India: A study of top 500 companies. *Global Business and Management Research: An International Journal*, 2(1), 41–56.

Geneletti, D., & Dawa, D. (2009). Environmental impact assessment of mountain tourism in developing regions: A study in Ladakh, Indian Himalaya. *Environmental Impact Assessment Review*, 29(4), 229–242.

Gioli, G., Thapa, G., Khan, F., Dasgupta, P., Nathan, D., Chhetri, N., Adhikari, L., Mohanty, S.K., Aurino, E., & Scott, L.M. (2019). Understanding and tackling poverty and vulnerability in mountain livelihoods in the Hindu Kush Himalaya. In P. Wester, A. Mishra, A. Mukherji, & A.B. Shrestha (Eds.), *The Hindu Kush Himalaya Assessment: Mountains, Climate Change, Sustainability and People* (pp. 421–455). Cham: Springer.

Godde, P.M., Price, M.F., & Zimmermann, F.M. (2000). Tourism and development in mountain regions: Moving forward into the new millennium. In P.M. Godde, M.F. Price, & F.M. Zimmermann (Eds.), *Tourism and Development in Mountain Regions* (pp. 1–25). Wallingford: CABI.

Grötzbach, E. (1985). Tourismus im Indischen Westhimalaya—Entwicklung und räumliche Struktur. In J. Steinbach (Ed.), *Beiträge zur Fremdenverkehrsgeographie* (pp. 27–47). Eichstätt: Fachgebiet Geographie der Katholischen Universität Eichstätt.

Grötzbach, E. (1994). Hindu-Heiligtümer als Pilgerziele im Hochhimalaya (Hindu Shrines as Pilgrim Destinations in the High Himalayas). *Erdkunde*, 48(3), 181–193.

Gupta, J. (2015). Tourism in Himachal Pradesh: A study of Kullu Manali. *International Journal of Engineering Research & Management Technology*, 2(3), 183–191.

Inoue, Y., & Lee, S. (2011). Effects of different dimensions of corporate social responsibility on corporate financial performance in tourism-related industries. *Tourism Management*, 32(4), 790–804.

Jiang, H., Yang, Y., & Bai, Y. (2018). Evaluation of all-for-one tourism in mountain areas using multi-source data. *Sustainability*, 10(11), 4065.

Kasim, A. (2006). The need for business environmental and social responsibility in the tourism industry. *International Journal of Hospitality & Tourism Administration*, 7(1), 1–22.

Kennedy, D. (1996). *The Magic Mountains*. Berkeley: University of California Press.

Khan, M., Lockhart, J.C., & Bathurst, R.J. (2018). Institutional impacts on corporate social responsibility: A comparative analysis of New Zealand and Pakistan. *International Journal of Corporate Social Responsibility*, 3(1), 1–13.

Kingdom of Bhutan (2021). *Tourism Policy of the Kingdom of Bhutan*. Thimphu: Tourism Council of Bhutan.

Köhler, P., de Boer, B., von der Heydt, A.S., Stap, L.B., & van de Wal, R.S. (2015). On the state-dependency of the equilibrium climate sensitivity during the last 5 million years. *Climate of the Past*, 11, 1801–1823.

Kolk, A., Hong, P., & Van Dolen, W. (2010). Corporate social responsibility in China: An analysis of domestic and foreign retailers' sustainability dimensions. *Business Strategy and the Environment*, 19(5), 289–303.

Kruk, E., & Banskota, K. (2007). Mountain tourism in Nepal: From impacts to sustainability. In *Tourism and Himalayan Biodiversity* (pp. 15–34). Srinagar: Transmedia.

Kunwar, R.R. (2012). *Tourists & Tourism: Science and Industry Interface*. Kathmandu: Ganga Sen Kunwar.

Lin, L.W. (2010). Corporate social responsibility in China: Window dressing or structural change. *Berkeley Journal of International Law*, 28(1), 64–100.

Lone, E.J., Ali, A., & Khan, I. (2016). Corporate governance and corporate social responsibility disclosure: Evidence from Pakistan. *Corporate Governance: The International Journal of Business in Society*, 16(5), 785–797.

Manente, M., Minghetti, V., & Mingotto, E. (2014). *Responsible Tourism and CSR: Assessment Systems for Sustainable Development of SMEs in Tourism*. Cham: Springer.

Manzoor, F., Wei, L., & Asif, M. (2019). The contribution of sustainable tourism to economic growth and employment in Pakistan. *International Journal of Environmental Research and Public Health*, 16(19), 3785.

Martínez, P., & del Bosque, I.R. (2013). CSR and customer loyalty: The roles of trust, customer identification with the company and satisfaction. *International Journal of Hospitality Management*, 35, 89–99.

Martínez, P., Pérez, A., & Rodriguez del Bosque, I. (2013). Measuring corporate social responsibility in tourism: Development and validation of an efficient measurement scale in the hospitality industry. *Journal of Travel & Tourism Marketing*, 30(4), 365–385.

Messerli, B., & Ives, J.D. (1997). *Mountains of the World: A Global Priority*. Nashville: Parthenon Publishing.

Ministry of Culture, Tourism and Civil Aviation (2020). *Nepal Tourism Statistics-2014*. Singha Durbar: Government of Nepal, Ministry of Culture, Tourism & Civil Aviation.

Ministry of Tourism, Government of India (2018). Indian tourism statistics at a glance 2018. Retrieved from http://tourism.gov.in/sites/default/files/Other/ITS_Glance_2018_Eng_Version_for_Mail.pdf

Mutana, S., & Mukwada, G. (2018). Mountain-route tourism and sustainability: A discourse analysis of literature and possible future research. *Journal of Outdoor Recreation and Tourism*, 24, 59–65.

Naitthani, P., & Kainthola, S. (2015). Impact of conservation and development on the vicinity of Nanda Devi national park in the north India. *Journal of Alpine Research/ Revue de Géographie Alpine*, 103(3), 1–20.

Nepal, S.K. (2000). Tourism in protected areas: The Nepalese Himalaya. *Annals of Tourism Research*, 27(3), 661–681.

Nepal, S.K. (2002). Tourism as a key to sustainable mountain development: The Nepalese Himalayas in retrospect. *Unasylva*, 53, 38–46.

Nepal, S.K. (2011). Mountain tourism and climate change: Implications for the Nepal Himalaya. *Nepal Tourism and Development Review*, 1(1), 1–14.

Nepal, S.K., & Chipeniuk, R. (2005). Mountain tourism: Toward a conceptual framework. *Tourism Geographies*, 7(3), 313–333.

Nepal, S.K., & Karst, H. (2017). Tourism in Bhutan and Nepal. In C.M. Hall & S. Page (Eds.), *The Routledge Handbook of Tourism in Asia* (pp. 307–318). London: Routledge.

Nyaupane, G.P., & Thapa, B. (2004). Evaluation of ecotourism: A comparative assessment in the Annapurna conservation area project, Nepal. *Journal of Ecotourism*, 3(1), 20–45.

Nyaupane, G.P., & Thapa, B. (2006). Perceptions of environmental impacts of tourism: A case study at ACAP, Nepal. *The International Journal of Sustainable Development and World Ecology*, 13(1), 51–61.

Nyaupane, G.P., & Timothy, D.J. (2010). Power, regionalism and tourism policy in Bhutan. *Annals of Tourism Research*, 37(4), 969–988.

Panapanaan, V.M., Linnanen, L., Karvonen, M.M., & Phan, V.T. (2003). Roadmapping corporate social responsibility in Finnish companies. *Journal of Business Ethics*, 44(2–3), 133–148.

Panwar, R., Rhine, T., Hansen, E., & Juslin, H. (2006). Corporate responsibility: Balancing economic, environmental, and social issues in the forest products industry. *Forest Products Journal*, 56(2), 4–12.

Paskova, M., & Zelenka, J. (2019). How crucial is the social responsibility for tourism sustainability? *Social Responsibility Journal*, 15(4), 534–552.

Patterson, C. (2015). *Sustainable Tourism: Business Development, Operations, and Management*. Champaign, IL: Human Kinetics.

Rai, D.B. (2017). Tourism development and economic and socio-cultural consequences in Everest region. *Geographical Journal of Nepal*, 10, 89–104.

Ray, S., & Raju, S.S. (Eds.). (2014). *Implementing Corporate Social Responsibility: Indian Perspectives*. New Delhi: Springer.

Raza, J., & Majid, A. (2016). Perceptions and practices of corporate social responsibility among SMEs in Pakistan. *Quality & Quantity*, 50(6), 2625–2650.

Rinzin, C., Vermeulen, W.J., & Glasbergen, P. (2007). Ecotourism as a mechanism for sustainable development: The case of Bhutan. *Environmental Sciences*, 4(2), 109–125.

Sacareau, I. (2009). Changes in environmental policy and mountain tourism in Nepal. *Journal of Alpine Research/Revue de Géographie Alpine*, 97(3), 1–11.

Sajjad, A., & Eweje, G. (2014). Corporate social responsibility in Pakistan: Current trends and future directions. *Corporate Social Responsibility and Sustainability: Emerging Trends in Developing Economies*, 8, 163–187.

Salisbury, R., & Hawley, E. (2011). *The Himalaya by the Numbers: A Statistical Analysis of Mountaineering in the Nepal Himalaya*. Frederick, MD: Vajra Publications.

Sebastian, I. (2015). Business and corporate social responsibility in a gross national happiness economy: Insights from Bhutan. *Proceedings of the International Association for Business and Society*, 26, 24–38.

Schroeder, K., & Sproule-Jones, M. (2012). Culture and policies for sustainable tourism: A South Asian comparison. *Journal of Comparative Policy Analysis: Research and Practice*, 14(4), 330–351.

Singh, J.S. (2006). Sustainable development of the Indian Himalayan region: Linking ecological and economic concerns. *Current Science*, 90(6), 784–788.

Singh, R.B., Mal, S., & Kala, C.P. (2009). Community responses to mountain tourism: A case in Bhyundar Valley, Indian Himalaya. *Journal of Mountain Science*, 6(4), 394–404.

Singh, R.B., & Mishra, D.K. (2004). Green tourism in mountain regions-reducing vulnerability and promoting people and place centric development in the Himalayas. *Journal of Mountain Science*, 1(1), 57–64.

Singh, S. (2002). Tourism in India: Policy pitfalls. *Asia Pacific Journal of Tourism Research*, 7(1), 45–59.

Shafer, W.E., Fukukawa, K., & Lee, G.M. (2007). Values and the perceived importance of ethics and social responsibility: The US versus China. *Journal of Business Ethics*, 70(3), 265–284.

Sharma, P. (2009). Sustainable mountain tourism development in Nepal: An historical perspective. In E. Kruk, H. Kreutzmann, & J. Richter (Eds.), *Proceedings of the Regional Workshop, Integrated Tourism Concepts to Contribute to Sustainable Mountain Development in Nepal. Section 2* (pp. 40–47). Kathmandu: Deutsche Gesellschaft für Internationale Zusammenarbeit.

Sharma, T. (2018). Tourism growth under India's new CSR regime. *Advances in Hospitality and Leisure*, 14, 161–171.

Sharpley, R. (2010). *The Myth of Sustainable Tourism*. Preston: Centre for Sustainable Development, University of Central Lancashire.

Sherpa, Y. (2012). *Mountain Tourism: A Boon or a Bane? Impacts of Tourism on Himalayan Women*. Kathmandu: Federation of Women Entrepreneurs Association of Nepal.

Shujahi, A.H., & Hussain, A. (2016). *Economic and Environmental Costs of Tourism: Evidence from District Abbottabad*. Islamabad: Pakistan Institute of Development Economics.

Slusariuc, G.C., & Bîcă, M.P. (2015). Mountain tourism—pleasure and necessity. *Ecoforum Journal*, 4(2), 119–126.

Sofield, T., & Li, S. (2011). Tourism governance and sustainable national development in China: A macro-level synthesis. *Journal of Sustainable Tourism*, 19(4–5), 501–534.

Soni, G., & Hussain, S. (2016). A study on the impact of tourism on the Himalayas. *E-Commerce for Future & Trends*, 3(1), 27–35.

Spiller, R. (2000). Ethical business and investment: A model for business and society. *Journal of Business Ethics*, 27(1–2), 149–160.

Tang, B. (2012). Contemporary corporate social responsibility (CSR) in China: A case study of a Chinese compliant. *Seven Pillars Institute Moral Cents*, 1(2), 13–22.

Teoh, S. (2012). The ethics platform in tourism research: A Western Australian perspective of Bhutan's GNH tourism model. *Journal of Bhutan Studies*, 27, 34–66.

Tham, A., Ruhanen, L., & Raciti, M. (2020). Tourism with and by Indigenous and ethnic communities in the Asia Pacific region: A bricolage of people, places and partnerships. *Journal of Heritage Tourism*, 15(3), 243–248.

Timothy, D.J. (1998). Cooperative tourism planning in a developing destination. *Journal of Sustainable Tourism*, 6(1), 52–68.

Timothy, D.J. (2021). *Cultural Heritage and Tourism: An Introduction* (2nd ed.). Bristol: Channel View Publications.

Tiwari, P.C., & Joshi, B. (2016). *Rapid Urban Growth in Mountainous Regions: The Case of Nainital, India.* Tempe: Global Institute of Sustainability, Arizona State University.

Tourism Council of Bhutan (2010). *Bhutan Tourism Monitor: Annual Report 2009.* Thimphu: Tourism Council of Bhutan.

Tourism Council of Bhutan (2018). *Bhutan Tourism Monitor: Annual Report 2018.* ThimphuL Tourism Council of Bhutan. Retrieved July 20, 2020, from www.tourism.gov.bt/resources/annual-reports

Tourism Operators Initiative & Center for Environmental Leadership in Business (2003). *Supply Chain Management for Tour Operators: A Handbook on Integrating Sustainability into the Tour Operators' Supply Chain Systems.* Paris: Tourism Operators Initiative and Center for Environmental Leadership in Business.

UNEP and UNWTO (2005). *Making Tourism More Sustainable—A Guide for Policy Makers.* New York: United Nations Environment Program and World Tourism Organization.

Upadhayaya, P. (2015). Sustainability threats to mountain tourism with tourist mechanizes mobility induced global warming: A case study of Nepal. *Journal of Tourism and Hospitality,* 4(2), 1000148.

us Saqib, N., Yaqub, A., Amin, G., Khan, I., Ajab, H., Zeb, I., & Ahmad, D. (2019). The impact of tourism on local communities and their environment in Gilgit Baltistan, Pakistan: A local community perspective. *Environmental & Socio-economic Studies,* 7(3), 24–37.

Walder, G. (2000). *Tourism Development and Environmental Management in Nepal: A Study of Sagarmatha National Park and the Annapurna Conservation Area Project, with Special Reference to Upper Mustang.* Unpublished, master's thesis, Bournemouth University, Bournemouth, UK.

Wang, Y., Wu, N., Kunze, C., Long, R., & Perlik, M. (2019). Drivers of change to mountain sustainability in the Hindu Kush Himalaya. In P. Wester, A. Mishra, A. Mukherji, & A.B. Shrestha (Eds.), *The Hindu Kush Himalaya Assessment: Mountains, Climate Change, Sustainability and People the Hindu Kush Himalaya Assessment* (pp. 17–56). Cham: Springer.

Welzel, C. (2006). *Corporate Social Responsibility in Nepal: A Chance for Peace and Prosperity?* Kathmandu: Report Based Upon a Mission to Nepal.

World Business Council for Sustainable Development (1999). *Corporate Social Responsibility: Meeting Changing Expectations.* Geneva: World Business Council for Sustainable Development.

World Economic Forum (2019). *The Travel & Tourism Competitive Report 2019.* Geneva: World Economic Forum.

WTTC (2017). *Travel & Tourism Economic Impact, 2017: Pakistan.* London: World Travel and Tourism Council.

Zhang, Y.L., Zhang, J., Zhang, H.O., Zhang, R.Y., Wang, Y., Guo, Y.R., & Wei, Z.C. (2017). Residents' environmental conservation behaviour in the mountain tourism destinations in China: Case studies of Jiuzhaigou and Mount Qingcheng. *Journal of Mountain Science,* 14(12), 2555–2567.

# 15 Social Enterprises and Sustainable Development Through Tourism in the Himalaya

*Jonathon Day*

## Introduction

Tourism is a tool for communities to achieve social, environmental, and economic goals. Indeed, tourism has the potential to help achieve the United Nations Sustainable Development Goals (SDGs), a set of 17 goals addressing the grand challenges facing humanity. While tourism is often associated with for-profit businesses, NGOs and social enterprises are leveraging tourism-related activities to achieve not only economic growth but also other social goals. The chapter examines the roles of socially responsible tourism enterprises, social enterprises, NGOs, and other tourism-related programs that encourage gender equality, improve health and well-being, reduce inequalities, create decent work, and several other SDGs.

The chapter explores some tourism-related business models that are currently being applied to bring about sustainable development in the Himalaya. With examples from rural communities in Nepal, the chapter explores both the potential and the limitations of traditional sustainable tourism businesses, as well as the role of social enterprises in tourism development—concepts and practices that can be utilized throughout the Himalayan region. The chapter provides empirical examples of social entrepreneurship and sustainable development in the Solukhumbu and Tanahu districts of Nepal, each using different approaches to achieve specific social outcomes. These cases demonstrate the power of tourism to support community-building activities over the long term. They extend the model of tourism for development beyond the immediate tourism supply chain and provide evidence that social enterprise-based tourism and hybrid NGO-traditional business model enterprises can stimulate programs that provide electricity, clean water, medical care, and education, in addition to jobs. The chapter also examines the power of tourism to overcome long-established social inequalities as exemplified through a community-based tourism enterprise in a Dalit (so-called untouchable caste) village.

DOI: 10.4324/9781003030126-19

## Sustainable Development and Tourism

### *Sustainable Development*

At present, the world is facing many socioeconomic and environmental challenges. To counter these challenges, the quest for sustainable development has been a global priority for almost four decades. The Brundtland Report, also known as *Our Common Future*, defines sustainable development as that which 'meets the needs of the present without compromising the ability of future generations to meet their own needs' (WCED, 1987, p. 8) and states the imperative of adopting sustainable practices. There has been growing awareness and concern over increased environmental degradation and continued social inequalities. Improving people's quality of life around the globe while addressing significant environmental issues, many of which are caused by unsustainable development practices, has become a defining challenge of the current generation. The world currently faces a variety of significant, even existential, issues caused by unsustainable development.

While the concept of sustainability is generally accepted, operationalizing sustainable development has been challenging. The complexity of the challenge can be overwhelming, so in an effort to streamline sustainable development, including many of the most pressing social and environmental issues, 17 Sustainable Development Goals (SDGs) were established by the United Nations and stakeholders from around the world to help stakeholders focus on specific areas of concern (Table 15.1).

*Table 15.1* The United Nations Sustainable Development Goals

| *The 17 UN Sustainable Development Goals* |
|---|
| 1. No Poverty |
| 2. Zero Hunger |
| 3. Good Health and Well-being |
| 4. Quality Education |
| 5. Gender Equality |
| 6. Clean Water and Sanitation |
| 7. Affordable and Clean Energy |
| 8. Decent Work and Economic Growth |
| 9. Industry, Innovation, and Infrastructure |
| 10. Reduced Inequalities |
| 11. Sustainable Cities and Communities |
| 12. Responsible Consumption and Production |
| 13. Climate Action |
| 14. Life Below Water |
| 15. Life on Land |
| 16. Peace, Justice, and Institutions |
| 17. Partnerships |
| *United Nations (2016)* |

The global focus on achieving these goals may obscure the local actions needed to realize the goals. The operationalization of sustainable development, including the achievement of the SDGs, can be considered wicked problems, which are both complex and complicated (Day, 2021), and while they may share common characteristics across locations or situations, each one is unique. Wicked problems are characterized by incomplete information and circumstances that are always changing. There are no simple solutions for these problems, and solutions are 'neither wholly right nor completely wrong' (Morrison et al., 2019, p. 6). There are many interdependencies in these complex systems, and solving one part of the problem may lead to unintended consequences in another area. As an example, sustainable development actions that may work well in one village may not be effective in another. Given the nature of the challenge, we must expect a large variety of actors working in different ways to achieve specific goals that contribute to the larger goals of sustainability. While government commitment to the achievement of SDGs is important, governments are only one of many actors that should be committed to achieving SDGs. It is appealing to consider that the problems of sustainable development, such as poverty, education, and climate change, will be solved from the top-down. In reality, however, solving these issues will require the contributions of individuals and organizations, and solutions will need to come from small villages, towns and cities, and states and nations.

Each of the Himalayan countries has committed to the SDGs. Pakistan was the first country to align its national development goals with the SDGs (Government of Pakistan, 2019). India played a crucial role in the development of the SDGs and has also aligned its national development goals with the SDGs (National Institution for Transforming India, 2017). China has indicated the importance of the SDGs and has incorporated actions to achieve them in their development plans, including the 13th Five-Year Plan for Economic and Social Development of China (Ministry of Foreign Affairs, 2019). Bhutan has adopted the SDGs and has incorporated the UN goals into its Gross National Happiness Development Framework (Royal Government of Bhutan, 2018). Nepal has also adopted the SDGs and incorporated them into its latest five-year plan (National Planning Commission, 2020). These commitments provide a strong foundation for actions to achieve the SGDs, but government resources are limited and the magnitude of the challenges is great.

While government policies and programs contribute to the achievement of sustainability goals, government action alone will not solve these issues. The accomplishment of sustainability goals requires effort by many actors, including policymakers and governments, as well as businesses, NGOs and the civil sector, and individuals. As challenges for sustainable development are significant, innovative responses are required. This chapter focuses on the contributions of socially responsible businesses, social enterprises, and NGOs directly associated with these organizations. While businesses are becoming more engaged in social issues and are becoming more socially responsible, many NGOs are adopting entrepreneurial activities. The traditional business model of NGOs and charities is changing as funding sources, both public and private, become less reliable. The limits of traditional funding sources and growing societal needs have unleashed considerable creativity in the social sector. Changemakers have embraced new forms of fundraising to achieve their objectives, including crowdsourcing and social-entrepreneurial activities. At the intersection of socially responsible businesses and social enterprises, tourism is providing an opportunity for both business and NGOs to contribute to SDGs.

## Tourism as a Driver of Sustainable Development

It is worthwhile to consider the size and impact of tourism to appreciate its ability to contribute to sustainable development. From an economic perspective, tourism is a significant contributor to global trade and employment. The World Tourism Organization (UNWTO) reports that 1.4 billion international trips were conducted in 2018, generating over $US1.7 trillion in tourism exports. While these numbers are impressive by themselves, it is vital to note that domestic travel within countries increases the socioeconomic impacts of tourism considerably. Tourism is an important generator of economic activity across the Himalayan region, and it is of particular importance to Bhutan, Nepal, India, and China. For example, tourism plays a vital role in economic life in Nepal. In 2019, before the impacts of the coronavirus pandemic, tourism contributed 6.7% of the Nepalese economy and employed over 1 million people (WTTC, 2020). While the pandemic has significantly reduced these numbers, it is clear that tourism educes a level of activity that creates opportunities for sustainable development in Nepal and other countries in the region.

The ability of tourism to contribute to achieving the SDGs has captured the attention of policymakers, academics, and the tourism industry. Tourism has been identified as a tool to achieve a number of positive outcomes, including poverty alleviation (Croes & Rivera, 2015; UNWTO, 2002), empowerment (Boley & McGehee, 2014; Yang et al., 2020), environmental conservation (Chen & Qiu, 2017; K C et al., 2015), and cultural heritage preservation, among others.

Nevertheless, the presence of tourism does not guarantee positive outcomes. To achieve these positive outcomes, tourism must adopt the principles of sustainable development. As such, the UNWTO (2005, pp. 11–12) characterizes sustainable tourism as that which

1. Makes optimal use of environmental resources that constitute a key element in tourism development, maintaining essential ecological processes and helping to conserve natural heritage and biodiversity.
2. Respects the socio-cultural authenticity of host communities, conserves their built and living cultural heritage and traditional values, and contributes to inter-cultural understanding and tolerance.
3. Ensures viable, long-term economic operations, providing socioeconomic benefits to all stakeholders that are fairly distributed, including stable employment and income-earning opportunities and social services to host communities, and contributing to poverty alleviation.

Sustainable tourism may be a mechanism for achieving the SDGs (Gössling & Hall, 2019; Tham et al., 2020). Tourism is identified explicitly as contributing to SDGs 8, 12, 14 (see Table 15.1), but the World Tourism Organization argues that tourism has the *potential* to help achieve all 17 goals (UNWTO, 2015). Sustainable development through tourism was the focus of the UNWTO's International Year of Sustainable Development through Tourism in 2017, which specifically aimed to advance inclusivity and sustainable economic growth; support social inclusiveness, employment, and poverty reduction; enhance resource efficiency, environmental protection, and climate change mitigation; promote cultural values, diversity, and heritage; and develop mutual understanding, peace, and security (UNWTO, 2017).

Despite the optimism of the ability of tourism to contribute to the SDGs, it is clear that tourism is not a panacea, and significant effort will be required to achieve these positive outcomes. Boluk et al. (2019) advocate for a critical view of tourism as a tool to achieve the SDGs. At least two obvious factors must be considered. The first is that the negative impacts of tourism have been well documented (Buckley, 2012). Unsustainable tourism can lead to negative social, economic, and environmental outcomes, including crime, increased health risks, exploitation of local residents, social friction, air pollution, depletion of natural resources, destruction of ecosystems and reduced biodiversity, economic dependence, lack of opportunity throughout the year because of seasonality, overtourism, and commercialization of cultural heritage (Cook et al., 2010; Day, 2016; Morrison et al., 2018). These negative impacts can only be avoided with careful and deliberate planning. The second factor is that achieving sustainable tourism, or more specifically, sustainable development through tourism (Butler, 1999), has proven challenging, and there is growing frustration that more significant progress has not been achieved in implementing sustainability practices in the tourism system (Buckley, 2012; Maxim, 2014; Ruhanen et al., 2015). It is clear that tourism-related activities must be carefully managed if the promise of sustainable tourism is to be achieved. Nevertheless, the growth of tourism has generated a wide range of business opportunities for individuals and organizations seeking to address specific social needs across the Himalaya.

### Responsible Travel Companies and Tourism Social Enterprises

The role of nongovernment participation in achieving sustainable development in tourism is driven by two interrelated trends. The first trend is the growing concern of for-profit businesses about corporate social responsibility and sustainability principles. In general, businesses exhibit different responses to corporate social responsibility, from 'social obligation' (the minimum required by law) to 'social responsibility' in which they proactively seek to support the cause of sustainability (Day & Mody, 2017). However, these generalizations obscure the complexity of social responsibility and the range of decisions each company must make to be socially responsible. The adoption of sustainable tourism practices by businesses is an important contribution to overall sustainability in the tourism system. These changes in the ways companies operate are driven by factors such as consumer demand, license to operate, and increased efficiencies (Porter & Kramer, 2008). There is a growing expectation by consumers and business leaders that businesses will contribute toward solving social and environmental issues. There is a small but growing trend for tourism businesses to adopt socially responsible tourism practices as concerns for sustainability and sustainable tourism increase.

*Social Entrepreneurs and Social Enterprises*

Another trend is the increasing importance of social enterprises and social entrepreneurship in the achievement of social goals. A simple definition of a social entrepreneur is 'one who uses business principles to solve social problems' (Sheldon et al., 2017, p. 5). Alvord et al. (2004) state that social entrepreneurship 'creates innovative solutions to immediate social problems and mobilizes the ideas, capacities, resources, and social arrangements required for sustainable social transformations' (Alvord et al., 2004, p. 262). In recent years, there has been a growing movement of thought leaders and organizations, including the Skoll Foundation, Ashoka, Kaufman Foundation, and Unlimited, highlighting the benefits of social entrepreneurship (Day & Mody, 2017). The application of social entrepreneurship in tourism, particularly in rural regions, is well documented (Kline et al., 2014; Mottiar et al., 2018; Peng & Lin, 2016; Sloan et al., 2014).

Social enterprises are businesses established by social entrepreneurs, whose purpose is to achieve a social or environmental goal, as opposed to generating a profit. They sell services to fund their missions and maintain their roles as viable businesses while simultaneously creating values that address social issues. While social enterprises may stand alone, it is common to see hybrid, mutually supportive business structures working in unison to achieve social goals. These may include combinations of for-profit socially responsible companies working in conjunction with more traditional NGOs. For example, the Guludo Beach Lodge in Mozambique operated as a socially responsible ecotourism resort with the Nema Foundation—funded by the operations of the resort and donors—undertaking social and economic development activities (Carter-James & Dowling, 2017; Daniele et al., 2017). Similar relationships exist between G Adventures (tour wholesaler) and Planeterra, an NGO that uses community-based tourism to help alleviate poverty.

One style of tourism that has embraced social enterprise is community-based tourism. This approach as a manifestation of social entrepreneurship supports infrastructure development and improves local employment opportunities. It is a grassroots endeavor that contributes to better income and employment opportunities, poverty reduction, improved quality of life, and increased education (Okazaki, 2008; Sloan et al., 2014; Timothy, 1999, 2007). Community-based social enterprises are characterized by their collective and bottom-up decision-making and collaboration. This business model is not without issues. Sarkar and Sinha (2015) note tensions between ecological sustainability, local culture, and financial matters. Determining the equitable distribution of direct benefits requires attention.

While the study of social enterprise is still developing, some studies have examined social entrepreneurship. Zahra et al. (2009, p. 523) identified three types of social entrepreneurs: social bricoleurs, social constructionists, and social engineers. Social constructionists and social engineers tend to work on larger-scale activities and have captured much attention. The work of Mohammed Yunus and his Grameen Bank (Day & Mody, 2017) in transforming banking as a means of overcoming poverty has unleashed considerable enthusiasm for social entrepreneurs—social constructionists and social engineers—to create scalable solutions to development problems. In contrast, the activities of social bricoleurs can be described as 'small scale and local in nature' as they 'perceive and act upon opportunities to address local social needs' (Zahra et al., 2009, p. 523). While there are some examples of organizations (e.g., Planeterra and the TreadRight Foundation) trying to scale back their social impact by supporting social entrepreneurship, most social entrepreneurs in tourism are social bricoluers (Day & Mody, 2017). These enterprises tend to focus on specific issues in certain geographic areas. The social entrepreneurs described in this chapter focus on local issues in specific Himalayan locations.

*Critical Roles in Socially Responsible, Socially*
*Entrepreneurial Businesses*

The role individuals play in the development of social enterprises and socially responsible organizations is crucial in understanding these organizations. Considerable focus has been directed to the role of individual social entrepreneurs, in much the same way as studies of entrepreneurship have focused on specific entrepreneurs. As the concept of social entrepreneurship has been popularized by organizations and protagonists (Day & Mody, 2017), there has been a focus on the 'heroic' social entrepreneur. From this perspective, a social entrepreneur is described as a mission leader, emotionally charged, an opinion leader, a social value creator, and a visionary (Abu-Saifan, 2012, p. 25). While the notion of the heroic social entrepreneur may be generally appealing, research provides a more nuanced understanding of the motivations of social entrepreneurs (Boluk & Mottiar, 2014; Mody et al., 2016). The focus on the driven individual, an 'unreasonable person' (Elkington, 2008), while appealing, tends to overlook the many roles and relationships required to implement programs that lead to social change. Reconciling the popular concept of the heroic social entrepreneur with distributed entrepreneurship (Spear, 2006) and distributed leadership, the hallmarks of social enterprises, is necessary.

Mottiar et al. (2018) identify three critical roles in the development of social entrepreneurship in rural destinations: the opportunist, the catalyst, and the network architect. Opportunists recognize opportunity and advocate for tourism as a means of achieving social goals within the community; catalysts play an important role in 'developing and implementing a collective vision' (Mottiar et al., 2018, p. 84); and network architects build and nurture networks of people to achieve the goals of the social enterprise. Mottiar et al. (2018) note that some, but not all, social entrepreneurs undertake all three roles, and situational and personal factors come into play.

## Social Entrepreneurship in the Himalaya

The Himalayan region is home to wide-ranging development endeavors, including many social enterprises that aim to improve the livelihoods of communities, women, youth, and ethnic minorities through tourism in India, Nepal, Pakistan, China, and Bhutan (Prasad, 2014; Praszkier & Nowak, 2011). Several homestay and handicraft enterprises have emerged to employ women and benefit their communities (Acharya & Halpenny, 2013; Peaty, 2009). In India, the Ladakhi Women's Travel Company was established to empower Ladakhi women financially and socially. The enterprise caters to women travelers and is owned and operated by women with only female guides (Bulsara et al., 2015). Likewise, certain elements of volunteer tourism and ecotourism have been connected to social entrepreneurship in the Himalaya (Rana et al., 2014; Singh, 2002).

The following sections examine two cases that highlight the potential of social entrepreneurship to support the achievement of sustainable tourism. While there are similarities in these cases, they also contrast in several ways. The case of Adventure Alternative, Moving Mountains, and the villages of Bunburri, Bupsa, and Kharikhola is based in a region that has a long history of welcoming tourists, whereas the villagers of Aapshawara, Tanahu, have little previous experience in tourism, either as hosts or as guests.

### *Moving Mountains and Adventure Alternative in the Solukhumbu*

The villages of Bupsa, Bunburri, and Kharikhola are on the traditional path to the Everest Base Camp. In recent years, these villages have been bypassed by many visitors who fly into Lukla and then embark from there on their trek to the Everest Base Camp and beyond. Tourism in this region is well established; trails are dotted with teahouses, complete with food and basic accommodations. This area is a traditional region of the Sherpa people. Many of the people in these villages participate in tourism to some degree, being engaged in mountain guiding, including Everest guides, porterage, and providing accommodation and hospitality services.

Despite relative prosperity, the three villages face a number of social and economic development issues. In 2001, when Moving Mountains (MM) (a development NGO) and Adventure Alternative (AA) (a socially responsible adventure travel company) first began working in these villages, the communities were isolated, and supplies were delivered by donkey train. They lacked electricity, health care, and employment opportunities, which led to a steady stream of outward migration. Many villages were depleted as people moved to Kathmandu and other cities. Educational opportunities were limited, and small villages struggled to attract and retain teachers. In the early 2000s, the region was embroiled in the Nepalese Civil War/Maoist conflict (1996–2006). The region continues to experience noteworthy changes today. For example, the completion of a new road connecting the villages to Kathmandu may have a significant impact on the region, creating economic opportunity but introducing new social pressures.

*Key Organizations and Key People*

The connections between Moving Mountains Trust and Adventure Alternative in Bupsa, Bunburri, and Kharikhola in Solukhumbu district provide insights into the relationship between sustainable tourism, sustainable development, social enterprise, and socioeconomic development.

Adventure Alternative (AA), established in 1991, is an award-winning tour operator that sells responsible adventure travel activities around the world. It received the World Responsible Tourism Award for 'Best for Poverty Alleviation' from the World Travel Markets in 2014. AA's founder, Gavin Bate, was recognized for his personal contribution to responsible tourism at the awards the following year (Daniele et al., 2017). AA and Moving Mountains Trust (MM) work synergistically to achieve social goals (Daniele et al., 2017). Gavin Bate, also the founder of MM, has built a small 'ecosystem' of mutually supportive organizations to achieve social and economic development goals in a number of countries, including Kenya, Tanzania, and Nepal. AA's UK operation provides some financial support to both AA and MM in each of these countries, supporting a strong portfolio of social enterprises (AA) and charities (MM). While this support provides a degree of continuity, AA works to build local markets in each country to achieve financial independence and support its local development projects. MM also seeks contributions directly from benefactors for development projects. In addition to these activities, Gavin, a world-renowned mountain climber, has raised funds for MM projects through climbing expeditions, many of which are mutually reinforcing. For example, satisfied AA customers learn of MM's activities and support its programs, and Bate's fundraising climbs for MM raise awareness of AA as a premium, responsible travel company. As Bate notes, "AA and MM are wedded at the waist, with both companies contributing to the other and both feeding back to the community. People go on AA trips to see Bumburi (Nepal) because of MM, and people support MM because they've been on an AA trip. When it works together it's a great balance of tourism and engagement" (personal communication with the author).

Central to the AA/MM web of organizations are long and deep relationships. Throughout the AA/MM network, strong personal ties built over 20 years of shared effort provide the foundation on which the work has developed. In Nepal, Ang Chhongba Sherpa, the chairman of Moving Mountains Nepal (MMN), and Gavin Bate have been friends and worked together to support the development of Bupsa, Bumburi, and Karikhola. Ang Chhongba was born in the region and is deeply involved in the region and its issues. Ang Chhomba left the village to pursue an education at the age of seven with the personal goal of returning to give back to his community. Today, he works to improve the social and economic development of the region through MMN and AA.

Ang Chhongba's role in the organizations illustrates the critical role of social entrepreneurs within the community. As a social entrepreneur, he acts as a catalyst, opportunist, and network developer. Social economic development for the region is based on his personal connection to the region. Early in the process, Ang Chhongba identified the key priorities of lighting and electricity, health, and education. His vision for the region is a driving force behind much of the activity that has taken place. Nevertheless, to achieve this vision, nurturing the networks of stakeholders and collaboration between groups in the community is critical. Although Ang Chhongba's vision for the long-term economic development of the region fits stereotypes of the sole social entrepreneur, the reality is that he relies on collaboration and deep relationships built over lifetimes. Development activities are determined by consulting with village leadership. The projects are complex collaborations requiring high levels of trust and social capital. The roles of Bate and Ang Chhongba show how social entrepreneurs assume different roles to achieve shared goals. Bate, a catalyst and opportunist influencing activity across the AA/MM network, assumes the role of enabler for Chhongba and MMN, which provides financial and human resources. Similarly, Ang Chhongba enables village leadership when working together to achieve local development. Village leaders are also social entrepreneurs in this example, identifying opportunities, building networks of participants, and gathering resources, including land, financial assets, and human capital.

The relationship between AA and MMN highlights two roles of sustainable tourism in economic development. The first role is sustainable tourism as a driver of development. AA practices a number of important activities that support these communities. AA has stimulated new business to the region through its promotional efforts and program offerings. They create employment by hiring local people as guides and porters and using teahouses for food and accommodation in their itineraries. AA applies fair trade principles in its supplier relationships. For example, it pays a premium to the porters and guarantees employment throughout the season. These types of business practices have significant social benefits as they provide good livelihoods in the village, thereby stemming the urban flight to Kathmandu. Other flow-on effects include reduced poverty and the maintenance of family structures as breadwinners are not forced to leave the village. As such, sustainable tourism generates significant direct and indirect benefits for the region, but there are limitations to the reach of these activities. Sustainable tourism has limited ability to impact many development issues, such as the provision of health, infrastructure, and education.

While AA provides benefits through tourism, MMN provides sustainable development from tourism and other funding sources. MMN is able to address issues directly with development and works closely with local village leaders to establish development priorities. Significant projects supported by the AA/MMN have included fuel-efficient stove replacements, annual health clinics and the establishment of a local health clinic, and the establishment of a micro-hydroelectric plant. Another critical element of the commitment to sustainable development has been a commitment to education. MMN recognized that both facilities and teachers would be required to achieve its long-term educational goals. After funding the construction of primary schools in Bunburri and Bupsa, MMN committed to funding teachers in the community; the organization now supports several teachers at local schools. While the Nepalese government provides teachers to local schools, it is hard to attract and retain teachers in remote rural communities. Additionally, many school districts face the inability to meet the full needs of their communities. Given these challenges, MMN supports both additional funding to provide existing teachers with salaries that will encourage them to stay, and funds to support additional teachers to meet local needs in both primary and secondary schools. Ensuring students do not need to leave the region to obtain a quality education helps reduce the pressures on families and individuals to leave.

Activities are typically developed collaboratively with local village members contributing 'sweat equity' in projects and the organizations providing resources and financial contribution to the projects. The engagement of all stakeholders—villagers and the MM network—is critical to the success of regional sustainable development projects. This is perhaps most dramatically exemplified by the construction of the micro-hydro project in 2005. The need for electricity in the villages was identified early by Ang Chhongba and Bate when they first began working together. The task was significant and multifaceted. Bate raised funds for the project through donations earned for the challenge of climbing Everest solo and from the proceeds of tours to the Everest Base Camp. Ang Chhongba worked with village leaders to secure support and approval for the project and overcame opposition from one particular key stakeholder group: the remnants of the Maoist insurgency. Building the power plant required a lot of materials and equipment on steep mountain paths from Lukla to the three villages. With no vehicular access, village volunteers undertook the laborious task of carrying pipes, concrete, and stone blocks. At a significant cost to each family, villagers also donated land for the project. Volunteers associated with AA and MM provided additional support. Today, the micro-hydro project provides electricity to approximately 60 homes. MMN continues to work toward training and capacity building among villagers to ensure the longevity of its projects. This process improves the rural workforce capacity by adding new skills and providing higher-wage skilled jobs. The micro-hydro project illustrates some of the opportunities that emerge from social entrepreneurship in these communities. The electricity enables new commercial opportunities, including corn and wheat milling and tea production and sales. The generated power also ensures village children can do homework at night after helping with family chores. This collaborative approach to development has led to several projects that satisfy the needs of the villages, including annual medical camps, the establishment and operation of a medical center in Bumburi, financial support for teachers at both primary and secondary schools, the installation of eco-friendly, efficient stoves for cooking, and water supply projects in Bumburi and Kinai. The community-based development described here also enriches the local social fabric by restoring Buddhist temples in Bunburri and Bupsa and the recent construction of a village community hall in Bupsa.

There are two other dimensions to the relationship between AA, MMN, and the Solukhumbu villages. The first of these is the scope of commitment. Under the leadership of Ang Chhongba, MMN has committed to the villages for the long term. In this way, MMN has become an integral part of the area's social fabric and is not seen as an external NGO with competing projects in other locations. The second dimension is the time horizon against which the success of the relationship is measured. As already noted, this is a long-term relationship, and its direct and indirect benefits are measured in decades or generations rather than individual years.

These community-based activities began before the SDGs were established and were not undertaken with the goal of addressing global challenges. Instead, they arose in response to the specific needs of small communities in the Solukhumbu District of Nepal and just happened to align with several of the UN's SDGs. In spite of their local focus, these efforts exemplify successful social enterprise-based development from which other Himalayan regions can learn important lessons about poverty reduction, improved health and education, clean energy, economic growth, responsible production and consumption, and cross-sectoral collaboration.

### Aapshawara and the Hands-On Institute

The Aapshawara Community Dalit Homestay project represents a different approach to solving social development challenges. This project began more recently than the MMN/AA activities in Solukhumbu.

### The Social Development Challenge: Touchability

Although officially banned in Nepal, the caste system remains an important part of Nepalese culture. Dalits, also known as Untouchables, have suffered discrimination in a variety of forms, including ostracism, domination, exclusion, vilification, and violence. Because of caste-based discrimination and social biases, Dalits tend to be trapped in poverty due to a lack of economic opportunities (UNDP, 2008). Aapshawara is a Dalit village that has faced significant challenges. It is close to the popular tourism destination Pokhara, which is a gateway to the Annapurna circuit. Aapshawara is a rural village, with views of the Himalaya and the Seti River, in a region where many Dalits work as blacksmiths. In the wake of the 2015 Nepal earthquake, Aapshawara faced the challenge of rebuilding and creating a new foundation on which to reinvigorate village life.

### Key Organizations and People

The Hands-On Institute (HOI) is a small social enterprise based in Kathmandu, whose principals, Samrat Katwal and Bijay Poudel, recognized the acute challenges of the Dalit people and identified the potential of tourism-related activities to address caste discrimination. In the immediate aftermath of the 2015 earthquake, HOI initiated a discussion with Aapshawara's leadership on how to develop strategies to support long-term, sustainable development in the village. After extended conversations, the community committed to developing a village-based homestay experience—the Aapshawara Community Dalit Homestay, which is a small-scale, community-based enterprise.

At first, community members were unsure about the idea of a homestay in their village. Village-based homestays are increasingly common (Walter et al., 2018), and the government has established operating standards for managing homestay programs and marketing them for tourism. Even so, no Dalit village had ever offered a homestay program before. Resham Bahadur Bishwakarma (BK) saw the opportunity and became an early advocate of the concept. When the village decided to proceed, a committee was convened to manage the homestay-related activities (Manandhar & Singh, 2019). The leadership was determined consensually, with Resham Bahadur BK taking the role of president of the committee. Resham plays the multiple roles of social entrepreneur, opportunist, and network architect. The Aapshawara Homestay Committee (AHC) and village leadership proactively manage the impact of tourism on their community.

The collaboration with HOI is worth noting. HOI has operated as both an opportunist and a catalyst in identifying the potential opportunity and presenting it to village leadership. The organization has been an enabler for the social enterprise by providing AHC with expertise, training, and support in fund-raising. HOI also attracted funds from outside the community. These two mutually supportive social enterprises—HOI and the village—worked synergistically to achieve sustainable development goals.

*Capacity Building and Experience Development*

Many Dalits rely on traditional skills, including metalwork, sewing, bamboo work, and playing musical instruments. Unfortunately, due to the pressures of modernization, these skills are endangered. HOI recognized unique cultural experiences available through the Dalit village experience and saw them as a way not only to attract visitors and support economic development but also to raise appreciation of the Dalit heritage. HOI provides a critical external perspective that is able to recognize both the uniqueness of the visitor experience and the market opportunity. Thus, for the villagers, HOI provided an important outsiders' view of the unique cultural life represented in the village. The people of Aapshawara, themselves immersed in the everyday nature of their work and conditioned by society that their activities were not worthy of interest, were unable to see the potential value of their cultural assets, which is commonly the case in many traditional societies (Timothy, 1999). Villagers did not recognize that their traditional crafts and work activities could be of interest to visitors and so HOI worked with them to map their tangible and intangible tourism assets.

The process of building skills to effectively host visitors was established over time. Because Aapshawara was new to tourism, building a solid foundation for growing their venture was important. Activities undertaken in preparation for guests included the following:

1. A 7-day training program for prospective hosts including cooking and hospitality skills.
2. Home improvement to ensure appropriate facilities.
3. Field visit to a successful homestay. Many of the hosts had never experienced being a paying guest. Appreciating that role was a critical step in developing the skills to effectively host visitors.
4. Registration as an official homestay program (Manandhar & Singh, 2019).

HOI worked with the villagers to identify additional capacity building and training needs. Community-based tourism encompasses management challenges that must be decided by the villagers (Okazaki, 2008; Timothy, 1999). These decisions include income distribution; environmental, social, and cultural mitigation plans; identification of a management committee; and decisions on the use of common resources. Aapshawara village members have learned new skills and reinvigorated traditional skills as a result of the homestay project. HOI has recognized that its role as an enabler and capacity builder should mean that the AHC assumes full operation of their program. Relinquishing some operational functions by HOI, such as booking procedure for guests, has resulted in some short-term inefficiencies as villagers learn new systems, but it provided a greater sense of efficacy and autonomy.

The project has also stimulated development activities that should have long-term benefits for the community. The needs of the homestay program led to changes in cooking fuel use, sanitation, and waste management. A multipurpose community hall was built to support homestay meetings and other social needs of the village. The community has been decorated with exhibits from Dalit culture and heritage and is a showcase for the community (Manandhar & Singh, 2019).

### Protecting the Community From Tourism

While key people are catalysts in the process, social-entrepreneurial activities are built on collaboration and collective decision-making. The village deliberated on the implementation of the homestay project and implemented it in a controlled manner. AHC is notable in the care it took to ensure that tourism would not adversely impact the community. The organization recognized that Aapshawara was vulnerable, so they carefully prepared and mitigated for any problematic issues.

Several factors came into play with the interaction between tourists and hosts. In Nepalese culture, the practice of 'guest is God' is widely accepted, giving guests power over their hosts. At the same time, Dalits who have been socially subjugated easily fall prey to being taken advantage of by guests. On the other hand, Dalit hosts, with little exposure to people from overseas, had unrealistic expectations about the 'donations' they may receive from their new guests (Katwal, 2017). AHC recognized the potential risk of the situation and developed interventions that ensured guests were carefully briefed on culturally appropriate behaviors, and hosts were prepared to provide quality hospitality without comprising their cultural identity.

The coronavirus pandemic has significantly impacted tourism and there has been a reduction of visitors, particularly international visitors, to the Aapshawara Homestay. Fortunately, the homestay operations represent a small part of the economic life of the community and the negative impacts of the pandemic have been limited. Nevertheless, the pandemic reinforces the importance of recognizing that tourism can be volatile and there is danger in relying too heavily on a single activity for economic and social development.

*Benefits*

Aapshawara began the project with modest expectations. The goal was to attract small numbers of international visitors on educational trips and in study groups. Despite the small scale of the operation, several immediate benefits have accrued from the project. AHC has hosted over 280 local and international guests with proceeds distributed to participating homes and the village collective (Manandhar & Singh, 2019). The indirect benefits of the program have been significant, including village participants gaining critical intercultural perspectives and an increasing sense of self-efficacy and self-worth, which Scheyvens (1999) refers to as psychological and social empowerment. Interaction with different social groups, including other groups in Nepal and international visitors, has broadened participants' worldview. The Dalits have also gained a greater appreciation for their own cultural heritage and contribution to society as a result of seeing the interest of outsiders. As Bishnu BK, one of the founders of the program, stated, "This has not just helped preserve our culture but it is also helping us live with dignity and self-respect, despite all societal norms" (Luitel & Rana, 2018, n.p.). The Aapshawara Homestay project—the first Dalit-owned homestay—has gained local, national, and international recognition. A feature article in one of Nepal's largest-selling English newspapers, the *Kathmandu Post*, titled "Dalit homestay blazes a trail in fighting 'untouchability'" (Luitel & Rana, 2018) highlights the impact the enterprise is having on perceptions of untouchability and tourism. The recognition the village has received has changed the perspectives of other Dalit communities. Community members have been invited to participate in local civic society and regional government. Finally, there is also some evidence that Dalit women who engaged in hosting duties have been empowered socially and economically by the experience (Manandhar & Singh, 2019). Again, while contributing to the SDGs was not an explicit goal of this project, it is abiding by many of the precepts of the UN goals, such as reducing inequalities, creating decent work and economic growth, reducing poverty, enhancing gender equality, providing clean water and sanitation, improving justice, and others.

## Conclusion

The challenge of achieving the UN's SDGs by 2030 is daunting. As a complex, wicked problem, we must recognize that meeting the goals will require the work of many actors—individuals, businesses, NGOs, and governments—and that there will be many different approaches to achieving the goals. Recognizing and learning from success stories, even on a local scale, can provide important insights for organizations seeking to contribute in effective ways. While it is understood that the examples highlighted here are not intended to be representative of all social enterprises in Nepal, or across the Himalaya, they do provide interesting insights into the value of socially responsible businesses and social enterprises in generating sustainable development. Perhaps most importantly, these examples show how sustainable development can be achieved through small-scale initiatives. As noted previously, wicked problems such as sustainable development require unique responses suited for specific circumstances. In both BBK and Aapshawara, the participants found ways to contribute to achieving broader sustainable development goals, even though the SDGs themselves did not drive these activities. Instead, the needs of the villages happed to align with the SDGs.

Tourism can be a tool for sustainable development if it is implemented deliberately and carefully. The Aapshawara case is an important lesson in recognizing the potential risks of tourism and managing them to achieve community goals. The Aapshawara villagers are a potentially vulnerable group for which poorly managed tourism may have negative consequences. Through careful planning and controlled growth, AHC is working to mitigate those risks.

While two success stories are presented here, there are limitations to the contributions of tourism activities—even responsible ones—in sustainable development. Economic and social benefits from tourism can contribute to the overall quality of life and stimulate infrastructure development, but direct intervention by NGOs, perhaps funded by tourism operations, is often required to address some development needs. Access to electricity, water, and education were all a result of MMN's direct support rather than the flow-down benefit of tourism operations. Sustainable development can be achieved both directly through tourism and with funds derived from tourism. It is also important to recognize that social, environmental, and economic benefits are not automatic from tourism. 'Trickle Down Tourism' is both ineffective and inefficient, and development activities need to be deliberately targeted to benefit specific groups.

It is time to reexamine the definition of social entrepreneurs. The common view of the social entrepreneur as a 'heroic messiah' (Mody & Day, 2017) implies single actors creating change through their personal characteristics. This definition clearly does not tell the whole story. The cases reported in this chapter reinforce the findings of Mottiar et al. (2018), that social entrepreneurs play many roles in the execution of these activities (e.g., opportunists, network architects, and catalysts) and that individual social entrepreneurs may take on some or all of these roles. It is clear from these Himalayan cases that, within a social enterprise, different people will play different roles, as noted by Spear (2006). It is proposed that 'enablers' should be added to Mottiar et al.'s (2018) list. Enablers are organizations or people who support the achievement of social goals by providing training, capacity building, and coaching. HOI and its principals, Samrat and Bijay, enable the AHC in developing the homestay program, while Moving Mountains Trust and Gavin Bate enable MMN to achieve its goals. In the examples provided here, the traditional view of a social entrepreneur does not accommodate the team approach these activities require, nor does it recognize complex collaborations between stakeholders.

In recent years, much of the literature on social entrepreneurship has focused on creating social enterprises that scale. Nevertheless, it is clear that the 'bricolage' social entrepreneur plays an important role in social development. Numerous factors contribute to the predominance of small-scale tourism projects in specific regions. These projects are characterized by strong social capital and rely on trust built over time, as well as deep grassroots understandings of the issues associated with each location. The cases presented here illustrate long-term commitments between organizations and individuals. These cannot be easily scaled, but they nevertheless make significant contributions to the world's social and environmental challenges. Although in much of the literature to date there is an assumption that social entrepreneurs establish social enterprises and entrepreneurs build for-profit businesses, this is not always the case. Sometimes, social entrepreneurs develop for-profit businesses to support NGOs, with the social enterprise being a hybrid between an NGO and a for-profit organization. Changemakers focused on improving social and economic conditions are applying a range of business models to achieve their goals. This creativity in addressing significant social and environmental issues justifies the enthusiasm for social entrepreneurship generated over the last few years, which is particularly notable in the Himalayan region.

As a final observation, the two Nepali cases also highlight the fragility of tourism as a driver of sustainable development. The impact of the COVID-19 pandemic on the tourism industry begs a critical evaluation of the industry as a driver of sustainable development. Tourism is sensitive to external issues, including natural disasters, climate change, and political unrest. Dependence on a single industry is always a risky strategy, but reliance on tourism may be particularly risky. The villages in both cases have the benefit of not being totally dependent on tourism, although the villages of Solukhumbu, which are more heavily dependent on tourism, have found 2020 to be a challenging year. In both cases, the local communities have traditional activities, such as farming, to rely on. Tourism is a source of economic and social development for many countries in the developing world, yet it is important to be mindful of its limitations.

## References

Abu-Saifan, S. (2012). Social entrepreneurship: Definitions and boundaries. *Technology Innovation Management Review* (February), 22–27.

Acharya, B.P., & Halpenny, E.A. (2013). Homestays as an alternative tourism product for sustainable community development: A case study of women-managed tourism product in rural Nepal. *Tourism Planning & Development*, 10(4), 367–387.

Alvord, S.H., Brown, L.D., & Letts, C.W. (2004). Social entrepreneurship and societal transformation: An exploratory study. *The Journal of Applied Behavioral Science*, 40(3), 260–282.

Boley, B.B., & McGehee, N.G. (2014). Measuring empowerment: Developing and validating the resident empowerment through tourism scale (RETS). *Tourism Management*, 45, 85–94.

Boluk, K., Cavaliere, C.T., & Higgins-Desbiolles, F. (2019). A critical framework for interrogating the United Nations sustainable development goals 2030 Agenda in tourism. *Journal of Sustainable Tourism*, 27(7), 847–864.

Boluk, K., & Mottiar, Z. (2014). Motivations of social entrepreneurs: Blurring the social contribution and profits dichotomy. *Social Enterprise Journal*, 10(1), 53–68.

Buckley, R. (2012). Sustainable tourism: Research and reality. *Annals of Tourism Research*, 39(2), 528–546.

Bulsara, H.P., Gandhi, S., & Chandwani, J. (2015). Social entrepreneurship in India: An exploratory study. *International Journal of Innovation*, 3(1), 7–16.

Butler, R.W. (1999). Sustainable tourism: A state-of-the-art review. *Tourism Geographies*, 1(1), 7–25.

Carter-James, A., & Dowling, R. (2017). Guludo Beach Lodge and the Nema Foundation, Mozambique. In P.J. Sheldon & R. Daniele (Eds.), *Social Entrepreneurship and Tourism: Philosophy and Practice* (pp. 221–235). Cham: Springer.

Chen, B.-X., & Qiu, Z.-M. (2017). Community attitudes toward ecotourism development and environmental conservation in nature reserve: A case of Fujian Wuyishan National Nature Reserve, China. *Journal of Mountain Science*, 14(7), 1405–1418.

Cook, R., Yale, L., & Marqua, J. (2010). *Tourism: The Business of Travel* (4th ed.). Hoboken, NJ: Prentice Hall.

Croes, R., & Rivera, M. (2015). *Poverty Alleviation Through Tourism Development*. Oakville: Apple Academic Press.

Daniele, R., Bate, G., & Quezada, I. (2017). Adventure alternative and moving mountain trust: A hybrid business model for social entrepreneurship in tourism In P.J. Sheldon & R. Daniele (Eds.), *Social Entrepreneurship and Tourism: Philosophy and Practice* (pp. 265–277). Cham: Springer.

Day, J. (2016). *An Introduction to Sustainable Tourism and Responsible Travel* (Beta ed.). West Lafayette, IN: Placemark Solutions.

Day, J. (2021). Sustainable tourism in cities. In A. Morrison & J.A. Coca-Stefaniak (Eds.), *Routledge Handbook of Tourism Cities* (pp. 53–64). London: Routledge

Day, J., & Mody, M. (2017). Social entrepreneurship typologies and tourism: Conceptual frameworks. In P.J. Sheldon & R. Daniele (Eds.), *Social Entrepreneurship and Tourism: Philosophy and Practice* (pp. 57–80). Cham: Springer.

Elkington, J. (2008). *The Power of Unreasonable People: How Social Entrepreneurs Create Markets That Change the World*. Boston: Harvard Business School Press.

Gössling, S., & Hall, C.M. (2019). Sharing versus collaborative economy: How to align ICT developments and the SDGs in tourism? *Journal of Sustainable Tourism*, 27(1), 74–96.

Government of Pakistan (2019). *Pakistan's Implementation of the 2030 Agenda for Sustainable Development: Voluntary National Review*. Islamabad: Government of Pakistan.

K C, A., Rijal, K., & Sapkota, R.P. (2015). Role of ecotourism in environmental conservation and socioeconomic development in Annapurna conservation area, Nepal. *International Journal of Sustainable Development & World Ecology*, 22(3), 251–258.

Katwal, S. (2017). *Aapshara Dalit Community Home-Stay—Project Update Report*. Kathmandu: Aapshara Homestay Project.

Kline, C., Shah, N., & Rubright, H. (2014). Applying the positive theory of social entrepreneurship to understand food entrepreneurs and their operations. *Tourism Planning & Development*, 11(3), 330–342.

Luitel, B., & Rana, P. (2018). Dalit homestay blazes a trail in fighting 'untouchability'. *Kathmandu Post*, December 29. Retrieved from https://kathmandupost.com/national/2018/12/29/dalit-homestay-blazes-a-trail-in-fighting-untouchability

Manandhar, R., & Singh, S. (2019). Sustainable community economy: A case of Aapshwara Community Dalit homestay, Tanahun, Nepal Paper presented at the IOE Graduate Conference, Nepal, May.

Maxim, C. (2014). Drivers of success in implementing sustainable tourism policies in urban areas. *Tourism Planning & Development*, 12(1), 1–11.

Ministry of Foreign Affairs (2019). *China's Progress Report on Implementation of the 2030 Agenda for Sustainable Development*. Beijing: Ministry of Foreign Affairs, People's Republic of China.

Mody, M., & Day, J. (2017). Heroic messiahs or everyday businessmen? The rhetoric and the reality of social entrepreneurship in India. In P.J. Sheldon & R. Daniele (Eds.), *Social Entrepreneurship and Tourism: Philosophy and Practice* (pp. 207–220). Cham: Springer.

Mody, M., Day, J., Sydnor, S., & Jaffe, W. (2016). Examining the motivations for social entrepreneurship using Max Weber's typology of rationality. *International Journal of Contemporary Hospitality Management*, 28(6), 1094–1114.

Morrison, A., Lehto, X., & Day, J. (2018). *The Tourism System* (8th ed.). Dubuque, IA: Kendall Hunt.

Morrison, E., Hutcheson, S., Nilsen, E., Fadden, J., & Franklin, N. (2019). *Strategic Doing: Ten Skills for Agile Leadership*. Hoboken, NJ: Wiley.

Mottiar, Z., Boluk, K., & Kline, C. (2018). The roles of social entrepreneurs in rural destination development. *Annals of Tourism Research*, 68, 77–88.

National Institution for Transforming India (2017). *Voluntary National Review Report on the Implementation of Sustainable Development Goals*. New York: United Nations High-Level Political Forum on Sustainable Development.

National Planning Commission (2020). *National Review of Sustainable Development Goals*. Kathmandu: National Planning Commission, Government of Nepal.

Okazaki, E. (2008). A community-based tourism model: Its conception and use. *Journal of Sustainable Tourism*, 16(5), 511–529.

Peaty, D. (2009). Community-based tourism in the Indian Himalaya: Homestays and lodges. *Journal of Ritsumeikan Social Sciences and Humanities*, 2, 25–44.

Peng, K.-L., & Lin, P.M.C. (2016). Social entrepreneurs: Innovating rural tourism through the activism of service science. *International Journal of Contemporary Hospitality Management*, 28(6), 1225–1244.

Porter, M., & Kramer, M. (2008). Strategy and society: The link between competitive advantage and corporate social responsibility. In M. Porter (Ed.), *On Competition: Updated and Expanded Edition* (pp. 451–479). Boston: Harvard Business Review Books.

Prasad, C.S. (2014). *Thinking Through Social Innovation and Social Entrepreneurship in India*. Ottawa: International Development Research Centre.

Praszkier, R., & Nowak, A. (2011). *Social Entrepreneurship: Theory and Practice.* Cambridge: Cambridge University Press.

Rana, P.S., P J, S., & Patnaik, S. (2014). Ecotourism in Himalayas: A case study of Jaunsar region (Uttarakhand, India). In H. Koçak, S. Tuzemen, & A.C. Gulluce (Eds.), *Proceedings of Eurasian Silk Road Universities Consortium, Investigating the Social, Cultural and Global Issues Along the Silk Road* (pp. 2–22). Saarbrücken: Lambert.

Royal Government of Bhutan (2018). *Sustainable Development and Happiness: Bhutan's Voluntary National Review Report on the Implementation of the 2030 Agenda for Sustainable Development.* Thimphu: Royal Government of Bhutan.

Ruhanen, L., Weiler, B., Moyle, B.D., & McLennan, C.-L.J. (2015). Trends and patterns in sustainable tourism research: A 25-year bibliometric analysis. *Journal of Sustainable Tourism, 23*(4), 517–535.

Sarkar, R., & Sinha, A. (2015). The village as a social entrepreneur: Balancing conservation and livelihoods. *Tourism Management Perspectives, 16,* 100–106.

Scheyvens, R. (1999). Ecotourism and the empowerment of local communities. *Tourism Management, 20*(2), 245–249.

Sheldon, P.J., Pollock, A., & Daniele, R. (2017). Social entrepreneurship and tourism: Setting the stage. In P.J. Sheldon & R. Daniele (Eds.), *Social Entrepreneurship and Tourism Philosophy and Practice* (pp. 1–18). New York: Springer.

Singh, T.V. (2002). Altruistic tourism: Another shade of sustainable tourism—the case of Kanda community. *Tourism, 50*(4), 361–370.

Sloan, P., Legrand, W., & Simons-Kaufmann, C. (2014). A survey of social entrepreneurial community-based hospitality and tourism initiatives in developing economies: A new business approach for industry. *Worldwide Hospitality and Tourism Themes, 6*(1), 51–61.

Spear, R. (2006). Social entrepreneurship: A different model? *International Journal of Social Economics, 33*(5/6), 399–410.

Tham, A., Ruhanen, L., & Raciti, M. (2020). Tourism with and by Indigenous and ethnic communities in the Asia Pacific region: A bricolage of people, places and partnerships. *Journal of Heritage Tourism, 15*(3), 243–248.

Timothy, D.J. (1999). Participatory planning: A view of tourism in Indonesia. *Annals of Tourism Research, 26*(2), 371–391.

Timothy, D.J. (2007). Empowerment and stakeholder participation in tourism destination communities. In A. Church & T. Coles (Eds.), *Tourism, Power and Space* (pp. 199–216). London: Routledge.

UNDP (2008). *The Dalits of Nepal and a New Constitution: A Resource on the Situation of Dalits in Nepal, Their Demand and the Implications for a New Constitution.* Kathmandu: United Nations Information Center.

United Nations (2016). Sustainable development goals. Retrieved from www.un.org/sustainabledevelopment/sustainable-development-goals/

UNWTO (2002). *Tourism and Poverty Alleviation.* Madrid: World Tourism Organization.

UNWTO (2005). *Making Tourism More Sustainable: A Guide for Policy Makers.* Madrid: World Tourism Organization.

UNWTO (2015). *Tourism and the Sustainable Development Goals.* Madrid: World Tourism Organization.

UNWTO (2017). *Why Tourism 2017 International Year of Sustainable Tourism for Development.* Madrid: World Tourism Organization.

Walter, P., Regmi, K.D., & Khanal, P.R. (2018). Host learning in community-based ecotourism in Nepal: The case of Sirubari and Ghalegaun homestays. *Tourism Management Perspectives, 26,* 49–58.

WCED (1987). *Our Common Future*. Oxford: Oxford University Press.

WTTC (2020). *Nepal: 2020 Annual Research: Key Highlights*. London. Retrieved from https://wttc.org/Research/Economic-Impact

Yang, J., Wang, J., Zhang, L., & Xiao, X. (2020). How to promote ethnic village residents' behavior participating in tourism poverty alleviation: A tourism empowerment perspective. *Frontiers in Psychology*, 11, 2064.

Zahra, S.A., Gedajlovic, E., Neubaum, D.O., & Shulman, J.M. (2009). A typology of social entrepreneurs: Motives, search processes and ethical challenges. *Journal of Business Venturing*, 24, 519–532.

# 16 Heritage Tourism and Sustainable Development

## Prospects for a Cultural Landscapes Framework

*Neel Kamal Chapagain and Jharna Joshi*

## Introduction

Tourism and heritage conservation can support each other if planned appropriately through a culturally sensitive and collaborative approach. This relationship between these two seemingly diverging perspectives (protection and use) is seen as increasingly compatible owing to their shared interest: the heritage values of a place or region. However, the reality in many regions, including South Asia and the Himalaya, is different (Timothy, 2021c; Timothy & Nyaupane, 2009). While heritage and cultural values have been the focus of preservation activities in the heritage sector, tourism may view these same values as commodifiable assets (Chapagain, 2016). This sometimes leads to dissonance between preservation and promotion (Tahan, 2020). It is therefore important that tourism and heritage conservation or planning policies are prepared to enable an integrated and collaborative approach.

Destinations known for spectacular cultural landscapes attract tourism, which often lies at the core of economic development (Graburn, 1995). Tourism requires certain infrastructure, such as hotels and restaurants, and these inevitably change the landscape. Landscape changes due to the uncontrolled growth of tourism is evident in many destinations, mainly through land-use change, infrastructure development, abandoning traditional livelihoods, and urbanization processes (Gkoltsiou et al., 2013; Urry, 2002). These changes include the growth of specific tourism-oriented built environments and a loss of environmental and cultural diversity when earlier landscapes are replaced by landscapes of tourism (Gkoltsiou et al., 2013).

Globalization and the recent development of communication technologies have increasingly homogenized the world, influencing people to conform to international standards rather than adhere to local cultural norms (Nepal et al., 2015). Tourism increases demand for modern facilities that are not part of traditional built environments and settlements. In developing countries that have weak building regulations and enforcement mechanisms, new building designs are determined by their owners, usually following contemporary trends (Nepal, 2005). With local governments adopting tourism for economic development, the demands of the sector often determine owners' design decisions and influence local development plans and policies. In sparsely populated and fragile Himalayan communities, the

DOI: 10.4324/9781003030126-20

visibility of these changes can be more dramatic. This chapter reviews tourism and cultural heritage policies and efforts in the Himalayan region, with special references to Upper Mustang in Nepal, to understand the dynamics and suggest a potential way forward.

## Tourism in the Himalaya

The Himalaya have captured imaginative and romantic titles such as 'way to heaven', 'Shangri La', and 'roof of the world'. These mystical nicknames have triggered many a tourist gaze (Urry, 2009; Chapagain, 2011). The Himalayan range has been subjected to academic studies ranging from their geophysical characteristics to their role as protectors of unique cultures. These associations have made the Himalaya a destination for spiritual journeys, exotic explorations, scientific research, nature appreciation, and recently, adventure and cultural tourism. Each of these perspectives relates to certain values which demand specific attention to support those values. For example, pilgrims would wish to maintain the sanctity of the mountains by espousing the stories and beliefs surrounding the sacred sites. Explorers of the exotic 'other' would wish to retain the unconventionality and mystique of the mountains, while geological or anthropological researchers would wish to preserve evidence of geomorphological change and cultural norms. Nature enthusiasts would desire to maintain the region's wildness, adventurers would wish to gain access to more challenging localities, and cultural and heritage tourists would appreciate the region's intangible and tangible heritage. Some of these ideals or values eventually connect to or contrast with the idea of conservation, whereas some of them may lead to exploitation. Within these multitudinous interests among 'outsiders', some may be appreciated by the local population while others may be contested.

The Himalaya are sacred for many communities, and pilgrims have been travelling to this region since ancient times. Modern tourism has emerged as a distinct sector, comprising a sizeable proportion of local and national economies in most of the Himalayan countries. All six Himalayan states have vast cultural heritage resources, many of which form a considerable portion of their tourism industries (Chand, 2013; Timothy, 2021c). Even though Afghanistan and Pakistan have significant heritage tourism potential, they have suffered from the impacts of war and image problems, which have hampered efforts to develop a cultural tourism sector (Arshad et al., 2018; Coulson et al., 2014; Nyaupane & Budruk, 2009). The region is particularly known for its Indigenous peoples and their living cultures (i.e., music, festivals, and food), rural villages, religious shrines and temples, hill station colonial outposts, agricultural landscapes, teahouses, and tea plantations (Chaudhuri, 2012; Dua, 2005; Jolliffe, 2007; Karar, 2010; Malodia & Singla, 2017; Nyaupane et al., 2015; Pradhan, 2014).

Along with an increase in tourism footfall, concerns about nature and heritage conservation and management have grown (Chapagain, 2016). The recent discussions on climate change and sustainable development have boosted people's interests in conservation and heritage management. To understand these issues, we need to consider local environments, cultures, and livelihoods prior to delving into tourism and heritage issues.

Sharma (2000) highlights some of the specific conditions of the Himalaya that require a critical understanding. These conditions include inaccessibility, fragility, diversity, niche tourisms, and marginality (Jodha, 1991). Sharma (2000) suggests three important concepts to guide tourism in the Himalaya: sustainability, carrying capacity, and participatory local development. The common threads across the Himalayan cases identified by Sharma (2000) that are relevant to the present discussion include policies on the role of tourism in local and regional development, the need for proactive planning and program interventions, concerns for carrying capacity, building local institutions, participatory tourism planning and local development, sharing revenue from mountain tourism, enhancing linkages with the local production base, human resource development at the local level, and conducting visitor surveys.

In Nepal and India, the replacing of traditional livelihoods (i.e., subsistence farming) with service industries, such as tourism and, more recently, road construction, is changing the built environment, especially along the main trekking routes (Pawson et al., 1984). With eight of the ten highest peaks in the world, mountaineering and adventure tourism dominates the tourism industry in Nepal. With limited potential for industrial development in the harsh mountain conditions, tourism has improved the living conditions of many residents. However, the changes to nature and culture wrought by the growing number of tourists is a salient concern for sustainable development in this region (Adams, 1992; Banskota, 2012; Brown et al., 1997; Pandey et al., 1995; Pawson et al., 1984; Sparrowhawk & Holden, 1999; Stevens, 1993a, 1993b; Wells, 1993; Zurick, 1992). Deforestation and a changing built environment have occurred even in remote destinations such as Lomanthang, which was closed to Western tourists until 1992 (Byers, 1987; Chapagain, 2011; Shackley, 1994, 1996; Stevens, 1993b). Even isolated Ladakh has faced the serious consequences of tourism. The only areas that have not seen extensive ecological and cultural changes are Sikkim and Bhutan, where state controls on tourism have helped mitigate the industry's negative consequences.

Transformations in the traditional functions of villages from agriculture to tourism are changing the sizes, characteristics, functions, and spatial distributions and dimensions of traditional settlements (Nepal, 2005; Joshi, 2019). Increasing tourism can also displace villagers as homes are purchased by non-permanent residents or other outside investors buy local hotels, restaurants, and shops. Locals are often pressured to sell their properties, and villages become unaffordable due to tourism-induced costs of living and gentrification in popular tourist destinations (Goulding et al., 2014).

In most developing countries, systematically weak policy, plan, and law enforcement, and a lack of coordination between government agencies result in conflicting programs and projects, such as the recent increase in road construction to popular trekking and pilgrimage destinations in Nepal without proper assessments of its impact on the environment, local economy, and culture (Lama & Job, 2014). Residents perceive the roads as 'development', providing easier access to cities and related services. Road connections are also considered by development agencies as a tool for poverty alleviation because they generate employment and provide access to markets. While there has not been a concomitantly significant increase in the export of local products, new roads tend to encourage the importation of non-traditional construction materials (e.g., cement) whose use alters the built landscape (Joshi, 2019). Other means of faster mobility have also raised concerns and drawn criticism, particularly because of rapid changes and unequal economic consequences that come with them. One example is the Manakamana cable car project in the Gorkha district of Nepal, which was introduced to facilitate pilgrims' access to a popular religious site; however, the project seems to have contributed to unsustainable and unequal development. The long-established/traditional use of local materials, which for centuries resulted in distinctive architectural characteristics and cultural landscapes, has gradually decreased as uniform designs and patterns of development have been adopted across different geographic regions (Phillips & Yannas, 1999). This trend endangers the unique and inimitable heritage of Himalayan traditional settlements, which ultimately may become uniform and undistinguishable from those in other parts of the world, creating standardized landscapes devoid of place meaning and identity (Relph, 1976). Ladakh, Tibet, and Karakoram have faced similar issues.

Nepal (2005) views the changing traditional landscapes on the trekking trails in Nepal positively as a result of economic, social, and cultural development. However, he emphasizes the importance of understanding the process of change to develop sustainable tourism strategies. As lodge owners' decisions are driven by earnings, their desire to cater to tourists seems to drive change in the vernacular landscape (Lim, 2007). Driven by the same concern, Bhutan's sustainable development policy has adopted a 'high value, low volume' tourism strategy "that was grounded on the firm belief that uncontrolled tourism would overburden Bhutan's limited facilities and threaten the traditional culture, values and the environment" (Rinzin et al., 2007, p. 109; see also Nyaupane & Timothy, 2010). Since its opening for tourism in 1974, and with private sector involvement since 1991, Bhutan's sustainable tourism focus emphasizes the country's 'Gross National Happiness' (GNH) index instead of the more commonly used Gross Domestic Product (GDP) (Leishipem, 2013; Rinzin et al., 2007). However, the success of Bhutan's sustainable tourism policy has been questioned for its inherent power dynamics and the fact that the policy is applied only to non-South Asians, whereas the country's main foreign markets are India and other countries of South Asia (Nyaupane & Timothy, 2010). Schroeder (2015) further suggests that the GNH framework alone cannot achieve the most desired outcomes. Instead, he suggests that the policy of environmental and cultural protection and its careful enforcement have been instrumental in ensuring the success of the Bhutanese tourism sector (Schroeder, 2015).

Geneletti and Dawa (2009) examine patterns of trekking tourism in Ladakh, India, using GIS and other spatial assessment methods to highlight the impact of tourism on mountain landscapes. Such impacts on landscapes correspond to existing trails and pilgrimage routes. However, they found that the mitigation of trekkers' landscape impacts has been achieved through the spiritual values associated with these landscapes. When pilgrimage blurs with non-pilgrimage tourism, the environmental impacts are more severe. Non-pilgrimage tourism demands more infrastructure, which invites more visits, thereby destabilizing the intricate relationship between pilgrims and the landscape, which is a core pilgrimage value in mountain regions.

Most of the literature on tourism-landscape relationships in the Himalaya focuses on the impact of tourism on the environment, which has been expanded in recent studies to include communities. Most studies on landscapes have yet to look beyond the natural elements to include cultural heritage. Recent work on landscape changes (Nepal, 2005; Nyaupane et al., 2014; Sur & Singh, 2020) and landscape aesthetics (Beza, 2010; Jutla, 2016) are limited in terms of how they conceptualize landscape; however, interest in the impact of tourism on the changing landscape is growing (Sundriyal et al., 2018; Zurick & Rose, 2009). There is a need to expand the research scope to include the potential of tourism in the Himalaya from different perspectives, as until now it has focused on the montane natural environment. Recently, international symposiums in the Himalaya have highlighted pertinent issues including the linkage between cultural heritage and tourism (Wickens et al., 2017). Different aspects of the landscape, including future changes, are inherently interconnected and require a more holistic examination under the cultural landscape framework to understand the inseparability of nature and culture in the region and to encourage protection and sustainable growth.

Countries in the Himalayan Karakoram range seem less prepared with robust policies. Pakistan, for example, lacks policies that focus on its northern mountain region, although the Northern Areas Tourism Development Board prepared a draft (Ahmed, 2003; Karim et al., 2012). That part of Pakistan lacks enabling processes (proper statistics, infrastructure, and attention to environmental and cultural heritage), creating a missed opportunity for sustainable development (Ahmed, 2003). Mountainous Pakistan hosts the world's second highest peak (K2) and many significant landscape features. Scholars have flagged concerns about the country's strict conservation approaches, particularly in the northern protected areas, for their potential failure to connect conservation to local people and their customary rights (Knudsen, 1999). This is a representative case indicating the problems with strict conservationist approaches to nature protection. As tensions heightened in some national parks in Pakistan during the 1990s, there were promising participatory developments taking place in Nepal. The next section highlights participatory development and communities in Upper Mustang, Nepal, as an empirical look at landscape evaluations.

## Upper Mustang, Nepal

Mustang is a district in northwestern Nepal, famous for its high-altitude flora, fauna, landscapes, settlements, and cultural heritage. On the basis of its geographic distinctions and ethnic groups, the district is broadly classified in two regions: lower and upper. A description of the socio-political dynamics between Lower Mustang and Upper Mustang is beyond the scope of this chapter, but both of them have received significant interest in both the tourism and heritage sectors (Bista & von der Heide, 1997; Kharel, 2017; Shackley, 1994). The example described here focuses on Upper Mustang, of which the walled settlement of Lomanthang is considered the cultural capital.

Lomanthang, a unique earthen-walled town in the Himalaya, was settled in the early fifteenth century as the capital of the then Kingdom of Lo (Dhungel, 2002). The city's establishment is intertwined with the religious and traditional beliefs about how the entire landscape of Upper Mustang was created by the subjugation of certain demons and demonesses by a Tibetan Buddhist sage named *Padmasambhav*. Such intricate intertwining of cultural beliefs and natural features makes the entire landscape a canvas of aesthetics, wonders of geomorphology, religious and spiritual connectedness, and overall a tradition of living settlements during the past few centuries amid a harsh environment due to the region's high altitude and arid climate.

With its historic importance as the capital of the emerging kingdom of Lo in the fifteenth century, Lomanthang was settled with the king's palace and other important religious-cultural institutions situated within a walled settlement. Much of fifteenth-century physical fabric remains today, along with its descendant population. In 2008, the Department of Archaeology of the government of Nepal nominated the 'Medieval Earthern Walled City of Lo Manthang' on the Tentative List of World Heritage Sites. Chapagain (2017) argues that the nomination of only the walled settlement separates it from its broader cultural landscape context, making it an island rather than being part of a larger historical, cultural, and geographical legacy within which Lomanthang has evolved over time. This is the basis for the argument of a cultural landscapes framework in this chapter. The focus here is two aspects of heritage and tourism management in this region: the Annapurna Conservation Area Project (primarily focusing on environmental conservation) and scattered efforts at cultural heritage conservation.

Although Nepal formally opened for foreign tourists in the 1950s, Upper Mustang largely remained a 'forbidden kingdom', closed mainly for political and security reasons. Upper Mustang was part of a larger Tibetan kingdom (Dhungel, 2002), but it joined the emerging Gorkha Kingdom (the forerunner of the unified Kingdom of Nepal) led by King Prithivi Narayan Shah in the eighteenth century. Upper Mustang was opened formally for foreign visitors in 1992, following the political change from an absolute monarchy to a constitutional monarchy in 1990. The opening of Upper Mustang had two fundamental policy premises. The first was to cater to the demands of both the tourism sector (mostly led from Kathmandu) and the locals who wanted to benefit from tourism as they had seen in Lower Mustang. The second policy was to use tourism as an economic means of supporting conservation and development. As part of the second policy, 70% of tourism revenue was to be channelled back to local development. For this, a very high fee and limited number of tourists per year were initiated, and the Annapurna

Conservation Area Project (ACAP) was entrusted as the key agency to manage these activities in the field.

The ACAP is a successful conservation model and has been a primary stakeholder on behalf of the Nepalese government, to monitor and manage tourism and conservation activities (Khadka & Nepal, 2010). Upper Mustang is part of ACAP's jurisdiction. The project was established as a protected area by the government of Nepal in 1986 with Upper Mustang being added in 1992. This new approach to protected area management by local communities was established after the original national park proposal with support of the World Wildlife Fund, even though it faced widespread local opposition (Stevens, 1997). ACAP aimed to "achieve sustained balance between nature conservation and socio-economic improvement in the Annapurna Conservation Area (ACA) thereby assist National Trust for Nature Conservation [NTNC] in achieving its goal" (NTNC, 2020, n.p.). ACAP was first tested as a pilot programme in the Ghandruk Village Development Committee (VDC) in 1986. After being notified in the Gazette as a 'conservation area' in 1992, ACAP's programme covered the entire area (NTNC, 2020, n.p.)

At its inception, ACAP recognized biodiversity and the mountain inhabitants and the need to integrate them. This resulted in the driving principle of ACAP: conservation for development. There are several features that make the Annapurna region a unique place in the world. It contains the world's deepest river gorge—Kali Gandaki Gorge, which is 3 miles long and 1.5 miles wide, a valley with fossils from the Tethys Sea dating 60 million years ago. The region contains the world's largest rhododendron forest. Tilicho Lake in Manang is the world's highest altitude freshwater lake.

The biological diversity of the Annapurna region is equally rivalled by cultural diversity. Gurung and Magar are the dominant groups in the south, whereas Thakali, Manange, and Loba are dominant in the north. Each of these groups speaks its own dialect and has unique cultures and traditions. There are also Brahmin, Chhetri, and other occupational castes although in comparatively smaller numbers. Hindu, Buddhist, and pre-Buddhist religions are prevalent across the region. The local people reside in the five districts of the 15 rural municipalities of the Annapurna Conservation Area (ACA). The natural and cultural features of ACA have made it the most popular trekking destination in the country, drawing more than 60% of the country's total trekkers (NTNC, 2020, n.p.).

The success and efforts of ACAP are a well-known model for community-based area conservation in Nepal and beyond (Nyaupane & Thapa, 2004; Stevens, 1997). Building on this model, the rest of this chapter introduces the cultural significance of the landscape to suggest a nature-culture symbiosis. Chapagain (2011) notes that although cultural heritage was recognized as one of the programme areas within ACAP, it was mostly limited to education and awareness of cultural heritage. The Department of Archaeology had initiated some repair works to Lomanthang's city wall in the 1980s, but the agency's active involvement has been limited. Only in 1997, when a conservation and restoration project of Thubchen Gompa in Lomanthang began, was cultural heritage seen as an important aspect of regional conservation and development (Chapagain, 2011). Interestingly, the heritage conservation project was also implemented through the then King Mahendra Trust for Nature Conservation (now National Trust for Nature Conservation), which is the parent organization for the ACAP. Yet, heritage protection was not well integrated with environmental and biodiversity conservation. While ACAP is appreciated for its community-based approach and has been held up as a role model for other regions, cultural heritage remains only a peripheral concern (Chapagain, 2011). On the other hand, in 2008, the Department of Archaeology nominated Lomanthang to UNESCO's Tentative List of World Heritage Sites. However, this was done without prepared policies and established local management systems for heritage and tourism (Chapagain, 2013). This resulted in conflict between the interests of outsiders (i.e., experts and the state) and local residents with regard to conservation and development. However, Lomanthang and Upper Mustang are not alone in this plight.

Along with its inclusion under the ACAP, as noted before, Upper Mustang was officially opened for tourism in 1992. Until then, foreign travellers were not permitted to visit for two reasons. First (and the formal explanation) was that the ecosystems were too fragile for tourist activities; hence, the idea of a conservation project as discussed earlier was promoted but tourism was not. The second reason (and informal but more realistic) was the region's politically sensitive location with regard to the Khampa insurgency against China and the associated CIA operations and support for the rebels in the 1970s (Chapagain, 2011).

Lower Mustang was already benefitting from tourism, which caused Upper Mustang's population to push for opening the region as well (Chapagain, 2011, 2017). Initially, the carrying capacity (quota) was set at 200 visitors annually with a trekking permit fee of USD $500 per person for seven days. Shackley (1994) warned that Lomanthang had already exceeded its carrying capacity two years after the region opened and had realized only limited financial benefits as tour operators preferred tented camps over local accommodations. Chapagain (2011) argues that the idea of carrying capacity was applied haphazardly, with the annual quota apparently going up from 200 to 500, and then over 1,000 in 2007. In 2008, the quota limit was lifted, resulting in more than 2,000 visits that year (Chapagain, 2011, 2017). Currently, there is no annual quota and the trekking permit fee has been lowered to US$500 per person for the first 10 days and US$50 per day for each additional day to increase tourism revenue.

The Upper Mustang case in the 1990s and the first decade of the new millennium (see Chapagain, 2011 for a detailed account on this) clearly show that the activities under local development, including the promotion of tourism and activities to preserve cultural and natural heritage, are almost entirely disjointed. Upper Mustang is thus a classic case for problematizing the heritage tourism approach. The plight of Upper Mustang resonates in Ladakh, where the matter is more complicated with the development of infrastructure, border security, and other national issues (Vogel & Field, 2020). Bhutan, on the other hand, has recently begun to exhibit similar symptoms where, despite the focus on high-end tourism by not allowing independent travel, recent reports on the adverse impacts of tourism have raised serious concerns. Bhutan's cultural preservation and tourism quality can be cautiously replicated elsewhere in the Himalaya.

The tourism assets of Upper Mustang include montane nature, a built environment, living cultures, and festivals. We suggest that it is the Himalayan 'wilderness' combined with exotic cultures (e.g., the 'lost' mystical Tibetan culture that appeals to the Western travel psyche) that drives tourism here. Figure 16.1 captures some of the key tourism assets that overlap with heritage conservation interests. Similarly, Figure 16.2 illustrates some of the key stakeholders in relation to their proximity and closeness with the region of Upper Mustang.

Overlaying Figures 16.1 and 16.2 can be informative about a desirable relationship among the larger contexts and features (tangible, intangible, natural, and cultural) emerging within the larger context, and various stakeholders. Experience from the past three decades of tourism in Upper Mustang has not been very encouraging in terms of tourism's contribution to both local development and conservation. Conservation goals seem to have been fairly well achieved primarily through two parallel activities: ACAP's focus on nature protection and a few ad hoc heritage conservation initiatives achieving some tangible outcomes in heritage restoration. However, the long-term viability of conservation initiatives has been a key concern. ACAP was originally meant to be handed over entirely to a local management system, but this has not been achieved. The cultural conservation efforts have primarily been driven by foreign funding; for example, the American Himalayan Foundation was the main donor in the 1990s and 2000s,

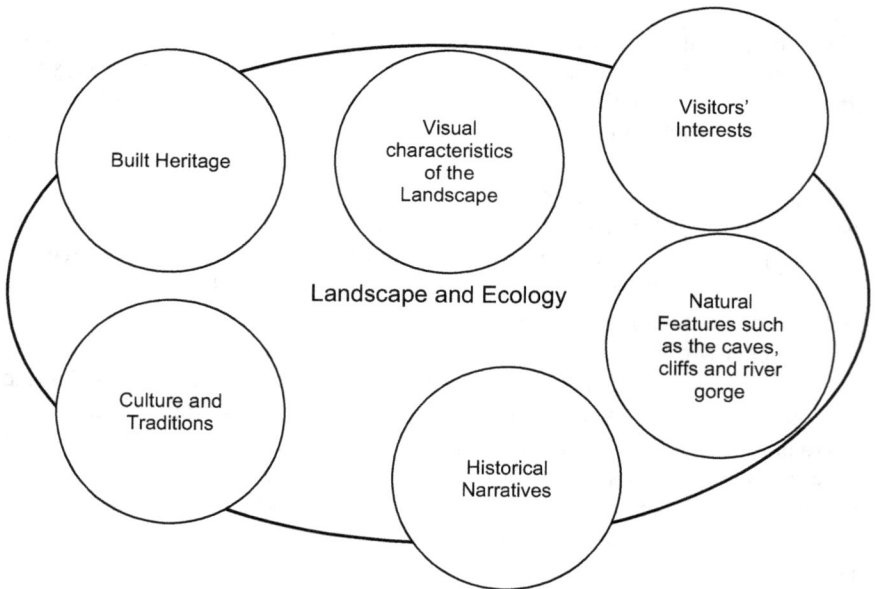

*Figure 16.1* Key tourism assets of Upper Mustang

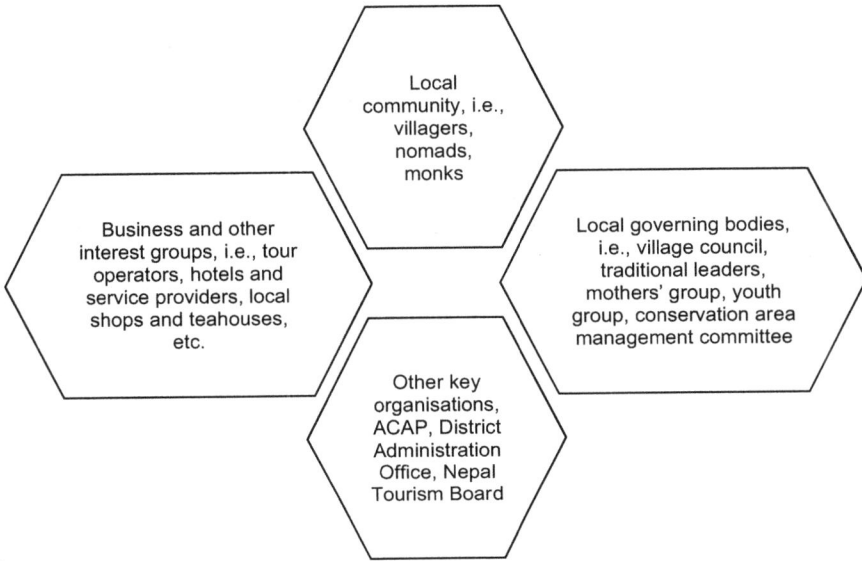

*Figure 16.2* Key stakeholders in Upper Mustang

whereas various embassies including those of India, China, and other states have played piecemeal roles as per their own convenience and interest in supporting monks' activities and monastery expansion projects. Compared to these foreign expenditures on sponsored preservation activities, there have been very few community-initiated and tourism-supported conservation efforts.

The functions of everyday life and daily survival are pre-eminent concerns for every community in the Global South. In most cases, nature and culture conservation are secondary considerations when people's basic survival is in question (Timothy, 2021a, 2021b). After the 1992 opening of tourism, when the inhabitants of Upper Mustang did not see any direct benefit from tourism, they resorted to closing down the area for visitors as a way of protesting the lack of shared tourism revenues between the central government and local administrations. There have been instances of conflict between outside experts and locals, triggered by outsiders' arguments for conservation or preservation without acknowledging and addressing the everyday needs of residents, such as health care and education. The epitome of these conflicts surfaced in 2001–2002 when residents built a dirt road from the Chinese border right to the main gate of the walled town of Lomanthang, with a plan to continue further. This happened after Lomanthang had received a great deal of funding for cultural heritage preservation since the mid-1990s, while the community's pressing concerns were ignored by both experts and the national government. The local government actually siphoned its budgets into the efforts to build the road. Despite the presence of ACAP at the grassroots level, it failed to facilitate meaningful negotiations between the local communities and policymakers to provide basic services such as roads. As a result of this conflict, today Upper Mustang is connected by a road, but its condition and sustainability, in terms of both infrastructure and ecology, are questionable. For some tourists, the road provides convenient access. For trekkers, however, the road is seen as a disturbance because it essentially destroyed the hiking trail, and therefore devastated the 'authentic' trekking experience because there are no alternate routes. For the locals, the economic viability of establishing local services (e.g., bed and breakfasts) is questionable except during the annual festival of Tenji, when local hospitality services are in short supply. In other words, not seeing Upper Mustang as part of a broader naturally and culturally symbiotic landscape, instead focusing only on Lomanthang as a heritage destination in isolation of its surroundings, has seriously disturbed the dynamics between the many stakeholders of this unique Himalayan destination.

This chapter now looks through a broader heritage lens and expands the 'specific-centric' nature conservation framework into a more holistic landscape perspective. While the ACAP's approach to community-based initiatives to interlink conservation with development is laudable, the original purpose of handing over ACAP management to the community within ten years of its inception could not take place even after 35 years of operation. This scenario calls for a rethinking of how nature and culture in a given landscape are understood.

## The Cultural Landscapes Framework in Heritage Tourism in the Himalaya

The term cultural landscape in this chapter has a holistic meaning to include the natural and built environments with their associated cultural dimensions. Landscapes are generally formed as a result of human interaction with nature, adapting people's history, traditions, and sociocultural contexts. Societies constantly rework their landscapes, creating and re-creating them, and the way in which people interact with and understand the landscape depends on specific time, place, and historical conditions. Traditional landscapes, especially in mountainous regions, generally integrate the natural and cultural environments utilizing local resources, materials, and labour and are often regarded as being sustainable based on Indigenous, local knowledge (Phillips, 2003). They are developed over generations and adapted to local conditions resulting in distinctive characteristics (Bodach et al., 2014; Phillips & Yannas, 1999; Relph, 1976), such as those heritagescapes of the Himalaya (Nigam, 2002; Nüsser, 2001).

With the growth of alternative and niche markets that desire experiential tourism, the potential for places that offer distinctive landscapes is also growing (Terkenli, 2002). The significance of the relationship between landscape and tourism lies in the interdependence between the two. While the tourism industry needs distinctive landscapes, places need tourism for economic development, especially Himalayan landscapes that have limited alternative development potential. However, as the tourism industry grows, it stimulates urbanization, which in turn boosts local economic activities and leads to changes, expanding settlements and transforming the distinctive character that attracted tourists in the first place (Dyer, 2006; Hoskins, 1953). Landscapes are also modified because of residents' perceptions of, and desire for, 'modern' lifestyles that regard traditional landscapes as too ordinary, outdated, and inadequate for contemporary life (Knapp, 1992; Timothy, 2021b).

Throughout history, the term 'landscape' has had a range of interpretations from nature-only to a social construct, where natural, cultural, social, and economic worlds overlap, interact, and integrate (Stoffelen & Vanneste, 2015). Early researchers viewed landscape as a way of seeing the world that was visual and detached (Setten, 2006). For example, Sauer (1963, p. 343) defined cultural landscape as a geographical area "fashioned from a natural landscape by a culture group" where "culture is the agent, the natural area is the medium, the cultural landscape the result". There is continuing discussion about the interpretation of cultural landscapes with additional concepts and terms such as tourism landscape, geotourism, nature, environment, region, place, and space used in landscape and tourism studies (Jones, 2003; Stoffelen & Vanneste, 2015). Despite the different terminologies, certain similarities seek to capture the meaning rather than view landscape as something to be understood in a detached, objective manner.

The concept of cultural landscape as heritage is widely used in the cultural geography, landscape planning and World Heritage literature, and UNESCO now lists properties under the cultural landscape category. UNESCO defines cultural landscapes as the cultural properties that represent the "combined works of nature and of man" (UNESCO World Heritage Centre, 2015, p. 11). The European Landscape Convention (ELC) expands on this definition of landscape as "a zone or area as perceived by local people or visitors, whose visual features and character are the result of the action of natural and/or cultural (that is, human) factors" (Council of Europe, 2000, n.p.). Unlike UNESCO, the ELC definition is not limited to remarkable and historic landscapes, but includes ordinary everyday landscapes (Antrop, 2013; Déjeant-Pons, 2006), which are also a vital, albeit often neglected, part of human heritage (Timothy, 2014a, 2018). The ELC recognizes the important role of landscapes in people's well-being and their participation in decision-making processes, which leads to greater attachment and ownership of landscapes. Unfortunately, this is still lagging behind in most Himalayan countries (Timothy, 2021c).

These definitions reflect the evolving character of landscapes and consider the natural and cultural components as intertwined continuums rather than separately. However, a nature-culture divide exists historically in the World Heritage domain. For example, the two advisory bodies of UNESCO's World Heritage Convention—the International Union for Conservation of Nature (IUCN) for nature and International Council on Monuments and Sites (ICOMOS) for culture—each has different agendas, guidelines, and categories. However, both advisory bodies and the third one, the International Centre for the Study of Preservation and Restoration of Cultural Property (ICCROM), have recently begun aggressively promoting nature-culture linkages.

The concept of landscape has evolved so that the interrelationships between nature and culture are now seen as essential characteristics of landscapes. The study of landscapes has also continuously evolved with the complexity, diversity, and dynamic nature of tourism and landscapes (Nepal, 2009). Tourism and geography researchers have studied landscape-tourism interactions and their ongoing changes; however, there is a lack of consensus on tourism landscape approaches and interpretations (Terkenli, 2008). Although it is well-established that landscapes are the fundamental assets of many tourist destinations, research into the crossover between landscape and tourism is either tourism-cantered or landscape-cantered (Stoffelen & Vanneste, 2015).

Cultural tourism is well-established, but research tends to focus separately on cultural and natural landscape-based tourism. At many destinations, the natural and cultural landscapes are packaged as two separate products, such as national parks and wilderness areas as natural attractions, while dance, art, architecture, food, and music are sold as cultural heritage. Only a few destinations, such as the Mongolian steppe, are promoted as an integrated cultural landscape (Buckley et al., 2008). This natural and cultural tourism divide is further separated into rural and urban tourism. While the interconnections between rural tourism and natural landscapes are widely accepted, especially with the potential of tourism for economic revival and protection of rural agricultural landscapes (Oliver & Jenkins, 2003; Saarinen, 2008), urban destinations are viewed generally as crucibles of culture.

Both natural and cultural landscapes have been conserved, restored, and modified for and by tourism. In the West, natural areas that were primarily wilderness were preserved for recreational and tourism purposes. Historic areas and villages have also been preserved and promoted for cultural exploration. However, pressure for economic development puts landscape resources and conservation efforts at risk in preference to tourism development (Han, 2006). This pressure is particularly acute in countries of the Global South where planning and implementation mechanisms are inadequate.

Conservation policies also tend to focus on one aspect of the landscape, compromising other aspects and affecting the communities living within it (Eagles & McCool, 2002). For example, when conservation policies focus on natural landscapes, biodiversity may be protected, but the costs for local people, especially in subsistence-based rural communities in the Himalaya may be substantial (Han, 2006; Joshi, 2019; Thing et al., 2017). Protected areas are often seen as playgrounds for foreigners and wealthy urban residents, and may be resented when benefits are not shared with local communities (Adams & Infield, 2003; Guha, 1997). Restricting traditional access to natural resources, such as wood for cooking, building and repairing houses, and establishing monuments to preserve biodiversity, not only creates conflicts between decision-makers and residents but can also impede conservation efforts (Eagles & McCool, 2002; Ghimire & Pimbert, 2009).

The protection of built cultural heritage is a complex and interdisciplinary field that should address not only physical structures but also the different narratives and memories of significant value in dynamic communities (Amar, 2017). Traditionally, the conservation of built landscapes focused on restoring and preserving buildings and monuments with historic, architectural, memorial, and national/international significance. In recent decades, built heritage increasingly includes the vernacular landscapes as well as their associated intangible values (Timothy, 2014b). However, there exists a contradiction between preserving the past for cultural values, historicity, or nostalgia and the need for development in response to changing societal values (Nasser, 2003). Those who study buildings tend to detach them from their surroundings and fail to link the influence of the context and the people living and working in it on the built environment (Dyer, 2006). Studies often detach urban sites from natural landscapes or rural heritagescapes, instead focusing on conserving the cities' physical and symbolic heritage values (Harrill & Potts, 2003; Scott, 2010).

The built landscape is multidimensional and includes settlement patterns, surrounding natural and agricultural areas, the vista, and the people who live in and use the landscape (Orbaşli, 2000). When there is a contest between the commercial values of tourism and the protective values of heritage conservation, tourism tends to win as local economies become more dependent on tourism (Nasser, 2003). The heritage tourism literature highlights this difficult relationship with two opposing views (Timothy, 2021a). On one hand, the cultural heritage sector sees conflicting interests where heritage values are compromised for commercial gain (Avieli, 2015; Harrison & Hitchcock, 2005; Joshi, 2014; Urry, 2002) and communities are adversely affected (Harrill & Potts, 2003). On the other hand, there are suggestions of mutual benefits where tourism strengthens people's identities and builds social solidarity. In developing countries, with the pressing demands of health, education, and sanitation combined with budget restrictions and weak legislation, heritage conservation becomes a predictable product for tourism development (Orbaşli, 2000). Both situations are exacerbated when villages and towns are declared national or world heritage with strict restrictions imposed on residents by governments as they target and promote tourism (Avieli, 2015).

Landscapes are vital to the tourism industry as the first medium of contact between the tourist and the destination, yet the industry receives constant criticism as an exploiter of landscapes (Terkenli, 2008). As one of the world's leading export earners, tourism creates landscapes and cultural industries that are staged for the purpose of tourist consumption (Terkenli, 2002), although this is not yet a major concern in most parts of the Himalaya. Certain Himalayan destinations are noted for specific landscapes that attract tourism, which necessitates tourism facilities and services. This results in modified landscapes, for example accommodations built specifically for tourists, which Jackson (1970) describes as other-directed architecture. Sometimes, these buildings follow traditional architecture, usually borrowing styles and decorations but built specifically for outsiders, thereby creating places that are detached from local communities (Relph, 1976). As the demand for accommodations rises, larger structures are built and may not conform to traditional forms, functions, or scales. The resulting 'touristified' landscapes homogenize the place and modify its local and regional character.

Rapid unplanned expansion results in strained resources such as water supply, which in turn can create conflict between residents and visitors, tourism entrepreneurs, and those not directly benefitting from the tourism industry (Joshi, 2019). An increase in the number of tourists may also bring about greater cultural exchange, in many cases diluting the cultures of both the host and the visitor. Some changes may be rapid and perceivable, such as urban growth, but most changes are slow and gradual.

This loss of traditional character and the homogenization of landscapes potentially create localities that lack distinctiveness, and may become placeless spaces (Relph, 1976). Tourism businesses are driven by profit-making, and their design preferences are often guided by the commodification of traditional architecture and culture (Weaver, 2009). This change endangers the sense of place for local residents, while the attractiveness of the landscape may decrease for tourists. As the interdependence between tourism and the destination increases, the distinction between organic and contrived landscapes becomes blurred, and over-staged landscapes become a permanent part of the destination (Cohen, 1995; Ritzer & Liska, 1997).

Change in both natural and cultural landscapes and tourism is inevitable. Landscape formation is an ongoing process that adds to the complexities of conserving cultural landscapes. At times, individual components change, even disappear, but the landscape as a whole may not necessarily appear to change in the same way or at the same rate. Therefore, a holistic approach to landscape, which considers that the overall landscape is more than the sum of its component parts is necessary (Antrop, 1998). While landscape changes are usually slow, natural disasters can change them quickly. For example, the 2015 Gorkha earthquake in Nepal flattened many settlements and caused landslides that swept away forested areas, which will take many years to regrow. However, today, many human-induced changes are consciously planned to achieve predetermined goals. While sudden and large changes and their impacts are easily visible, the impacts of slow and moderate change are gradual and may take a generation or more to become visible (Lew, 2014).

Policies initiated to conserve natural or cultural landscapes institute requirements that make it unaffordable for the local community to remain in their traditional homes; they are often displaced by tourism investors and second-home markets (Goulding et al., 2014; Kaltenborn et al., 2008; Phillips, 2003). This is beginning to happen in the Himalaya with regard to trekking tourism. Landscapes are living and dynamic organisms that encompass both natural and cultural values. It is important to acknowledge the crucial role of Indigenous communities and native knowledge to sustainably manage Himalayan resources. The strict division between culture and nature in policy formation may be detrimental, as these are not mutually exclusive, and such actions can force false boundaries between the inseparable concepts of nature and culture. The consequences of emphasizing one over the other sacrifices one in favour of the other. Tourism has the potential to provide economic opportunities that can contribute to holistic landscape conservation by integrating planning frameworks that adhere to the tenets of sustainable development.

## Conclusion: The Way Forward

Revising the current Tentative World Heritage nomination for the walled settlement of Lomanthang, using a cultural landscapes or even a biocultural landscapes perspective, is warranted (Chapagain, 2017). A broader framework that includes the diverse natural and cultural aspects of the landscape, as well as the people who inhabit it, would achieve inclusive and sustainable results.

Analysing various issues, community concerns, and the possible integration of contemporary heritage and tourism discourses, a cultural landscape-based heritage tourism framework for the Upper Mustang region and other parts of the Himalaya can be better conceptualized and operationalized. The landscape within this region corresponds to various cultural and historical landmarks. In contemporary terms, Lo-Tsho-Dun, the seven villages of Lo, are somehow distinct from the other villages in the rest of the Mustang district. Hence, it makes sense to consider this area a distinct Upper Mustang cultural landscape. The second pragmatic option would be to consider the existing Annapurna Conservation Area as a biocultural landscape. This would also be an easy option to explore, in terms of both assessing the values and putting together a managerial framework, taking advantage of the success story of ACAP in the past few decades. This can be developed onto the established management frameworks that ACAP has, but expanding it to integrate cultural landscape elements including the specific pockets of historically important cultural heritage sites.

This broad approach incorporates heritage tourism as a vehicle to facilitate heritage management and address some of the practical concerns and aspirations of many Himalayan communities (Singh, 2021). Here, one may recall the two historic turning points: setting up of the Annapurna Conservation Area Project and its expansion to include Upper Mustang with an emphasis on conservation for development. The term development relates to community development, which is pertinent to the second historic event—that of opening up the restricted area of Upper Mustang for controlled tourism. Although there have been issues of tourism revenue and carrying capacity, these two historic events in principle had the intent of balancing conservation and local development through monitored responsible tourism. It is therefore crucial to enhance the discourse and practice of heritage tourism in the Himalaya with a broader framework like that of cultural landscapes so that tangible, intangible, natural, and cultural heritage; conservation values; and community aspirations are integrated in a holistic framework. This is useful in establishing policies that matter to both the land and its people in identifying and promoting their values while also contributing to economic and environmental sustainability.

## References

Adams, V. (1992). Tourism and sherpas, Nepal: Reconstruction of reciprocity. *Annals of Tourism Research*, 19(3), 534–554.

Adams, W., & Infield, M. (2003). Who is on the gorilla's payroll? Claims on tourist revenue from a Ugandan national park. *World Development*, 31(1), 177–190.

Ahmed, B. (2003). *Sustainable Tourism and Cultural Heritage*. Gilgit, Pakistan: IUCN, Northern Areas Progamme.

Amar, J.H.N. (2017). *Conservation of Cultural Built Heritage: An Investigation of Stakeholder Perceptions in Australia and Tanzania*. Unpublished doctoral dissertation, Bond University.

Antrop, M. (1998). Landscape change: Plan or chaos? *Landscape and Urban Planning*, 41(3–4), 155–161.

Antrop, M. (2013). A brief history of landscape research. In P. Howard, I. Thompson, & E. Waterton (Eds.), *The Routledge Companion to Landscape Studies* (pp. 12–22). London: Routledge.

Arshad, M.I., Iqbal, M.A., & Shahbaz, M. (2018). Pakistan tourism industry and challenges: A review. *Asia Pacific Journal of Tourism Research*, 23(2), 121–132.

Avieli, N. (2015). The rise and fall of Hội An, a UNESCO World Heritage Site in Vietnam. *Journal of Social Issues in Southeast Asia*, 30(1), 35–71.

Banskota, K. (2012). Impact of tourism on local employment and incomes in three selected destinations: Case studies of Sauraha, Nagarkot and Bhaktapur. *Nepal Tourism and Development Review*, 2(1), 1–31.

Beza, B.B. (2010). The aesthetic value of a mountain landscape: A study of the Mt. Everest trek. *Landscape and Urban Planning*, 97(4), 306–317.

Bista, H.J.S., & von der Heide, S. (1997). An account of cultural heritage and nature conservation in Mustang, Nepal. *International Journal of Heritage Studies*, 3(3), 168–173.

Bodach, S., Lang, W., & Hamhaber, J. (2014). Climate responsive building design strategies of vernacular architecture in Nepal. *Energy & Buildings*, 81, 227–242.

Brown, K., Turner, R.K., Hameed, H., & Bateman, I. (1997). Environmental carrying capacity and tourism development in the Maldives and Nepal. *Environmental Conservation*, 24(4), 316–325.

Buckley, R., Ollenburg, C., & Zhong, L. (2008). Cultural landscape in Mongolian tourism. *Annals of Tourism Research*, 35(1), 47–61.

Byers, A. (1987). An assessment of landscape change in the Khumbu region of Nepal using repeat photography. *Mountain Research and Development*, 7(1), 77–81.

Chand, M. (2013). Residents' perceived benefits of heritage and support for tourism development in Pragpur, India. *Tourism*, 61(4), 379–394.

Chapagain, N.K. (2011). *Rethinking cultural heritage conservation at historic settlements: The case of Lomanthang, Nepal*. Unpublished Doctoral Dissertation. University of Wisconsin-Milwaukee.

Chapagain, N.K. (2013). World Heritage in Nepal: Critical reflections on Lomanthang's potential nomination. In N.K. Chapagain (Ed.), *Reflections on the Built Environment and Associated Practices: Essays in Honor of Professor Sudarshan Raj Tiwari* (pp. 37–53). Kathmandu: Department of Architecture, Institute of Engineering, Tribhuvan University.

Chapagain, N.K. (2016). Contextual approach to the question of authenticity in heritage management and tourism. *Journal of Heritage Management*, 1(2), 160–169.

Chapagain, N.K. (2017). Upper Mustang: A Cultural Landscape Framework. *Context – Built, Living and Natural*, XIII, 37–44.

Chaudhuri, T. (2012). Cultural tourism: A critical building block towards sustainable environment on Sikkim. *International Interdisciplinary Research Journal*, 2(3), 133–139.

Cohen, E. (1995). Contemporary tourism—trends and challenges: Sustainable authenticity or contrived post-modernity? In R.W. Butler & D.G. Pearce (Eds.), *Change in Tourism: People, Places, Processes* (pp. 12–29). London: Routledge.

Coulson, A.B., MacLaren, A.C., McKenzie, S., & O'Gorman, K.D. (2014). Hospitality codes and social exchange theory: The Pashtunwali and tourism in Afghanistan. *Tourism Management*, 45, 134–141.

Council of Europe (2000). European landscape convention. *Report and Convention Florence, ETS No.* 17(176), 8. Retrieved from https://doi.org/http://conventions.coe.int/Treaty/en/Treaties/Html/176.htm

Déjeant-Pons, M. (2006). The European landscape convention. *Landscape Research*, 31(4), 363–384.

Dhungel, R.K. (2002). *The Kingdom of Lo (Mustang): A Historical Study*. Kathmandu: Jigme S.P. Bista for Tashi Gephel Foundation.

Dua, J.C. (2005). Tourism and heritage resources in Garhwal Himalaya, an approach to planning and management. *Indian Historical Review*, 32(2), 294–298.

Dyer, C. (2006). Vernacular architecture and landscape history: The legacy of "The rebuilding of rural England" and "The making of the English landscape". *Vernacular Architecture*, 37(1), 24–32.

Eagles, P.F.J., & McCool, S.F. (2002). *Tourism in National Parks and Protected Areas: Planning and Management*. Wallingford: CABI.

Geneletti, D., & Dawa, D. (2009). Environmental impact assessment of mountain tourism in developing regions: A study in Ladakh, Indian Himalaya. *Environmental Impact Assessment Review*, 29(4), 229–242.

Ghimire, K.B.B., & Pimbert, M.P. (Eds.). (2009). *Social Change and Conservation*. London: Earthscan.

Gkoltsiou, A., Terkenli, T.S., & Koukoulas, S. (2013). Landscape indicators for the evaluation of tourist landscape structure. *International Journal of Sustainable Development & World Ecology*, 20(5), 461–475.

Goulding, R., Horan, E., & Tozzi, L. (2014). The importance of sustainable tourism in reversing the trend in the economic downturn and population decline of rural communities. *PASOS: Revista de Turismo y Patrimonio Cultural*, 12, 549–563.

Graburn, N. (1995). The past in the present in Japan: Nostalgia and neo-traditionalism in contemporary Japanese domestic tourism. In R.W. Butler & D.G. Pearce (Eds.), *Change in Tourism: People, Places, Processes* (pp. 47–70). London: Routledge.

Guha, R. (1997). Radical American environmentalism and wilderness preservation. In L. Robin, S. Sörlin, & P. Warde (Eds.), *The Future of Nature: Documents of Global Change* (pp. 409–431). New Haven: Yale University Press.

Han, F. (2006). *The Chinese View of Nature: Tourism in China's Scenic and Historic Interest Areas*. Unpublished doctoral dissertation, Queensland University of Technology.

Harrill, R., & Potts, T.D. (2003). Tourism planning in historic districts: Attitudes toward tourism development in Charleston. *Journal of the American Planning Association*, 69(3), 233–244.

Harrison, D., & Hitchcock, M. (2005). *The Politics of World Heritage: Negotiating Tourism and Conservation*. Clevedon: Channel View Publications.

Hoskins, W.G. (1953). The rebuilding of rural England, 1570–1640. *Past & Present*, 4(November), 44–59.

Jackson, J.B. (1970). Other-directed houses. In *Landscapes: Selected Writings of J. B. Jackson* (pp. 55–72). Amherst: University of Massachusetts Press.

Jodha, N.S. (1991). Mountain perspective and sustainability: A framework for development strategies. In M. Banskota, N.S. Jodha, & U. Pratap (Eds.), *Sustainable Mountain Agriculture: Perspectives and Issues* (Vol. 1, pp. 41–82). New Delhi: Oxford & IBH.

Jolliffe, L. (Ed.). (2007). *Tea and Tourism: Tourists, Traditions and Transformations*. Bristol: Channel View Publications.

Jones, M. (2003). The concept of cultural landscape: Discourse and narratives. In H. Palang & G. Fry (Eds.), *Landscape Interfaces: Cultural Heritage in Changing Landscapes* (pp. 21–52). Dordrecht: Kluwer.

Joshi, J. (2014). The Buddhist stupa at Bauddhanath: A world heritage site under pressure. In T.R. Gensheimer & C.L. Guicchard (Eds.), *World Heritage and National Registers: Stewardship in Perspective* (pp. 83–90). Piscataway, NJ: Transaction Publishers.

Joshi, J. (2019). *Landscape Aesthetics, Tourism, and Change: Case Studies in Nepal* Unpublished Doctoral Dissertation. Victoria University of Wellington.

Jutla, R.S. (2016). Understanding the Himalayan townscape of Shimla through resident and tourist perception. In S.F. McCool & K. Bosak (Eds.), *Reframing Sustainable Tourism* (pp. 137–159). Dordrecht: Springer.

Kaltenborn, B.P., Andersen, O., Nellemann, C., Bjerke, T., & Thrane, C. (2008). Resident attitudes towards mountain second-home tourism development in Norway: The effects of environmental attitudes. *Journal of Sustainable Tourism*, 16(6), 664–680.

Karar, A. (2010). Impact of pilgrim tourism at Haridwar. *The Anthropologist*, 12(2), 99–105

Karim, R., Durrani, S.A., & Hussain, A. (2012). Review of issues related to tourism policies regarding environmental management and customary practices of tourism in Gilgit-Baltistan, Pakistan. *Journal of Environmental Science and Engineering. B*, 1, 1087–1093.

Khadka, D., & Nepal, S.K. (2010). Local responses to participatory conservation in Annapurna Conservation Area, Nepal. *Environmental Management*, 45(2), 351–362.

Kharel, S. (2017). *Developing Rural Tourism Business in Nepal: Case Study of Mustang District, Nepal*. Unpublished thesis, Centria University of Applied Sciences, Kokkola, Finland.

Knapp, R.G. (1992). *Chinese Landscapes: The Village as a Place*. Hawaii: University of Hawaii Press.

Knudsen, A. (1999). Conservation and controversy in the Karakoram: Khunjerab National Park, Pakistan. *Journal of Political Ecology*, 6(1), 1–30.

Lama, A.K., & Job, H. (2014). Protected areas and road development: Sustainable development discourses in the Annapurna Conservation Area, Nepal. *Erdkunde*, 68(4), 229–250.

Leishipem, K. (2013). Modernisation, globalisation and development in Bhutan: Tourism as a catalyst. *Journal of Management & Public Policy*, 5(1), 5–11.

Lew, A.A. (2014). Scale, change and resilience in community tourism planning. *Tourism Geographies*, 16(1), 14–22.

Lim, F.K.G. (2007). Hotels as sites of power: Tourism, status, and politics in Nepal Himalaya. *Journal of the Royal Anthropological Institute*, 13, 721–738.

Malodia, S., & Singla, H. (2017). Using HOLSAT to evaluate satisfaction of religious tourists at sacred destinations: The case of religious travellers visiting sacred destinations in the Himalayas, India. *International Journal of Culture, Tourism and Hospitality Research*, 11(2), 255–270.

Nasser, N. (2003). Planning for urban heritage places: Reconciling conservation, Tourism, and sustainable development. *Journal of Planning Literature*, 17(4), 468–479.

Nepal, S.K. (2005). Tourism and remote mountain settlements: Spatial and temporal development of tourist infrastructure in the Mt Everest region, Nepal. *Tourism Geographies*, 7(2), 205–227.

Nepal, S.K. (2009). Traditions and trends: A review of geographical scholarship in tourism. *Tourism Geographies*, 11(1), 2–22.

Nepal, S.K., Verkoeyen, S., & Karrow, T. (2015). The end of sustainable tourism? Reorienting the debate. In M. Hughes, D.B. Weaver, & C. Pforr (Eds.), *The Business of Sustainable Tourism: Resolving the Paradox* (pp. 52–65). New York: Routledge.

Nigam, D. (2002). *Tourism, Environment, and Development of Garhwal Himalaya*. New Delhi: Mittal Publications.

NTNC (2020). Annapurna conservation area project (ACAP). Retrieved from https://ntnc. org.np/project/annapurna-conservation-area-project-acap

Nüsser, M. (2001). Understanding cultural landscape transformation: A re-photographic survey in Chitral, eastern Hindukush, Pakistan. *Landscape and Urban Planning*, 57(3–4), 241–255.

Nyaupane, G.P., & Budruk, M. (2009). South Asian heritage tourism: Conflict, colonialism, and cooperation. In D.J. Timothy, & Nyaupane, G.P. (Eds.), *Cultural Heritage and Tourism in the Developing World: A Regional Perspective* (pp. 127–145). London: Routledge.

Nyaupane, G.P., Lew, A.A., & Tatsugawa, K. (2014). Perceptions of trekking tourism and social and environmental change in Nepal's Himalayas. *Tourism Geographies*, 16(3), 415–437.

Nyaupane, G.P., & Thapa, B. (2004). Evaluation of ecotourism: A comparative assessment in the Annapurna Conservation Area Project, Nepal. *Journal of Ecotourism*, 3(1), 20–45.

Nyaupane, G.P., & Timothy, D.J. (2010). Power, regionalism and tourism policy in Bhutan. *Annals of Tourism Research*, 37(4), 969–988.

Nyaupane, G.P., Timothy, D.J., & Poudel, S. (2015). Understanding tourists in religious destinations: A social distance perspective. *Tourism Management*, 48, 343–353.

Oliver, T., & Jenkins, T. (2003). Sustaining rural landscapes: The role of integrated tourism. *Landscape Research*, 28(3), 293–307.

Orbaşli, A. (2000). *Tourists in Historic Towns: Urban Conservation and Heritage Management*. London: E & FN Spon.

Pandey, R.N., Chettri, P., Kunwar, R.R., & Ghimire, G. (1995). *Case Study on the Effects of Tourism on Culture and the Environment: Nepal—Chitwan-Sauraha and Pokhara-Ghandruk*. Bangkok: UNESCO.

Pawson, I.G., Stanford, D.D., Adams, V., & Nurbu, M. (1984). Growth of tourism in Nepal's Everest Region: Impact on the physical environment and structure of human settlements. *Mountain Research and Development*, 4(3), 237–246.

Phillips, C. (2003). *Sustainable Place: A Place of Sustainable Development*. Chichester: Wiley.

Phillips, C., & Yannas, S. (1999). Sustainable place: A place of sustainable development. *AA Files*, (39), 77–79.

Pradhan, K.M. (2014). Cultural tourism and Nepal. *Transnational Corporations Review*, 6(3), 238–247.

Relph, E. (1976). *Place and Placelessness*. London: Pion.

Rinzin, C., Vermeulen, W.J., & Glasbergen, P. (2007). Public perceptions of Bhutan's approach to sustainable development in practice. *Sustainable Development*, 15(1), 52–68.

Ritzer, G., & Liska, A. (1997). "McDisneyization" and "Post-tourism". Complementary perspectives on contemporary tourism. In C. Rojek & J. Urry (Eds.), *Touring Cultures: Transformations of Travel and Theory* (pp. 96–109). London: Routledge.

Saarinen, J. (2008). Tourism and touristic representations of nature. In A A. Lew, C.M. Hall, & A.M. Williams (Eds.), *A Companion to Tourism* (pp. 438–449). Oxford: Wiley-Blackwell.

Sauer, C.O. (1963). *Land and Life: A Selection from the Writings of Carl Ortwin Sauer* (J. Leighly, Ed.). Berkeley: University of California Press.

Schroeder, K. (2015). Cultural values and sustainable tourism governance in Bhutan. *Sustainability*, 7(12), 16616–16630.

Scott, A.J. (2010). The cultural economy of landscape and prospects for peripheral development in the twenty-first century: The case of the English Lake District. *European Planning Studies*, 18(10).

Setten, G. (2006). Fusion or exclusion? Reflections on conceptual practices of landscape and place in human geography. *Norwegian Journal of Geography*, 60(1), 32–45.

Shackley, M. (1994). The land of Lo, Nepal/Tibet: The first eight months of tourism. *Tourism Management*, 15(1), 17–26.

Shackley, M. (1996). Too much room at the inn? *Annals of Tourism Research*, 23(2), 449–162.

Sharma, P. (2000). *Tourism as Development: Case Studies From the Himalaya*. Lalitpur: Himal Books.

Singh, S. (2021). Time, tourism area 'life-cycle,' evolution and heritage. *Journal of Heritage Tourism*, 16(2), 218–229.

Sparrowhawk, J., & Holden, A. (1999). Human development: The role of tourism-based NGOs in Nepal. *Tourism Recreation Research*, 24(2), 37–43.

Stevens, S.F. (1993a). *Claiming the High Ground: Sherpas, Subsistence, and Environmental Change in the Highest Himalaya*. Berkeley: University of California Press.

Stevens, S.F. (1993b). Tourism, change and continuity in the Mount Everest Region, Nepal. *American Geographical Society*, 83(4), 410–427.

Stevens, S.F. (1997). Annapurna conservation area: Empowerment, conservation, and development in Nepal. In S. Stevens (Ed.), *Conservation through Cultural Survival—Indigenous Peoples and Protected Areas* (pp. 237–361). Washington, DC: Island Press.

Stoffelen, A., & Vanneste, D. (2015). An integrative geotourism approach: Bridging conflicts in tourism landscape research. *Tourism Geographies*, 17(4), 544–560.

Sundriyal, S., Shridhar, V., Madhwal, S., Pandey, K., & Sharma, V. (2018). Impacts of tourism development on the physical environment of Mussoorie, a hill station in the lower Himalayan range of India. *Journal of Mountain Science*, 15(10), 2276–2291.

Sur, U., & Singh, P. (2020). Assessment of landscape change of lesser Himalayan road corridor of Uttarakhand, India. *Journal of Landscape Ecology*, 13(3), 1–22.

Tahan, L.G. (2020). Archaeologists and tourism: Symbiosis or contestation? In D.J. Timothy & L.G. Tahan (Eds.), *Archaeology and Tourism: Touring the Past* (pp. 26–40). Bristol: Channel View Publications.

Terkenli, T.S. (2002). Landscapes of tourism: Towards a global cultural economy of space? *Tourism Geographies*, 4(3), 227–254.

Terkenli, T.S. (2008). Tourism and landscape. In A.A. Lew, C.M. Hall, & M. Williams (Eds.), *A Companion to Tourism* (pp. 339–348). Oxford: Wiley-Blackwell.

Thing, S.J., Jones, R., & Jones, C.B. (2017). The politics of conservation: Sonaha, riverscape in the Bardia National Park and buffer zone, Nepal. *Conservation and Society*, 15(3), 292–303.

Timothy, D.J. (2014a). Contemporary cultural heritage and tourism: Development issues and emerging trends. *Public Archaeology*, 13(3), 30–47.

Timothy, D.J. (2014b). Views of the vernacular: Tourism and heritage of the ordinary. In J. Kaminski, A. Benson, & D. Arnold (Eds.), *Contemporary Issues in Cultural Heritage Tourism* (pp. 32–44). London: Routledge.

Timothy, D.J. (2018). Making sense of heritage tourism: Research trends in a maturing field of study. *Tourism Management Perspectives*, 25: 177–180.

Timothy, D.J. (2021a). *Cultural Heritage and Tourism: An Introduction* (2nd ed.). Bristol: Channel View Publications.

Timothy, D.J. (2021b). Democratizing the cultural past: Western values, the Global South and cross-cultural perspective in heritage tourism. In J. Saarinen & J. Rogerson (Eds.), *Tourism, Change and the Global South*. London: Routledge.

Timothy, D.J. (2021c). Heritage and tourism: Alternative perspectives from South Asia. *South Asian Journal of Tourism and Hospitality*, 1(1), 35–57.

Timothy, D.J., & Nyaupane, G.P. (2009). *Cultural Heritage and Tourism in the Developing World: A Regional Perspective*. London: Routledge.

UNESCO World Heritage Centre (2015). *Operational Guidelines for the Implementation of the World Heritage Convention*. Paris: UNESCO.

Urry, J. (2002). *The Tourist Gaze* (2nd ed.). London: Sage.

Urry, J. (2009). *The Tourist Gaze: Leisure and Travel in Contemporary Societies*. London: Sage.

Vogel, B., & Field, J. (2020). (Re)constructing borders through the governance of tourism and trade in Ladakh, India. *Political Geography*, 82, 102226.

Weaver, A. (2009). Tourism and aesthetic design: Enchantment, style, and commerce. *Journal of Tourism and Cultural Change*, 7(3), 179–189.

Wells, M.P. (1993). Neglect of biological riches: The economics of nature tourism in Nepal. *Biodiversity and Conservation*, 2(4), 445–464.

Wickens, E., Bakir, A., & Avgeli, V. (2017). Sustainable tourism. *Sustainable Tourism Development: Issues Challenges & Debates*, 140–166.

Zurick, D.N. (1992). Adventure travel and sustainable tourism in the peripheral economy of Nepal. *Annals of the Association of American Geographers*, 82(4), 608–628.

Zurick, D.N., & Rose, A. (2009). Landscape change in Kathmandu valley. *Focus on Geography*, 51(4), 7–16.

# Conclusion

## The Experience of Himalayan Tourism

*Dallen J. Timothy and Gyan P. Nyaupane*

Mountains provide one of the most attractive ecosystems in the world for tourists. High peaks, sheer cliffs, glaciers and glacial lakes, snow-capped summits, river valleys, caves, grassy meadows adorned with wildflowers and fauna, evergreen vegetation, barren alpine landscapes, and switchback trails provide a natural backdrop and create an allure unequalled by any other natural system on the planet (Godde et al., 2000; Kala & Bagri, 2020; Singh, 2016). Besides their spectacular natural landscapes, mountains are home to some of the world's most unique and best-preserved cultures—preserved well because of their montane isolation and absence of contact with the outside world until quite recently. For millennia, people have survived the hardships of living at extreme altitudes, which has resulted not only in cultural preservation and unique subsistence livelihoods but also in physiological corporeal adaptations that have enabled mountain populations to remain and thrive at even very high elevations (Funnell & Price, 2003; Huggel et al., 2015). Most of the highest mountains in the world are still tectonically active, continuing in their formation through natural geophysical processes. By means of earthquakes, river erosion, glacial runoff, flooding, and volcanic activity, mountains are in a constant state of evolution. They are in fact a perfect laboratory for understanding geophysical processes and humankind's ability to adapt to terrestrial change.

In addition to their exquisite physical and cultural landscapes, high elevations provide cool retreats in otherwise very hot climates, such as the Atlas Mountains against the Sahara Desert in Morocco and Algeria, the White Mountains against the Sonoran Desert in the US Southwest, and the Himalayan hill stations against the heat of the North Indian River Plain (Lew & Wilkerson, 1997; Popp & El Fasskaoui, 2013; Sacareau, 2007). The touristic appeal of mountain cultures, ecosystems, topography, and moderated climatic conditions is unmistakable. While mountains on every continent provide a backdrop for many types of tourism, the Himalaya are particularly exemplary in their connections to tourism. Famous for their sheer height and dramatic natural features, the tales and folklore they have educed for generations, their sacred and spiritual aura, and the distinctiveness of their cultures have created a region where tourism continues to grow and flourish. Despite the Himalayan region's late arrival in the global tourism marketplace, especially certain areas of China (Tibet), Nepal, and Bhutan, it is now one of the

DOI: 10.4324/9781003030126-21

most desired destinations and, with the exceptions of Afghanistan, Pakistan, and certain parts of Bangladesh and Myanmar, tourism has exploded into a leading economic sector.

The chapters in this book highlight many of the types of tourism that have developed in the Himalaya, including ecotourism, adventure tourism, pilgrimage, volunteer tourism, heritage tourism, tea tourism, rural homestay tourism, and indigenous tourism, as well as the market characteristics, challenges, and opportunities associated with each one. Other contributors have focused on the general challenges of living and working in the Himalaya in relation to tourism, such as vulnerabilities to disasters, climate change, and external economic forces; high-altitude illnesses, which affect tourists, service providers, and destination residents themselves; political instability and conflict; social exclusion and the moral dilemmas it creates among certain classes of people employed in tourism; and the negative ecological impacts of tourism. Other chapters have focused on solutions approaches to understanding Himalayan tourism, including perspectives on corporate social responsibility, social enterprises, and sustainable development in the context of heritage and nature-based tourism. A lot of territory, both geographic and metaphoric, has been covered in the contents of this book, as it provides the most up-to-date and heretofore comprehensive examination of tourism-related issues in this dynamic mountain region. The main thrust of this concluding chapter is not reconversing about what has already been covered in the previous 16 chapters but examining some of the underlying issues that characterize Himalayan conditions and how these might be fruitfully brought to bear in future research activities.

## Future perspectives

Despite the breadth and depth of discussions in this volume, there are many issues that remain unresolved and understudied. The sections that follow attempt to reconcile some of this dearth and highlight crucial issues and underscore the opportunities at our doorstep to understand them better through additional research.

### *Overtourism and sustainability*

One of the biggest challenges facing the world of tourism today is 'overtourism', or overcrowded conditions where mass tourism has grown unabated to the point where it deteriorates the experience for the tourists themselves and makes the destination less liveable for its inhabitants (Dodds & Butler, 2019). Such things as urban gentrification, overcrowded public and private spaces, and negative social and ecological impacts are a direct result of overtourism, which frequently creates animosity between residents and visitors.

To date, there have been relatively few places in the Himalaya that might be considered 'overtouristified'. Yet, increased access in the past 20 years, the growing fame and reputation of the region as a desirable destination, greater technological advancements, and strong marketing efforts by several countries in the region, including India's long-time and successful 'Incredible India' campaign, Bhutan's effectual 'Happiness is a Place' slogan, and Nepal's 'Nepal: Back on Top of the World' promotional campaign following the devastating 2015 earthquakes and its current tagline, 'Lifetime Experiences', have worked together to create a region that was on a steep upward tourism trajectory before the onset of the COVID-19 pandemic in 2020.

Several Himalayan destinations have begun to experience overtourism. In India, Shimla's growing popularity as an ecotourism and mountain tourism destination has created a level of mass tourism that has contributed considerably to air pollution, excess litter, improper waste disposal, overuse of water resources, and housing shortages (Gupta & Chomplay, 2021). Likewise, Singh (2021) surmises that Haridwar, India, on the Ganges River experiences overtourism during pilgrimage times but that the city has not yet reached or exceeded its carrying capacity. Similarly, both Pokhara and Chitwan in Nepal have demonstrated early signs of overtourism (Phuyal, 2020), but not as much as Kathmandu. There, certain parts of the city were, until the pandemic, experiencing considerable overcrowding and deteriorating environments, such as the Thamel tourist neighbourhood (Morimoto, 2015). In an effort to de-escalate some of the crowding and its polluting effects, local authorities recently banned motorized vehicles in the area. According to one travel writer (Adhikari, 2018, n.p.), "Most visitors to Kathmandu end up in Thamel, a tourist district made famous by the hippies who started coming to the city in the 1960s and 70s. [The Thamel district] has everything an international visitor could need. But over the years, Thamel has become crowded and congested, and therefore, unappealing".

Aside from a handful of Himalayan cities, with the increasing popularity of mountain pilgrimages, trekking, and mountaineering, many prevalent religious sites and trekking routes have become overtouristified, resulting in severe deforestation, waste management problems, and crowded conditions on the trails and at sacred sites. This not only has crucial ecological implications but it also deteriorates the ambience of the landscape and experiences of the trekkers, as well as the communities that serve them (Timothy & Boyd, 2015). The most notorious overtourism phenomenon in the past few years, however, took place in the most isolated and one of the most fragile localities in the region. Nowhere is overtourism more obvious than, of all places, on the summit of Mount Everest (Butler, 2020; Wheeler, 2019). The Everest Base Camp in Nepal (and other mountain base camps) is increasingly busy with tourists each climbing season, but in 2019, a theretofore record-breaking 381 Everest permits were issued by the Nepali government for the short April–May climbing season.

Recent images in the popular media of hundreds of climbers queuing and crowding on the narrow precipices leading to Everest's summit have alarmed many observers (O'Grady, 2019). Besides the obvious ecological implications, there are deep concerns about climbers' health and safety. As of the end of May 2019, an unusually high number of climbers had perished due in large part to their inexperience and crowded conditions on the last stretch before the summit (Miller & Blumhardt, 2019; O'Grady, 2019). Overcommercialization and scale and geographic vulnerabilities have much to do with this problem (Nyaupane, 2015). The number of people encroaching upon the peak of Everest would not be a problem on a beach or at a major archaeological site. Yet, on a delicate and dangerous spot such as the summit of Mount Everest, which has been described as the 'size of a dining room table', personal safety and environmental security become a significant challenge (Wengel, 2021). The 2020 climbing season was cancelled because of the pandemic, but it opened again in 2021 with a record-breaking 408 Everest permits having been issued that year (Agence France-Presse, 2021). As this example illustrates, overtourism is not always about raw numbers alone. Some localities are able to absorb higher numbers of tourists than others. Overtourism, therefore, is not a problem itself. Instead, it is a symptom of deeper problems inherent in tourism practices (Clark & Nyaupane, 2020). In addition, some of the problems mentioned have also been sensationalized by the media to gain public attention.

In the Himalaya, there are many popular pilgrimage and trekking locations that are on the verge of irreversible damage to the fragile mountainscapes and ecosystems because of poor and haphazard tourism development and practices. The tourism successes are mostly gauged by the number of tourists, not by their positive impacts to the host communities and host environments. Himalayan tourism policies are often geared toward the volume of tourism instead of quality and long-term sustainability.

Although carrying capacity as a concept can be a tool to address overtourism, it is not a magic number. The carrying capacity of an entire destination is not as simple as a museum or stadium, where the capacity can be determined based on the available seats. Carrying capacity is also a perception and function of management. For example, a crowded mountain trail can be managed through the temporal or spatial distribution of tourists by making the trail one-way, or by reducing its impacts on the communities and environments. Other solutions to overtourism are a pre-booking online reservation system, educating visitors on proper conduct, engaging stakeholders in planning future developments, limiting tourist numbers at peak seasons, improving infrastructure capacity, promoting off-season travel, marketing lesser-known attractions, and demarketing popular tourist destinations (Clark & Nyaupane, 2020). Tourism stakeholders should consider some of these tools and strategies in the Himalaya going forward.

### *New forms of tourism and new niche markets*

Tourism scholars, managers, and marketers are continuously evaluating the tourism marketplace, identifying emerging niche markets that might be targeted for promotional efforts or research assessments. Several chapters in this volume have identified traditional and current niche tourisms, including ecotourism, heritage tourism, volunteer tourism, mountaineering, tea tourism, trekking, and pilgrimage. Unfortunately, many of these traditional types of tourism, especially trekking, mountaineering, and pilgrimage, have overshadowed the potential of other tourism types to thrive in the region (Apollo & Andreychouk, 2020), making it difficult to expand the image of the Himalaya beyond that of mountaineering and trekking (Choegyal, 2019). Yet, like most destinations throughout the world, the Himalaya have a lot of untapped potential for developing tourism products that have heretofore been largely ignored in the tourism marketplace (Choegyal, 2011). For example, agritourism is becoming a salient form of tourism, especially in rural parts of industrialized countries, where getaways to the countryside are seen as a way of getting back to people's cultural roots and a means of understanding past farming practices and rural lifestyles (Timothy, 2020). Although tea tourism is an important manifestation of agritourism in the region (Jolliffe, 2007), many other sorts of agricultural patterns and practices are unique to the Himalaya and have potential to become a salient part of rural tourism development (Timothy, 2021). There are some nascent efforts to develop agritourism, but these have only scratched the surface in empowering agrarian communities to utilize their farming heritage as a tourism product to help enhance their livelihoods beyond farming alone (Bhatta & Ohe, 2019; Bhatta et al., 2019).

Related to agritourism is the notion of food tourism or culinary tourism. This form of tourism is increasing dramatically throughout the world, particularly in countries whose culinary traditions have proliferated the world and become anchored as 'ethnic food' attractions beyond the borders of their origins. For example, Indian, Chinese, Thai, and Mexican food are now ethnic staples in many parts of the world, and there is a growing food tourism market surrounding their origins in India, China, Thailand, and Mexico. The Himalayan region has potential to become a unique and highly specialized food tourism destination, yet the countries of the region have not yet capitalized on this potentially lucrative, albeit likely small-scale, form of tourism.

The culinary traditions of the Himalayan parts of India, Pakistan, Afghanistan, and China have been heavily influenced by foods from the more populated, lower parts of those countries. For the more isolated states, in particular Bhutan and Nepal (and Tibet prior to the 1950s), however, their isolation and lack of colonization helped them preserve inimitable gustatory traditions that remain today. Although each region has its own specialty culinary items, there are some standard trends throughout the Himalayan highlands. Most of the food traditions of Bhutan, Nepal, and Tibet derive from cattle herding, subsistence agriculture, and foraging. Yak and water buffalo products are special and unique to the region, especially milk, cheese, and meat products, and these feature prominently in many regional and national dishes.

Tibetan and Nepali *momo*s are one of the most popular food items in the region, and its popularity has spread beyond the highlands to other parts of South Asia. These steamed Himalayan dumplings are usually stuffed with meat (chicken, yak, or water buffalo), vegetables, and sometimes cheese. Bhutan's traditional cuisines are perhaps the best preserved and representative of the Himalayan food traditions. Bhutanese red rice is unique in that it is cultivated at elevations between 2,500 and 2,800 metres and is grown with ancient glacier waters that are rich in minerals and nutrients. These conditions have enabled Bhutanese red rice to thrive for thousands of years. Almost all of the country's red rice production is done for subsistence purposes, but in recent years, it has been exported as a specialty item at very high prices (Colombari Filho et al., 2020). Although cheaper white rice from India has begun to replace some of the traditional red rice in Bhutanese meals, the local rice product continues to be an important part of the local heritage foodscape. Likewise, the national dish of Bhutan, *ema datshi*, is the foundational dish in Bhutanese cuisine. It is made of hot chilies and yak cheese but now has several variants using cow's milk and potatoes or other vegetables. Unique culinary traditions that tell stories of the past and humankind's adaptations to the natural limitations where they live are the essence of culinary tourism. Unique foods and the exclusive stories of surviving the harsh Himalayan environment and the evolution of food there could become a foundational element of food tourism in the region.

Another area of opportunity, which many mountain areas have begun adopting is the heritage trail. Although trekking routes are one of the main focal areas of tourism in the Himalaya, the region has a dearth of themed cultural/heritage trails. The advent of the Great Himalaya Trail, which focuses on nature and culture, is a huge undertaking that spans Bhutan, Nepal, and northwestern India (Choegyal, 2011). Smaller-scale heritage trails are becoming commonplace throughout the world with the job of linking together places of similar interest. The Guerilla Trail in western Nepal is one such effort, established to connect sites and localities associated with the Maoist rebellion of the early 2000s. Efforts to establish themed trails on a smaller scale have been known to help connect towns, villages, and rural areas into linear spaces to share common assets and spread the economic impact of tourism to isolated areas. Such efforts might help dissipate the negative impacts of tourism and broaden the positive economic impacts (Timothy & Boyd, 2015).

Heritage tourism and cultural tourism have with only a few exceptions been largely neglected in tourism research in the Himalaya and throughout South Asia in general (see Chapagain, 2016, 2020; Suntikul, 2016; Timothy, 2021). Although there is significant interest in the tangible and intangible cultural features of the region, in tourism practice, the cultural heritage characteristics of the Himalaya tend to be dwarfed by their physical geography. Thus, there is much untapped potential for cultural tourism, although we should note that pilgrimage is an important form of heritage tourism that appeals to millions of people in the Himalaya. Nonetheless, there is room for the expansion of other types of heritage tourism surrounding Indigenous cultures, village lifestyles, music and dance, and as noted before, culinary and agricultural heritage.

With remoteness and less human interference, the Himalaya have the potential to become a popular astrotourism (a form of tourism that involves visiting places for the night sky) destination. Many photographers and travel writers describe the clear skies of the High Himalaya as being among the world's best dark skies. The Himalayan region should be able to capitalize on its recently realized dark skies as an important natural and cultural resource for tourism and education. There is a global movement to designate dark sky places, such as communities, national parks and reserves, through the International Dark-Sky Association. Governments,

NGOs, communities, and the other tourism stakeholders should collaborate to protect, develop, and market astrotourism opportunities in the region.

The region also has a great deal of potential for growing spiritual and wellness tourism, including yoga, both spiritual and postural, and meditation. There has been a growing recognition of holistic well-being that goes beyond the physical needs of food, water, shelter, and physical health. The region has been known for thousands of years for its traditions of spiritual wellness, which has recently been expanded throughout the world. Many Himalayan destinations, including Rishikesh and Mysore, now advertise and offer wellness tourism packages for Western tourists who want to learn and practice Eastern philosophies of spiritual well-being, and experience related activities, including meditation, yoga, teaching about one's self, and zen retreats (Buzinde, 2020). As the world continues to grapple with COVID-19 and its resulting mental health issues, a growing awareness and appreciation of meaningful life experiences and self-care will create more demand for wellness tourism, which should be planned, managed, and marketed carefully.

### *Globalization and supranationalism*

The effects of globalization are felt everywhere. Neoliberal trade, technological development, international corporations buying locally owned hospitality services, increased tourism and human mobility, and changing food traditions are all manifestations of globalization (Timothy, 2019a). One of the most pertinent issues with regard to globalization in the Himalaya is supranationalism, or the ordering of the world into large-scale, multinational alliances that aim to reduce the barrier effects of international borders to create more equitable trade relationships and improve human mobility. The South Asian Association for Regional Cooperation (SAARC) is the dominant supranational alliance in the region. All Himalayan states except China and Myanmar are members of SAARC. The organization's main goal is economic integration, and to accomplish this, SAARC commenced the South Asian Free Trade Area in 2006 in preparation for an eventual customs union or common market patterned after the European Union. In 1992, the region's Visa Exemption Scheme was initiated to enable increased cross-border travel, but so far it only applies to certain dignitaries and people of influence (Shaheen, 2013; Taneja & Bimal, 2020). The SAARC Tourism Working Group has also initiated programs to promote the entire region as a nature and culture tourism destination through travel guides and documentaries (Nyaupane & Budruk, 2009).

Although the goals of SAARC include increased transfrontier collaboration and eventual integration, there are many obstacles standing in its way (Batra, 2013; Shaheen, 2013). Not least of these are the sour relations between some of its Himalayan member states, including India and Pakistan, and security concerns in others (e.g., Afghanistan and Pakistan). In the context of tourism, bringing the goals of SAARC to fruition has been a major challenge (Pillai, 2017; Timothy, 2019b), yet there are many opportunities for the Himalayan states to work together in bringing about some of the supranational goals of SAARC. Bhutan and Nepal have integrated certain tenets into their tourism policies. For example, citizens of SAARC countries get preferential treatment at many museums, historic sites, and protected areas in terms of admission fees. Likewise, a handful of SAARC states have adjusted their visa policies to reflect freer mobility among citizens of SAARC member countries, although this is not universal given current tensions and security concerns with some members states (Table 17.1). Bhutan has initiated a no-visa policy for citizens of Bangladesh, India, and the Maldives, although this is not applied across the board for SAARC citizens, as Afghanis, Sri Lankans, Pakistanis, and Nepalis are required to obtain a visa before arriving. The Maldives has the most liberal visa policy in all of SAARC, which affects arrivals from the Himalaya states. Most of the Himalayan members of the alliance have relatively open access to other member states' nationals, but this is dependent upon bilateral relations rather than the overall SAARC framework of free movement.

*Table 17.1* SAARC member states' tourist visa policies with regard to SAARC citizens

| SAARC member state | Visa-free travel for citizens of: | Visa on arrival for citizens of: | Visa required in advance for citizens of: |
|---|---|---|---|
| Afghanistan | – | – | Bangladesh, Bhutan, India, Maldives, Nepal, Pakistan, Sri Lanka |
| Bangladesh | Bhutan, Maldives | – | Afghanistan, India, Nepal, Pakistan, Sri Lanka |
| Bhutan | India, Bangladesh, Maldives | – | Afghanistan, Nepal, Pakistan, Sri Lanka |
| India | Bhutan, Maldives, Nepal | Bangladesh, Pakistan, Sri Lanka | Afghanistan |
| Maldives | India | Afghanistan, Bangladesh, Bhutan, Nepal, Pakistan, Sri Lanka | – |
| Nepal | India | Bangladesh, Bhutan, Maldives, Pakistan, Sri Lanka | Afghanistan |
| Pakistan | Maldives, Nepal | Sri Lanka | Afghanistan, Bangladesh, Bhutan, India |
| Sri Lanka | – | Bangladesh, Bhutan, India, Maldives, Nepal | Afghanistan, Pakistan |

Source: Compiled from various sources

Research to understand the effects of supranationalism in the Himalaya at the regional level and bilateral cross-border cooperation between state pairings is needed. Research has shown that cross-border cooperation in areas of environmental protection, infrastructure development, human resource mobility, tourist mobility, and tourism development has many geopolitical and socioeconomic advantages (Blasco et al., 2014; Prokkola, 2008; Stoffelen & Vanneste, 2017). These same approaches have the potential to help tourism thrive better in the Himalaya, especially in assisting the poorer countries to mobilize their financial and human resources better for tourism. This is a rather idealistic perspective, however, as many political and cultural hurdles remain to deter cross-border collaboration (Javed, 2019; Medhekar & Haq, 2020; Naazer, 2018). Nonetheless, understanding the successes and failures of supranationalism in the Himalaya would help planners and policy-makers better realize the potential benefits of this kind of transfrontier networking.

### Disasters and other vulnerabilities

Some of the authors in this book have underscored the importance of the physical and cultural vulnerabilities of the region. These vulnerabilities cannot be overstated; they dictate how people live their lives and the forms of tourism that can develop. When disasters happen, they affect thousands of people at a time, including tourists. While some disasters are human-induced, many are purely natural through tectonic and other geomorphological processes, but both are equally devastating. National authorities, community leaders, community members, and tourism providers have a responsibility to prepare disaster prevention plans for natural events to plan and devise alternative strategies for their management (Kala, 2014). Developing resilient communities that can withstand and adapt to rapidly changing conditions is an important part of the goals of sustainable development.

### Climate change and resource depletion

A few of the chapters in this volume discuss the problems associated with climate change and how it is currently (and will in the future) affecting the physical and social environments of the Himalaya, particularly how it will impact tourism. The biggest challenge of climate change in the Himalaya is its contribution to increasing natural disasters, particularly flooding, erosion, and landslides, as well as the need for endemic species to adapt to changing temperatures (Nyaupane & Chhetri, 2009). This has obvious implications for tourism (Nepal, 2011). Problems associated with climate change include increasing isolation in certain places when access is blocked, irregular precipitation, glacial lake outburst flooding, increasing danger levels on some trails and cliffs faces, environmental damage to ecosystems, forced migration of fauna species, increased need to import food products from outside the region, and potential changes to seasonal activities in the future (i.e., mountaineering, trekking, and pilgrimages).

Much has been written about the effects of global warming on tourism in small islands, low-lying coastal areas, and in polar regions, yet we know relatively little about its effects on tourism in mountain areas, with the exception of a growing number of studies that discuss decreasing snowfall and skiing. More work is needed to understand how Indigenous and marginalized people and other agriculturists and herders are having to adapt to changing climates in the Asian highlands and how these changes are manifested in tourism. If it is not already affecting service quality, hospitality, and the supply chain, then it certainly will in the future. Likewise, questions need to be raised regarding how climate change will affect cultures and social practices as people continue to adapt to changing circumstances.

As early as the 1990s, concerns were loudly voiced by academic observers about the resource depletion problems brought about by tourism, especially deforestation and reductions in wildwood supplies in popular trekking areas. This problem has become an especially sticky discussion point between rural service providers and environmentalists, but such actions continue to have gross environmental and economic implications through increased erosion and mudslides, deteriorated landscape aesthetics, as well as the economic downturns at the local level that come with depleted resources.

Research needs to address these important issues to help marginalized, rural populations adjust to their 'new normal' in ways that continue to sustain their livelihoods and protect their cultures. While the world community of states is trying to establish higher environmental standards of production and consumption, people and places on the global periphery remain most vulnerable. Special interest tourism has the potential to be a piece of the toolbox that will help in these efforts to adjust to environmental challenges and provide replacements to resource overuse, not least through proper training, education, and proper law enforcement (Timothy, 2012).

### *Geopolitical challenges—debordering and rebordering*

The tenets of the SAARC alliance and various binational agreements aim to bring about debordering processes, or the reduction in barrier effects of state frontiers. For the most part, this goal has yet to be fulfilled. As noted earlier, there are a few minor examples of transfrontier networking having filtered down to the ground in practice with regard to some aspects of tourism, but each member state can do much more to realize the goals of improving people's lives through bilateral and multilateral cooperation.

Instead of opening, as so many borders are doing in other parts of the world, the Himalayan state borders are continuing in a cycle of bordering and rebordering where, antithetical to the goals of SAARC and other alliances, international frontiers are becoming less permeable. This is particularly acute between India and Pakistan, India and China, Afghanistan and Pakistan, and India and Bangladesh. Many places with significant tourism potential remain closed to tourism in many respects. For example, access to Sikkim and the states of northeastern India is still heavily regulated, frustrating some people's desires to visit. Current Chinese restrictions on travel to Tibet for international and domestic tourists deter tourism to one of China's most scenic and culturally rich destinations. The borderlands of India and Pakistan in Jammu and Kashmir have considerable tourism potential, but the heavily militarized and conflicted area is essentially off limits to most would-be visitors. And of course, most parts of Afghanistan today are off limits to tourists owing to security concerns. With the Taliban's retaking of the country in August 2021, travel conditions there are even more volatile and unpredictable than they were during the US occupation.

## The COVID-19 pandemic

The COVID-19 pandemic has affected tourism more than other industries in the world. In 2020, it essentially halted global travel, and many countries closed their borders to international arrivals. In some cases (e.g., Australia), even subnational borders were closed, restricting domestic tourism even further. Although 2021 saw a certain level of recovery as countries began opening up again to outside visitors, forecasters predict that a full recovery from the 2020–2022 pandemic may take years, if not decades (Fotiadis et al., 2021; Hall et al., 2020).

Most Himalayan countries were caught off guard just as other countries in the world were. They immediately limited tourism and began requiring quarantines, mask-wearing, and vaccines (KC, 2021; Nepal, 2020). Although the 2020 climbing season on Mount Everest and on many other peaks was cancelled due to the virus, they reopened for the 2021 season. Unfortunately, however, carelessly enforced health regulations led to many people at the Everest Base Camp in Nepal (5,364 metres) falling ill to COVID-19, affecting not only other tourists but also their Sherpa and other native guides. Many climbers and support staff became seriously ill in May 2021, which not only strained the capabilities of medical staff at the Base Camp but also led China to declare its intentions to erect a border barrier or 'line of separation' on the tiny summit of Everest to prevent climbers on the Chinese side from intermingling with climbers from the Nepal side (Adhikari, 2021). At the time of writing, this Chinese action had not yet come to fruition; instead, China cancelled the 38 Everest permits it had issued for the 2021 season. Nonetheless, a new border fence is something Chinese officials are considering not only to prevent co-mingling across the international border but ostensibly also to exercise an increased level of control over its portion of Mount Everest.

## Final word

The goal of most academic work in the field of tourism and development today is to understand and facilitate the goals of sustainable tourism and its principles, which center on environmental and cultural protection, community empowerment, social equity, and economic development. In the Global South, most of the benefits of tourism go to the rich and powerful elites in a population. Instead of perpetuating social and economic inequity, tourism can be a powerful tool for social and economic transformation. Of particular importance is the protection of natural resources and the empowerment of Indigenous and disadvantaged groups of people, who have traditionally been marginalized in national economic development. If planned and managed properly, tourism in all its forms has the potential to protect the natural environment and empower disenfranchised people and tribal minorities socially, politically, economically, and psychologically (Scheyvens, 1999; Timothy, 2007). However, to be successful, tourism operators, community members and local entrepreneurs, government agents, and the tourists themselves must accept their own level of responsibility for ensuring the health and longevity of tourism and the natural and cultural assets on which it is founded (Baniya & Thapa, 2021; Karst & Nepal, in press; Timothy, 2012). Although tourism in the Himalaya is becoming advanced in age and geographic reach, it is not too late to implement policies that will support the UN Sustainable Development Goals and maintain a healthy environment that will outlive the tourists who currently visit (Richins & Hull, 2016). Some efforts, including homestay and cultural tourism, have been implemented in the Himalayan region to bring disadvantaged groups into the mainstream tourism industry, but more has to be done.

Most forms of tourism in the Himalayan region, except pilgrimage, are geared towards international tourism. Tourism development is typically demand driven, and the policies and infrastructure are developed based on the needs of international tourists (Nyaupane et al., 2020). Colonization has significantly changed the meaning and values of the Himalaya from rich '*dev bhumi*' spiritual land to unproductive wasteland with some aesthetic value (Hall, 1992). This trend continues as government and non-government sectors follow Western development paradigms, including international tourism.

Domestic and regional tourists in the Himalaya have distinct needs, norms, patterns, and behaviours guided by their collectivist values. Tourism resources, including forests, national parks and reserves, have been protected and managed using Western models of transactional relationships with nature. Indigenous values and practices guided by principles of moral and harmonious relationships with nature, rather than being separated from nature, have been proven to be more sustainable, equitable, and fair. The Himalaya should not be developed as any other tourism destination, but as a special destination where tourism activities are more conducive to protecting the ecosystems, cultures, and traditions of the region.

Resilience, adaptability, and change are the operative words of today in the Himalaya. However, if one overarching word could be used to describe the Himalaya and the tourism that occurs there, it would be 'dynamic'. The region is physically and culturally dynamic, and tourism is dynamic, organically as well as by design. Many forces come to play in the dynamicity of the region, including colonialism and poverty, plate tectonics and global warming, indigeneity and Indigenous knowledge, social responsibility and irresponsibility, vulnerability and strength, tourism and its impacts, and social and economic inequities. Although the Himalayan region is a place of unremitting transformation, one thing remains constant: Its charm and natural beauty will continue to attract visitors from all over the world who come to see its everlasting summits, its distinctive cultures, and its unmatched landscapes.

## References

Adhikari, D. (2018). Kathmandu city guide: 16 things to know before you go to Kathmandu. *Roads and Kingdoms*, July 19. Retrieved from https://roadsandkingdoms.com/travel-guide/kathmandu/know-before-you-go-to-kathmandu/

Adhikari, R. (2021). The COVID-19 pandemic may be the hardest mountain Nepal's Sherpas have ever had to climb. *Time*, May 21. Retrieved from https://time.com/6048857/covid-nepal-everest-sherpa/

Agence France-Press (2021). Everest climbing season was like no other. *VOA News*, June 5. Retrieved from www.voanews.com/south-central-asia/everest-climbing-season-was-no-other

Apollo, M., & Andreychouk, V. (2020). Mountaineering and the natural environment in developing countries: An insight to a comprehensive approach. *International Journal of Environmental Studies*, 77(6), 942–953.

Baniya, R., & Thapa, B. (2021). CSR communication among tourism SMEs through their websites. *Tourism and Hospitality*, 2(3), 319–326.

Batra, A. (2013). *Regional Economic Integration in South Asia: Trapped in Conflict?* London: Routledge.

Bhatta, K., Itagaki, K., & Ohe, Y. (2019). Determinant factors of farmers' willingness to start agritourism in rural Nepal. *Open Agriculture*, 4(1), 431–445.

Bhatta, K., & Ohe, Y. (2019). Farmers' willingness to establish community-based agritourism: Evidence from Phikuri village, Nepal. *International Journal of Tourism Sciences*, 19(2), 128–144.

Blasco, D., Guia, J., & Prats, L. (2014). Heritage tourism clusters along the borders of Mexico. *Journal of Heritage Tourism*, 9(1), 51–67.

Butler, R.W. (2020). Overtourism in rural areas. In H. Séraphin, T. Gladkikh, & T.V. Thanh (Eds.), *Overtourism: Causes, Implications and Solutions* (pp. 27–44). Cham, Switzerland: Springer.

Buzinde, C.N. (2020). Theoretical linkages between well-being and tourism: The case of self-determination theory and spiritual tourism. *Annals of Tourism Research*, 83, 102920.

Chapagain, N.K. (2016). Blurring boundaries and moving beyond the tangible/intangible and the natural/cultural classifications of heritage: Cases from Nepal. In K.D. Silva, & A. Sinha (Eds.), *Cultural Landscapes of South Asia* (pp. 44–58). London: Routledge.

Chapagain, N.K. (2020). Public archaeology in Nepal: Now and in the next 10 years. *Online Journal of Public Archaeology*, 10, 73–74.

Choegyal, L. (2011). The Great Himalaya Trail: A new Nepal tourism product with both trek marketing and development rationale. *Nepal Tourism & Development Review*, 1(1), 71–76.

Choegyal, L. (2019). Tourism in the Himalayas. *Druk Journal*, 5(2), n.p. (online journal)

Clark, C., & Nyaupane, G.P. (2020). Overtourism: An analysis of its coverage in the media by using framing theory. *Tourism Review International*, 24(2–3), 75–90.

Colombari Filho, J.M., Abreu, A.G.D., & Pereira, J.A. (2020). Red rice. In A. Costa de Oliveira, C. Pegoraro, & V. Ebeling Viana (Eds.), *The Future of Rice Demand: Quality Beyond Productivity* (pp. 283–296). Cham, Switzerland: Springer.

Dodds, R., & Butler, R.W. (Eds.). (2019). *Overtourism: Issues, Realities and Solutions*. Berlin: De Gruyter.

Fotiadis, A., Polyzos, S., & Huan, T.C.T. (2021). The good, the bad and the ugly on COVID-19 tourism recovery. *Annals of Tourism Research*, 87, 103117.

Funnell, D.C., & Price, M.F. (2003). Mountain geography: A review. *Geographical Journal*, 169(3), 183–190.

Godde, P.M., Price, M.F., & Zimmermann, F.M. (Eds.). (2000). *Tourism and Development in Mountain Regions*. Wallingford: CABI.

Gupta, V., & Chomplay, P. (2021). Local residents' perceptions regarding the negative impacts of overtourism: A case of Shimla. In A. Sharma, & A. Hassan (Eds.), *Overtourism as Destination Risk: Impacts and Solutions* (pp. 69–80). Bingley: Emerald.

Hall, C.M. (1992). *Wasteland to World Heritage: Preserving Australia's Wilderness*. Melbourne: Melbourne University Press.

Hall, C.M., Scott, D., & Gössling, S. (2020). Pandemics, transformations and tourism: Be careful what you wish for. *Tourism Geographies*, 22(3), 577–598.

Huggel, C., Carey, M., & Clague, J.J. (Eds.). (2015). *The High-Mountain Cryosphere*. Cambridge: Cambridge University Press.

Javed, A. (2019). South Asia's services trade: Barriers and prospects for integration. *International Journal of Management, Accounting and Economics*, 6(10), 752–760.

Jolliffe, L. (Ed.). (2007). *Tea and Tourism: Tourists, Traditions and Transformations*. Bristol: Channel View Publications.

Kala, C.P. (2014). Deluge, disaster and development in Uttarakhand Himalayan region of India: Challenges and lessons for disaster management. *International Journal of Disaster Risk Reduction*, 8, 143–152.

Kala, D., & Bagri, S.C. (2020). *Global Opportunities and Challenges for Rural and Mountain Tourism*. Hershey, PA: IGI Global.

Karst, H.E., & Nepal, S.K. (in press). Social-ecological wellbeing of communities engaged in ecotourism: Perspectives from Sakteng Wildlife Sanctuary, Bhutan. *Journal of Sustainable Tourism*, 1–23.

KC, A. (2021). Role of policymakers and operators towards tourism revival in the era of COVID-19 in Nepal. *Journal of Tourism Quarterly*, 3(2), 98–112.

Lew, A.A., & Wilkerson, J.S. (1997). Settlement history and tourism development in two Arizona mountain communities. *Yearbook of the Association of Pacific Coast Geographers*, 59(1), 87–100.

Medhekar, A., & Haq, F. (2020). Cross-border cooperation for bilateral trade, travel, and tourism: A challenge for India and Pakistan. In R.A. Castanho (Ed.), *Cross-Border Cooperation (CBC) Strategies for Sustainable Development* (pp. 168–191). Hershey, PA: IGI Global.

Miller, R.W., & Blumhardt, M. (2019). What's causing Mount Everest's deadly season? Overcrowding, inexperience and a long line to the top. *USA Today*, May 29. Retrieved from www.usatoday.com/story/news/world/2019/05/29/mount-everest-deaths-traffic-jam-blame-why-such-deadly-season/1258092001/

Morimoto, I. (2015). Tourism, consumption and the transformation of Thamel, Kathmandu. In C. Bates & M. Mio (Eds.), *Cities in South Asia* (pp. 325–341). London: Routledge.

Naazer, M.A. (2018). Politics, tourism and regional cooperation in South Asia. *Pakistan Journal*, 54(1), 20–42.

Nepal, S.K. (2011). Mountain tourism and climate change: Implications for the Nepal Himalaya. *Nepal Tourism and Development Review*, 1(1), 1–14.

Nepal, S.K. (2020). Adventure travel and tourism after COVID-19—business as usual or opportunity to reset? *Tourism Geographies*, 22(3), 646–650.

Nyaupane, G.P. (2015). Mountaineering on Mount Everest: Evolution, economy, ecology, and ethics. In G. Musa, A. Thompson-Carr, & J. Higham (Eds.), *Mountaineering Tourism* (pp. 265–271). London: Routledge.

Nyaupane, G.P., & Budruk, M. (2009). Heritage and tourism in South Asia: Conflict, colonialism, and cooperation. In D.J. Timothy, & G.P. Nyaupane (Eds.), *Cultural Heritage and Tourism in the Developing World: A Regional Perspective* (pp. 127–145). London: Routledge.

Nyaupane, G.P., & Chhetri, N. (2009). Vulnerability to climate change of nature-based tourism in the Nepalese Himalayas. *Tourism Geographies*, 11(1), 95–119.

Nyaupane, G.P., Paris, C.M., & Li, X.R. (2020). Domestic tourism in Asia. *Tourism Review International*, 24(1), 1–4.

O'Grady, S. (2019). Mount Everest has gotten so crowded that climbers are perishing in the traffic jams. *Washington Post*, May 25. Retrieved from www.washingtonpost.com/world/2019/05/24/mount-everest-has-gotten-so-crowded-that-climbers-are-perishing-traffic-jams/

Phuyal, S. (2020). *Developing a Tourism Opportunity Index Regarding the Prospective of Overtourism in Nepal*. Unpublished master's thesis, Missouri State University, Springfield, Missouri.

Pillai, K.R. (2017). Tourism in South Asia: An economic leverage to India. *Asia Pacific Journal of Tourism Research*, 22(7), 709–719.

Popp, H., & El Fasskaoui, B. (2013). Some observations on tourism developments in a peripheral region and the validity of global value chain theory: The Anti-Atlas Mountains in Morocco. *Erdkunde*, 67(3), 265–276.

Prokkola, E-K. (2008). Resources and barriers in tourism development: Cross-border cooperation, regionalization and destination building at the Finnish-Swedish border. *Fennia*, 186(1), 31–46.

Richins, H., & Hull, J.S. (Eds.). (2016). *Mountain Tourism: Experiences, Communities, Environments and Sustainable Futures*. Wallingford: CABI.

Sacareau, I. (2007). Himalayan hill stations from the British Raj to Indian tourism. *European Bulletin of Himalayan Research*, 31, 30–47.

Scheyvens, R. (1999). Ecotourism and the empowerment of local communities. *Tourism Management*, 20(2), 245–249.

Shaheen, I. (2013). South Asian association for regional cooperation (SAARC): Its role, hurdles and prospects. *IOSR Journal of Humanities and Social Science*, 15(6), 1–9.

Singh, L. (2021). Overtourism and its impacts in Haridwar from residents' perspective. In A. Sharma & A. Hassan (Eds.), *Overtourism as Destination Risk: Impacts and Solutions* (pp. 221–233). Bingley: Emerald.

Singh, S. (2016). Devising an electronically supported heritage conservation method for the valley of flowers in the Indian Himalayas. *Journal of Heritage Tourism*, 11(4), 411–419.

Stoffelen, A., & Vanneste, D. (2017). Tourism and cross-border regional development: Insights in European contexts. *European Planning Studies*, 25(6), 1013–1033.

Suntikul, W. (2016). Tourism, liquid modernity and Bhutanese traditional festivals. In K.D. Silva & A. Singha (Eds.), *Cultural Landscapes of South Asia: Studies in Heritage Conservation and Management* (pp. 191–202). London: Routledge.

Taneja, N., & Bimal, S. (2020). Informal trade in the SAARC region. In S. Raihan & P. De (Eds.), *Trade and Regional Integration in South Asia: A Tribute to Salam Kelegama* (pp. 267–277). Singapore: Springer.

Timothy, D.J. (2007). Empowerment and stakeholder participation in tourism destination communities. In A. Church & T. Coles (Eds.), *Tourism, Power and Space* (pp. 199–216). London: Routledge.

Timothy, D.J. (2012). Destination communities and responsible tourism. In D. Leslie (Ed.), *Responsible Tourism: Concepts, Theory and Practice* (pp. 72–81). Wallingford: CABI.

Timothy, D.J. (2019a). Globalisation: The shrinking world of tourism. In D.J. Timothy (Ed.), *Handbook of Globclisation and Tourism* (pp. 323–332). Cheltenham: Edward Elgar.

Timothy, D.J. (2019b). Tourism, border disputes and claims to territorial sovereignty. In R.K. Isaac, E. Çakmak, & R. Butler (Eds.), *Tourism and Hospitality in Conflict-Ridden Destinations* (pp. 25–38). London: Routledge.

Timothy, D.J. (2020). Heritage consumption, new tourism and the experience economy. In M. Gravari-Barbas (Ed), *A Research Agenda for Heritage Tourism* (pp. 203–217). Cheltenham: Edward Elgar.

Timothy, D.J. (2021). Heritage and tourism: Alternative perspectives from South Asia. *South Asian Journal of Tourism and Hospitality*, 1(1), 35–57.

Timothy, D.J., & Boyd, S.W. (2015). *Tourism and Trails: Cultural, Ecological and Management Issues*. Bristol: Channel View Publications.

Wengel, Y. (2021). The micro-trends of emerging adventure tourism activities in Nepal. *Journal of Tourism Futures*, 7(2), 209–215.

Wheeler, T. (2019). Foreword. In C. Milano, J.M. Cheer, & M. Novelli (Eds.), *Overtourism: Excesses, Discontents and Measures in Travel and Tourism* (pp. xv–xvii). Wallingford: CABI.

# Index

For Product Safety Concerns and Information please contact our EU
representative  GPSR@taylorandfrancis.com
Taylor & Francis Verlag GmbH, Kaufingerstraße 24, 80331 München, Germany

www.ingramcontent.com/pod-product-compliance
Lightning Source LLC
Chambersburg PA
CBHW061623220326
41598CB00026BA/3859